Tarek Zaki

Short-Channel Organic Thin-Film Transistors

Fabrication, Characterization, Modeling and Circuit Demonstration

Doctoral Thesis accepted by
the University of Stuttgart, Germany

Author
Dr. Tarek Zaki
Institute for Microelectronics Stuttgart
Stuttgart
Germany

Supervisor
Prof. Joachim N. Burghartz
Institute for Microelectronics Stuttgart
Stuttgart
Germany

ISSN 2190-5053 ISSN 2190-5061 (electronic)
Springer Theses
ISBN 978-3-319-18895-9 ISBN 978-3-319-18896-6 (eBook)
DOI 10.1007/978-3-319-18896-6

Library of Congress Control Number: 2015939661

Springer Cham Heidelberg New York Dordrecht London

Printed on acid-free paper

Springer International Publishing AG Switzerland is part of Springer Science+Business Media
(www.springer.com)

Parts of this thesis have been published in the following journal articles:

1. J. Milvich, **T. Zaki**, M. Aghamohammadi, R. Röodel, U. Kraft, H. Klauk, and J. N. Burghartz, "Flexible low-voltage organic phototransistors based on air-stable dinaphtho[2,3-b:2′,3′-f]thieno[3,2-b]thiophene," *Organic Electronics*, vol. 20, pp. 63–68, May 2015.
2. U. Kraft, M. Sejfić, M. J. Kang, K. Takimiya, **T. Zaki**, F. Letzkus, J. N. Burghartz, E. Weber, and H. Klauk, "Flexible low-voltage organic complementary circuits: Finding the optimum combination of semiconductors and monolayer gate dielectrics," *Advanced Materials*, vol. 27, no. 2, pp. 207–214, Jan. 2015.
3. S. Scheinert, **T. Zaki**, R. Rödel, I. Hörselmann, H. Klauk, and J. N. Burghartz, "Numerical analysis of capacitance compact models for organic thin-film transistors," *Organic Electronics*, vol. 15, no. 7, pp. 1503–1508, Jul. 2014.
4. **T. Zaki**, S. Scheinert, I. Hörselmann, R. Rödel, F. Letzkus, H. Richter, U. Zschieschang, H. Klauk, and J. N. Burghartz, "Accurate capacitance modeling and characterization of organic thin-film transistors," *IEEE Transactions on Electron Devices*, vol. 61, no. 1, pp. 98–104, Jan. 2014.
5. R. Rödel, F. Letzkus, **T. Zaki**, J. N. Burghartz, U. Kraft, U. Zschieschang, K. Kern, and H. Klauk, "Contact properties of high-mobility, air-stable, low-voltage organic n-channel thin-film transistors based on a naphthalene tetracarboxylic diimide," *Applied Physics Letters*, vol. 102, no. 233303, pp. 233303-1–233303-5, Jun. 2013.
6. U. Zschieschang, R. Hoffmockel, R. Rödel, U. Kraft, M. J. Kang, K. Takimiya, **T. Zaki**, F. Letzkus, J. Butschke, H. Richter, J. N. Burghartz, and H. Klauk, "Megahertz operation of flexible low-voltage organic thin-film transistors," *Organic Electronics*, vol. 14, no. 6, pp. 1516–1520, Jun. 2013.
7. **T. Zaki**, R. Rödel, F. Letzkus, H. Richter, U. Zschieschang, H. Klauk, and J. N. Burghartz, "AC characterization of organic thin-film transistors with asymmetric gate-to-source and gate-to-drain overlaps," *Organic Electronics*, vol. 14, no. 5, pp. 1318–1322, May 2013.
8. **T. Zaki**, R. Rödel, F. Letzkus, H. Richter, U. Zschieschang, H. Klauk, and J. N. Burghartz, "S-parameter characterization of submicrometer low-voltage organic thin-film transistors," *IEEE Electron Device Letters*, vol. 34, no. 4, pp. 520–522, Apr. 2013.
9. **T. Zaki**, F. Ante, U. Zschieschang, J. Butschke, F. Letzkus, H. Richter, H. Klauk, and J. N. Burghartz, "A 3.3 V 6-bit 100 kS/s current-steering D/A converter using p-type organic thin-film transistors on glass," *IEEE Journal of Solid-State Circuits*, vol. 47, no. 1, pp. 292–300, Jan. 2012.
10. F. Ante, D. Kälblein, **T. Zaki**, U. Zschieschang, K. Takimiya, M. Ikeda, T. Sekitani, T. Someya, J. N. Burghartz, K. Kern, and H. Klauk, "Contact resistance and megahertz operation of aggressively scaled organic transistors," *Small*, vol. 8, no. 1, pp. 73–79, Jan. 2012.

Dedicated to my beloved parents

Supervisor's Foreword

The year 2015 marks the 50th anniversary of Gordon Moore's 1965 prediction that miniaturization of microelectronics would follow an exponential growth path with a doubled density on chip every 24 months, widely known as Moore's law. This has fostered tremendous growth in consumer and industrial electronics applications. Electronic chips have become increasingly complex and are commonly realized while minimizing chip area and, thus, cost. Chips are, therefore, miniature size products by nature and are unsuitable for use in any large-area and mechanically flexible electronic applications, such as flexible or curved displays, photovoltaics, and certain sensor as well as life style devices.

It was not surprising that after the discovery of conductive polymers in the 1970s, researchers in academia and industry envisioned a tremendous opportunity in developing electronic products that are macrosize by nature. The Nobel Prize awarded to Heeger, MacDiarmid, and Shirakawa in 2000 for their work on conductive polymers has even more ignited enthusiasm in this field. Organic photovoltaics and organic displays, in particular, are manufactured in large volumes today and feature superiority to other technological solutions in terms of performance and cost. However, organic electronic circuits are still far from industrial applications, which is mainly due to the low carrier mobility of organic semiconductors, falling two orders of magnitudes behind the silicon value. Also, operating voltages for organic field-effect transistors (OTFTs) are in the tens of volts and OTFTs are, thus, not well suitable for battery-powered devices which are the basis for entering the promising industrial markets. Therefore, there is high need for low-voltage OTFT technology which will also enable flexible electronic systems that utilize both high-performance ultra-thin silicon chips and large-area organic electronics, which we recently proposed as hybrid systems-in-foil (HySiF).

This thesis addresses the main shortcomings of current OTFT technology by aiming at increasing the performance level of OTFTs at a voltage supply that is compatible with that of modern silicon chips. The use of stencil masks for patterning OTFTs at submicrometer channel lengths not only greatly advances OTFT performance but also offers an insight at the parasitic effects that become apparent when OTFT technology will advance to that level of miniaturization in the future. Stencil mask patterning also leads to far better device matching which allows for exploiting circuit topologies known from silicon electronics. S-parameter characterization on individual OTFTs, which is first demonstrated in this thesis, will help to advance OTFT compact modeling and, thus, circuit design to pave the way to viable electronic products realized in organic or HySiF technologies.

Stuttgart
January 2015

Prof. Joachim N. Burghartz

Abstract

Plastic electronics based on organic thin-film transistors (OTFTs) pave the way for cheap, flexible, and large-area products. Over the past few years, OTFTs have undergone remarkable progress in terms of reliability, performance and scale of integration. This work takes advantage of high-resolution silicon stencil masks to build air-stable complementary OTFTs using a low-temperature fabrication process. Many factors contribute to the allure of this technology; the masks exhibit excellent stiffness and stability, thus allowing to pattern the OTFTs with submicrometer channel lengths and superb device uniformity. Furthermore, the OTFTs employ an ultra-thin gate dielectric that provides a sufficiently high capacitance of the order of 1 $\mu F/cm^2$ to enable the transistors to operate at voltages as low as 3 V.

The critical challenges in this development are the subtle mechanisms that govern the properties of the aggressively scaled OTFTs. These mechanisms, dictated by device physics, have to be described and implemented into circuit design tools to ensure adequate simulation accuracy. This is particularly beneficial to gain deeper insight into materials-related limitations. The primary objective of this work is to bridge the gap between device modeling and mixed-signal circuits by establishing an OTFT compact model, together with realizing the world's fastest organic digital-to-analog converter (DAC).

A unified model that captures the essence in the static/dynamic behavior of the OTFTs is derived. Approaches to incorporate the implicit bias-dependent parasitic effects in the model are elucidated and accordingly a reliable fit to experimental data of OTFTs with different dimensions is obtained. It is demonstrated that the charge storage behavior in the intrinsic OTFTs agrees very well with Meyer's capacitance model. Moreover, the first comprehensive study of the frequency response of OTFTs using S-parameter characterization is presented. In view of the low supply voltage and air stability, a record cutoff frequency of 3.7 MHz for a channel length of 0.6 μm and a gate overlap of 5 μm is accomplished. Finally, a 6-bit current-steering DAC, comprising as many as 129 OTFTs, is designed. The converter achieves a 1000-fold faster update rate (100 kS/s) than prior state of the art.

Acknowledgments

The past 4 years of my research journey at the University of Stuttgart and the Institute for Microelectronics Stuttgart (IMS CHIPS) have been challenging, where certainly the rewards are inspirational, sensational, and powerful. In the same way, I am in awe of the German University in Cairo (GUC) for setting the primary bricks of that journey and for that I am very grateful. It has given me immense honor to learn from and work with many distinguished scholars and colleagues. I am very fortunate to have been endowed with a true passion for knowledge and cherish all these unforgettable experiences. With the deepest gratitude, I wish to express my thankfulness here to all those who supported me in making this work possible.

First and foremost, I am very grateful to my advisor, Prof. Dr.-Ing. Joachim N. Burghartz, Director of IMS CHIPS, for his invaluable guidance and encouragement. His rich knowledge and insightful suggestions from the initial conception to the end of this work are highly appreciated. I am also greatly indebted to his assistance in matters that were beyond the academic concerns, which have helped me to endure all kinds of difficulties during the course of my graduate studies.

Furthermore, I would like to sincerely thank my supervisor, Dr. Harald Richter, Head of the ASIC Department at IMS CHIPS. He enlightened me with the first glance of research in the field of organic electronics and kept me motivated since then. He has been truly a constant inspiration to me owing to his vast amount of skills and expertize in many areas. It is no doubt that his precious support has contributed enormously to the success of this work and I could not have imagined of a better supervisor.

I would also like to acknowledge with much appreciation the essential role of our partners and collaborators, who have shared generously their materials, experiences, and ideas. This includes Dr. Hagen Klauk and all his group at the Max Planck Institute for Solid State Research; particularly Dr. Ute Zschieschang, Dr. Frederik Ante, Mr. Reinhold Rödel, and Ms. Ulrike Kraft. No enough words could possibly convey my thankfulness fully to them for their endless support in fabricating the organic samples for us. Moreover, I owe special thanks to PD Dr.-Ing. habil. Susanne Scheinert and Mr. Ingo Hörselmann of the Institute of Solid

State Electronics, Technical University of Ilmenau, who introduced me to two-dimensional device simulations of our organic transistors. I have been delighted by their kind hospitality, which made my stay in the city of Ilmenau absolutely wonderful. I still remember quite well the stunning view of the city from the top of Kickelhahn hill (861 m) where Goethe has written one of his most famous and admirable poems (Wanderer's Night Song II) onto a wall of a wooden lodge:

Over all of the hills
Peace comes anew,
The woodland stills
All through;
The birds make no sound on the bough.
Wait a while,
Soon now
Peace comes to you.

—Goethe (1780) and translated by John Whaley (1999)

These are truly thoughtful words that I will never forget. Furthermore, my deep gratitude goes to Prof. Dr.-Ing. Manfred Berroth, Dr. Sandra Klinger, and Mr. Johannes Digel of the Institute of Electrical and Optical Communications Engineering, University of Stuttgart, for their assistance in high-frequency characterization. Finally, I extend my sincere thankfulness and appreciation to Prof. Dr. Boris Murmann and Mr. Alex Omid-Zohoor, Stanford University, for their helpful and constructive discussions.

This work is partially funded by the German Research Foundation (DFG) under Grant BU 1962/4-1, thereby I would like to take this opportunity to thank them for their generous support. Likewise, I am grateful to the IEEE Electron Device Society (IEEE-EDS) for selecting me as the recipient of the 2013 EDS Ph.D. Student Fellowship award. It is an achievement that has brought pride to my family, colleagues, friends, and myself.

I am indebted to all my colleagues at IMS CHIPS for making a very friendly and receptive work atmosphere, especially Dr. Florian Letzkus for fabricating stencil masks, Dr. Horst Rempp, Dr. Stefan Endler, and Mr. Alexander Frank for being the best supportive office-mates, Mr. Thomas Deuble for his help with translating the abstract, Ir. Cor Scherjon for always sharing his wise advice, Ms. Helga Schmidtke for managing all administrative issues, Mr. Mahadi Ul-Hassan for helping me in the laboratory, and Mr. Jürgen Ade for taking professional photographs, not to mention the emotional support and pleasurable company of all my friends in Germany and Egypt.

Last but not least, I must save my most intense gratitude to my beloved parents, for the most treasured and heartfelt of gifts; their wonderful love, endless support, and priceless encouragement.

Stuttgart
January 2015

Tarek Zaki

Contents

Nomenclature

Abbreviations

a-Si:H	Hydrogenated Amorphous Silicon
AC	Alternating Current
ADC	Analog-to-Digital Converter
AFM	Atomic Force Microscopy
AM-LCD	Active-Matrix Liquid Crystal Display
BHJ	Bulk Heterojunction
BJT	Bipolar Junction Transistor
CD	Critical Dimension
CRT	Cathode Ray Tube
CSA	Current-Source Array
CVC	Current-to-Voltage Converter
DAC	Digital-to-Analog Converter
DC	Direct Current
DFM	Design For Manufacturability
DNL	Differential Nonlinearity
DNTT	Dinaphtho-Thieno-Thiophene
DRC	Design Rule Check
DRIE	Deep Reactive Ion Etching
DSSC	Dye-Sensitized Solar Cell
DUT	Device Under Test
EL	Electroluminescence
$F_{16}CuPc$	Hexadecafluorocopperphthalocyanine
FET	Field-Effect Transistor
GDS	Graphic Database System
GSG	Ground-Signal-Ground
HBT	Heterojunction Bipolar Transistor
HOMO	Highest Occupied Molecular Orbital
IC	Integrated Circuit

ICC	Integrated Circuit Card
IGFET	Insulated-Gate Field-Effect Transistor
IMS CHIPS	Institute for Microelectronics Stuttgart
INL	Integral Nonlinearity
LCR	Inductance-Capacitance-Resistance
LSB	Least Significant Bit
LUMO	Lowest Unoccupied Molecular Orbital
MFP	Membrane Flow Process
MIM	Metal-Insulator-Metal
MIS	Metal-Insulator-Semiconductor
MOS	Metal-Oxide-Semiconductor
MPI-SSR	Max Planck Institute for Solid-State Research
MSB	Most Significant Bit
NV-RAM	Nonvolatile Random Access Memory
OHJ	Ordered Heterojunction
OLED	Organic Light-Emitting Diode
OPVC	Organic Photovoltaic Cell
OTFT	Organic Thin-Film Transistor
OVPD	Organic Vapor Phase Deposition
PANI	Polyaniline
PCell	Parameterized Cell
PEL	Parameter Extraction Language
PEN	Poly(ethylene 2,6-naphthalate)
PET	Poly(ethylene terephthalate)
PHJ	Planar Heterojunction
PLED	Polymer Light-Emitting Diode
PoC	Point-of-Care
PTC	Positive Temperature Coefficient
RC	Resistance-Capacitance
RF	Radio Frequency
RFID	Radiofrequency Identification
rMSE	Relative Mean Square Error
ROM	Read Only Memory
SAM	Self-Assembled Monolayer
SAP	Self-Aligned Inkjet Printing
SEM	Scanning Electron Microscope
SFDR	Spurious-Free Dynamic Range
SiF	System-in-Foil
SIJ	Subfemtoliter Inkjet Printing
SMU	Source Measure Unit
SOI	Silicon-on-Insulator
SSL	Solid-State Lighting
SWA	Switch Array
TEOS	Tetraethyl Orthosilicate
TFT	Thin-Film Transistor

TLM	Transmission Line Method
TMAH	Tetramethylammonium Hydroxide
VNA	Vector Network Analyzer
VRH	Variable Range Hopping
VTC	Voltage Transfer Characteristic
VTE	Vacuum Thermal Evaporation
WFP	Wafer Flow Process
WORM	Write Once Read Many

Symbols

β	Short-circuit current gain (–)
μ	Effective charge carrier mobility (cm^2/Vs)
μ_o	Intrinsic charge carrier mobility (cm^2/Vs)
ω	Angular frequency (rad/s)
ρ_c	Normalized contact resistivity (Ωm^2)
τ	Effective charge response time (s)
τ_d	Signal delay per ring oscillator stage (s)
ε_i	Insulator effective permittivity (–)
ε_o	Vacuum permittivity (F/m)
C_{ch}	Channel capacitance (F)
C_{gd}	Gate-drain capacitance (F)
C_{gs}	Gate-source capacitance (F)
C_G	Total gate capacitance (F)
C_I	Insulator capacitance per unit area (nF/cm^2)
c_i	Dielectric capacitance per unit length (F/cm)
C_j	Junction capacitance at the contact (nF/cm^2)
C_{ov}	Gate-overlap capacitance (F)
c_s	Semiconductor capacitance per unit length (F/cm)
d	Semiconductor layer thickness (nm)
d_{ox}	Oxide layer thickness (nm)
d_{SAM}	Self-assembled monolayer thickness (nm)
d_i	Insulator layer thickness (nm)
E_G	Band gap (eV)
E_{ch}	Electric field in the channel (V/m)
E_C	Conduction band minimum (eV)
E_F	Fermi energy level (eV)
E_V	Valence band maximum (eV)
f	Frequency (Hz)
f_{rio}	Oscillation frequency of a ring oscillator (Hz)
f_T	Current-gain cutoff frequency (Hz)
G_{ch}	Channel conductance (A/V)

g_{ch}	Channel conductance (A/V)
g_m	Transconductance (A/V)
I_D	Drain current (A)
I_{on}/I_{off}	On/off current ratio (A/A)
I_{ref}	Reference current (A)
I_{SS}	Reverse bias saturation current (A)
k_B	Boltzmann constant (eV/K)
L	Channel length (μm)
L_{ov}	Gate-overlap length (μm)
L_{gd}	Gate-drain overlap length (μm)
L_{gs}	Gate-source overlap length (μm)
L_T	Transfer length (μm)
N_A	Doping concentration (cm^{-3})
N_{if}	Fixed interface charges concentration (cm^{-2})
n_o	Total conducting charge per unit area (C/cm^2)
q	Absolute electron charge (C)
$q\phi_{Al}$	Aluminum work function (eV)
$q\phi_{Au}$	Gold work function (eV)
$q\phi_s$	Semiconductor work function (eV)
Q_{ch}	Sheet charge density of the channel (C/cm^2)
Q_G	Total intrinsic gate charges (C)
Q_{if}	Interface-trapped charge concentration (C/cm^2)
Q_i	Insulator charge concentration (C/cm^2)
R_{ch}	Channel resistance (Ω)
r_{ch}	Channel resistance per unit length (Ω/cm)
R_C	Contact resistance (Ω)
R_D	Drain contact resistance (Ω)
r_j	Junction resistance at the contact (Ω cm)
r_o	Output resistance (Ω cm)
R_{sheet}	Sheet resistance (Ω/□)
R_S	Source contact resistance (Ω)
r_s	Series resistance at the contact (Ω cm)
R_{tot}	Total transistor resistance (Ω)
$S_{s\text{-}th}$	Subthreshold slope (mV/dec)
T	Temperature (K)
V_{bias}	Bias voltage (V)
V_{DD}	Supply voltage (V)
V_D	Drain voltage (V)
v_d	Drift velocity (m/s)
V_{FB}	Flat-band voltage (V)
V_{GT}	Gate-overdrive voltage (V)
V_G	Gate voltage (V)
V_{OH}	Inverter output high voltage (V)
V_{OL}	Inverter output low voltage (V)

V_{ON}	Turn-on voltage (V)
$V_{sw,on}$	Switch input ON voltage (V)
V_{swing}	Swing voltage (V)
V_S	Source voltage (V)
V_{TH}	Threshold voltage (V)
W	Channel width (μm)
Z_S	Source impedance (Ω)
Z_L	Load impedance (Ω)
Z_o	Characteristic impedance (Ω)

Chapter 1
Motivation

The rapid evolution of semiconductor technology is fueled by an unending demand for better performance and more functionality at reduced manufacturing costs [1]. A new generation of thin, flexible electronics based on organic semiconductors arises [2]. The organic materials are all the more appealing as they can be deposited over large areas at or near room temperature, thus providing compatibility with a wide range of unconventional substrates such as glass, plastic, fabric and paper. Furthermore, the ability to process the organic materials from solution opens a plethora of alternative high-throughput, low-cost patterning techniques that are adapted from the graphic art printing industry [3]. These strengths are currently unfolding in the production of organic thin-film transistors (OTFTs), light-emitting diodes (OLEDs) and photovoltaic cells (OPVCs) [4]. The ensuing applications are manifold; they result in what some see as visually stimulating objects, such as flexible displays, panel lighting and transparent solar cells, and others see as a promotion to ubiquitous sensing in the form of low-end radio-frequency identification, smart food packaging, electronic nose and skin for robotics, and implantable or disposable health monitoring devices [5].

The field of flexible electronics is progressing quite dynamically and many other candidate materials, such as metal oxides, graphene and carbon nanotubes, are coming into sight [6, 7]. Also monolithic silicon chips can already be fabricated with a thickness of 20 μm and below, making them as flexible as foils [8–11]. Organic materials, on the contrary, display a notable advantage that they can be synthesized into seemingly limitless compound variations. Hence, their properties, including electrical, optical, appearance, chemical interactivity and biocompatibility, can be optimized depending on the target application. The question is not which technology will win this development race; instead, the aim should be to combine the best aspects of the different technologies in a hetero-integrated system approach [2]. The global market for printed, flexible and organic electronics reached $16.04 billion in 2013 and is projected to cross $75 billion in 2023, as forecasted by IDTechEx [12]. This is dominated by displays, while emerging integrated circuits and sensors are much smaller segments though with a huge growth potential.

T. Zaki, *Short-Channel Organic Thin-Film Transistors*, Springer Theses,
DOI 10.1007/978-3-319-18896-6_1

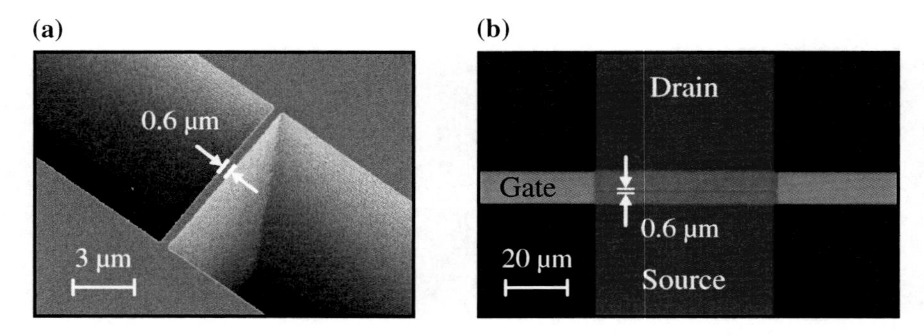

Fig. 1.1 **a** Photograph of a silicon stencil mask used to pattern the source and drain contacts of an OTFT. **b** *Top-view* photograph of an inverted-staggered (*bottom*-gate, *top*-contacts) OTFT with a channel length of 0.6 μm fabricated on a glass substrate. The organic semiconductor layer is transparent [18]

This work is devoted to the development of short-channel OTFTs, the performance and matching of which rivals that of the low-voltage, air-stable OTFTs reported to date. In cooperation with the Institute for Microelectronics Stuttgart (IMS CHIPS) and the Max Planck Institute for Solid-State Research (MPI-SSR), a new OTFT fabrication process based on high-resolution silicon stencil masks is established [13–17]. Figure 1.1 shows close-up micrographs of a stencil mask and an OTFT with a feature size of 0.6 μm, demonstrating that submicrometer channels with a very smooth edge roughness is feasible. This is, to the best of our knowledge, the shortest channel OTFT fabricated so far through stencil mask lithography. The process intends not only to provide a solution for fast individual OTFTs, but also to enable the realization of large-scale organic mixed-signal integrated circuits that can operate—for the first time—in the high kilohertz range.

In order to boost the combination of high-functionality and compactness, the durability of the stencil masks has to be characterized. Moreover, the downscaling of OTFTs tends to augment crucial non-ideal properties, which have to be implemented in a device model for use in circuit simulation. For example, the increasing impact of device parasitics in short-channel OTFTs now hampers further progress towards the attained intrinsic performance. In view of these aspects, this work targets the following four key objectives:

- **Technology**—Investigate the reliability of the silicon stencil masks with respect to minimum dimensions, refurbishment, manufacturability, defects and uniformity. This is to ensure proper yield and to further push miniaturization and integration.
- **Characterization**—Carry out a comprehensive experimental study of the static/dynamic response of the OTFTs using different methods. For the first time, the cutoff frequency of stand-alone OTFTs is extracted by means of scattering parameter measurements. The correspondence between the different characterization techniques is elucidated and a performance benchmark for the state-of-the-art OTFTs is presented.

- **Modeling**—Derive a unified static/dynamic compact model and a small-signal equivalent circuit that describe accurately the intrinsic as well as the extrinsic effects of the OTFTs. The model parameters are physically justified; thus, they can be extracted from measurements in a consistent and convenient way. Experimentally-validated recommendations are also given to accomplish a breakthrough in the performance of organic integrated circuits.
- **Demonstrator**—Design fast organic mixed-signal integrated circuits to demonstrate the reproducibility and matching capability of the new manufacturing process. This leads to the first successful realization of organic current-steering digital-to-analog converters (DACs).

This work, therefore, offers a promising technology platform that extends the favorable advancement of OTFTs and opens up the prospect of many novel applications in fields ranging from communication and robotics to healthcare.

This thesis is organized in nine chapters, including this motivation. The materials, devices and applications in organic electronics are broad. For this reason, Chap. 2 gives a historical perspective and a general overview of the field. Chapter 3 focuses on the architecture and operation principle of an OTFT. Making use of the gradual channel approximation, the important electrical characteristics and extracted device parameters are described. In addition, the outstanding air stability, uniformity and performance (using ring oscillators) of the utilized low-voltage, inverted-staggered OTFTs are illustrated. Chapter 4 presents the new OTFT fabrication process, which employs silicon stencil masks for parallel patterning of submicrometer features without the need of chemicals or elevated temperatures. Issues concerning mask pattern distortion and membrane deformation, and methods to mitigate these effects, are also discussed. Chapter 5 is devoted to the static characterization and modeling of the OTFT, primarily using the transmission line method. The model takes into account not only the intrinsic steady-state behavior of the OTFT but also the boundary conditions imposed by the device geometry, particularly the contact resistances which are not necessarily linear. The charge storage dynamics of the OTFT are analyzed by means of admittance characterization and compact modeling, as explained in Chap. 6. Furthermore, a small-signal equivalent circuit is built to describe the conduction mechanism and quantitatively evaluate the parasitic impedance. Chapter 7 provides a study of the frequency response of aggressively-scaled and asymmetric OTFTs using scattering parameter characterization. Chapter 8 introduces the design, simulation and test results of a 6-bit binary and a 3-bit unary current-steering DAC, comprising unipolar and complementary OTFTs, respectively. Finally, in Chap. 9, conclusions and an outlook to future work are presented.

References

1. J.N. Burghartz (ed.), *Guide to State-of-the-Art Electron Devices* (Wiley-IEEE Press, Chichester, 2013)
2. K. Bock, Polymer electronics systems–Polytronics. Proc. IEEE **93**(8), 1400–1406 (2005)

3. H. Klauk (ed.), *Organic Electronics: Materials, Manufacturing and Applications*, 2nd edn. (Wiley-VCH, Weinheim, 2008)
4. G. Meller, T. Grasser (eds.), *Organic Electronics*, ser. Advances in Polymer Science, vol. 223. Springer, Berlin, 2010)
5. A. Moliton, R.C. Hiorns, The origin and development of (plastic) organic electronics. Polym. Int. **61**(3), 337–341 (2012)
6. S.R. Thomas, P. Pattanasattayavong, T.D. Anthopoulos, Solution-processable metal oxide semiconductors for thin-film transistor applications. Chem. Soc. Rev. **42**(16), 6910–6923 (2013)
7. D.-M. Sun, C. Liu, W.-C. Ren, H.-M. Cheng, A review of carbon nanotube- and graphene-based flexible thin-film transistors. Small **9**(8), 1188–1205 (2013)
8. J.N. Burghartz (ed.), *Ultra-Thin Chip Technology and Applications* (Springer, Berlin, 2011)
9. J.N. Burghartz, You can't be too thin or too flexible. IEEE Spectr. **50**(3), 38–61 (2013)
10. J.N. Burghartz, W. Appel, H.D. Rempp, M. Zimmermann, A new fabrication and assembly process for ultrathin chips. IEEE Trans. Electron Devices **56**(2), 321–327 (2009)
11. K. De Munck, T. Chiarella, P. De Moor, B. Swinnen, C. Van Hoof, Influence of extreme thinning on 130-nm standard CMOS devices for 3-D integration. IEEE Electron Device Lett. **29**(4), 322–324 (2008)
12. R. Das, P. Harrop, Printed, organic & flexible electronics: forecasts, players & opportunities 2013–2023. Technical report, IDTechEx (2013)
13. F. Ante, F. Letzkus, J. Butschke, U. Zschieschang, J.N. Burghartz, K. Kern, H. Klauk, Top-contact organic transistors and complementary circuits fabricated using high-resolution silicon stencil masks, in *Device Research Conference, June 2010*, pp. 175–176
14. F. Ante, F. Letzkus, J. Butschke, U. Zschieschang, K. Kern, J.N. Burghartz, H. Klauk, Submicron low-voltage organic transistors and circuits enabled by high-resolution silicon stencil masks, in *IEEE International Solid-State Circuits Conference Technical Digest, Dec. 2010*, pp. 21.6.1–21.6.4
15. F. Letzkus, T. Zaki, F. Ante, J. Butschke, H. Richter, H. Klauk, J.N. Burghartz, Si stencil masks for organic thin film transistor fabrication, in *Proceedings of the SPIE Photomask Technology, Sept 2011*, pp. 81662B-1–81662B-12
16. F. Letzkus, T. Zaki, F. Ante, J. Butschke, H. Richter, H. Klauk, J.N. Burghartz, Si Stencil-Masken für die Herstellung organischer Dünnschichttransistoren, in *MikroSystemTechnik Kongress*, pp. 181–184 (2011) (in German)
17. U. Zschieschang, R. Hofmockel, R. Rödel, U. Kraft, M.J. Kang, K. Takimiya, T. Zaki, F. Letzkus, J. Butschke, H. Richter, J.N. Burghartz, H. Klauk, Megahertz operation of flexible low-voltage organic thin-film transistors. Org. Electron. **14**(6), 1516–1520 (2013)
18. T. Zaki, R. Rödel, F. Letzkus, H. Richter, U. Zschieschang, H. Klauk, J.N. Burghartz, S-parameter characterization of submicrometer low-voltage organic thin-film transistors. IEEE Electron Device Lett. **34**(4), 520–522 (2013)

Chapter 2
Introduction to Organic Electronics

Organic electronics is forming a new basis for low-cost microelectronic technology on thin, lightweight and mechanically flexible substrates. The purpose of this chapter is to provide a general overview of the topic. The chapter describes the historical development of microelectronics starting from the evolution of vacuum tubes until the rise of organic devices, namely transistors, light-emitting diodes and photovoltaic cells. Emphasis is also made on recent leading advances in terms of materials, devices and applications.

2.1 History

Integrated circuits are often said to be the most important invention of the twentieth century. They have become ubiquitous in modern technology, finding their way into nearly all industries available including telecommunications, automotive, consumer electronics and many others [1]. An integrated circuit, also referred to as microchip, is a tiny electronic circuit in which all the components, such as resistors, capacitors and transistors, are housed on a single chip. Transistors, however, are the main building blocks of an electronic circuit; they are solid-state active devices, which act as switches or amplifiers [2]. Early transistors were discrete and large in size (few square centimeters), but the rapid pace of progress of integrated circuits has enabled packing of more than billion transistors onto a single chip (also with few square centimeters of area).

Before transistors and microchips, the electronic circuits were based on bulky devices and had to be painstakingly assembled piece by piece. The main family of such bulky devices was vacuum tubes (also called thermionic valves), which were used for rectification, amplification, switching or similar processing of electrical signals (Fig. 2.1). It was German scientist Heinrich Geissler who is credited with building the first vacuum tubes (also called Geissler tubes; they were almost completely evacuated) in 1855. He also noticed their strange glow of colored light when applying an electric field across the vacuum between two electrodes in 1857 [3].

T. Zaki, *Short-Channel Organic Thin-Film Transistors*, Springer Theses,
DOI 10.1007/978-3-319-18896-6_2

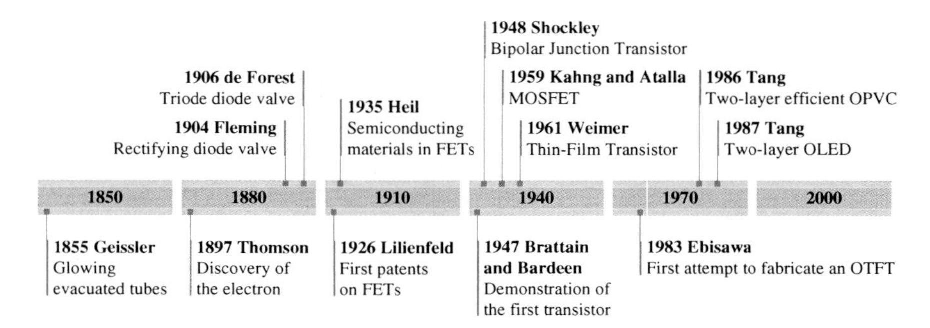

Fig. 2.1 Historical development of the electronic devices from the introduction of vacuum tubes to the beginning of using of organic compounds in transistors, light-emitting diodes and photovoltaic cells. The illustration is adapted from [5]

It was described later that the colored light is caused by rays, called cathode rays, which are carrying negative electric charges and hitting the air molecules inside the tube to produce light. It was only 40 years later, in 1897, that the nature of the cathode rays was completely understood by the British physicist Sir Joseph John Thomson; he used the vacuum tubes to calculate the mass of the negative electric charges, which were suggested before to be particles and observed later to have properties of both particles and waves, and finally discovered the electron [4].

A series of subsequent configurations of the vacuum tube valves were essential in building early computers and marked the beginning of the electronics industry. German scientist Karl Ferdinand Braun invented in 1897 the cathode ray tube (CRT), which was then used to realize screens for television sets, oscilloscopes and radars [6]; British electrical engineer and physicist Sir John Ambrose Fleming invented in 1904 the diode valve, which was used for the rectification of electrical signals [7]; American engineer Lee de Forest invented in 1906 the triode vacuum tube (also called Audion tube or triode), which was used for the switching or amplification of electrical signals [8]. The vacuum tube valves, however, had many limitations; they were bulky, fragile, rather slow, difficult to miniaturize, consumed too much energy and produced too much heat [9]. For instance, the first general-purpose computer that was built in 1946, the ENIAC (Electronic Numerical Integrator And Computer), comprised more than 17 thousand vacuum tubes to perform operations and calculations, weighed about 30 Tons and filled an entire room. In addition, there was in average one tube of the ENIAC damaged every two days [10].

The idea of replacing these thermionic valves with more promising and reliable solid-state devices that are based on semiconducting materials can be traced back to the late-1920s and early-1930s. Three patents are considered as the foundation of the principles of solid-state devices those from American (formerly Austro-Hungarian) physicist Julius Edgar Lilienfeld (Author of the two patents [11] and [12] filed in 1926 and 1928, respectively) and German electrical engineer Oskar Heil (Author of the patent [13] filed in 1935). Both described in their patents different variations of a method, aperture or device to control the flow of an electric current between two terminals of an electronically active material by means of a third potential applied to an insulated third terminal. Lilienfeld suggested in [12] that the active layer could be either pure metallic such as copper (Cu), compound such as cuprous oxide (Cu_2O), or preferably, a mixture of both. Although Cu_2O is a semiconductor, there was absolutely no definite indication in his claims of this class of solid materials or the necessity of using it for the active layer. Heil, on the other hand, is believed to be the first one who stated clearly in his patent that the active material should be made of thin layer of semiconductor such as tellurium (Te) or cuprous oxide (Cu_2O) [14, 15]. He also described that the semiconducting layer changes its resistance depending on the potential applied at the controlling metal terminal.

The presented inventions of Lilienfeld and Heil only embody concepts of solid-state devices as possible substitutes for the thermionic valves, with no indication of any reduction to practice. This class of solid-state devices was later named Field-Effect Transistors (FETs), where the name *transistor* is a shortened version of the original term *transfer resistor*, which conveys the operation principal of the device. Subsequent to these key conceptual inventions, many electronic device physicists and engineers pursued research on realizing semiconductor replacements for the unreliable vacuum tubes. More than a decade later, in 1947, the first successful semiconductor transistor was demonstrated [16–18]. This has marked the beginning of a series of milestones, at which different device configurations and materials were introduced. Four of the most important milestones are listed below:

- In 1947, American physicists Walter Brattain and John Bardeen demonstrated the first transistor action in a germanium point-contact device [18]. The transistor could amplify an input power up to 40 times [19], but had delicate mechanical configuration and was difficult to manufacture in high volume with sufficient reliability.
- In 1948, American physicist William Shockley invented the concept of Bipolar Junction Transistors (BJTs) [20, 21]. At that time, Shockley was actually leading a solid-state physics group at Bell Labs that included both Brattain and Bardeen. All the three scientists were awarded the 1956 Nobel Prize in Physics for their invention. For about three decades, the BJT was the main device of choice in the design of discrete and integrated circuits. BJTs are nowadays mostly made from silicon germanium (SiGe) and their use is often limited to very high-speed applications such as radio-frequency circuits for wireless systems [1].
- In 1959, South-Korean physicist Dawon Kahng and Egyptian engineer Martin M. Atalla demonstrated the first Metal-Oxide-Semiconductor Field-Effect Transistor (MOSFET), which had been long anticipated by Lilienfeld and Heil [22–24]. The device is mainly constructed by a stack of three layers: (i) silicon semiconductor as a base material, (ii) thermally-grown native oxide as an insulator, and (iii) metallic gate electrode as a controlling terminal. The MOSFET had largely superseded the BJT, owing to many favorable properties of silicon, especially due to the controllable and stable surface oxide. With the extraordinary progress and continued miniaturization of the silicon-based technology, MOSFETs became the dominant

device used in integrated circuits and electronics industry until today (with the exception of large-area applications).

- In 1961, American physicist Paul Weimer developed the first thin-film transistor (TFT), which is a FET-like device similar to the ones proposed by Lilienfeld and Heil [25–27]. The TFTs are fabricated on an insulating substrate such as glass. In a typical process, TFT would comprise polycrystalline or amorphous semiconducting materials (only in some few cases crystalline semiconductors are used as well). For example, Weimer used a semiconducting film of polycrystalline cadmium sulfide (CdSe) in his first demonstration [25]. Unlike MOSFETs, TFTs are supreme for large-area applications that do not require significantly high speeds such as active-matrix liquid crystal displays (AM-LCDs). The most common semiconducting material used today for TFTs is hydrogenated amorphous silicon (a-Si:H), owing to its low fabrication cost and easy processability [28].

Ever since, the evolution and development of the semiconductor technology have been following two main approaches, namely silicon monolithic circuits and thin-film circuits. The advancement of both technologies has been aiming of shrinking device geometries, increasing production throughput and yield, improving circuit reliability and fault tolerance, minimizing mismatch effects and reducing manufacturing costs. The rapid pace of innovation for both technologies has been pursuing certain trends [29]. For example, one of the major trends that enabled the proliferation of silicon monolithic circuits until they became omnipresent today in our daily life is commonly known as *Moore's Law* [30, 31]:

> *Transistor density on integrated circuit*
> *doubles about every two years.*
> —Gordon Moore (1965; updated 1975)

Increasing the density of transistors on chip implies more speed, complexity and functionality. Emerging applications, though, do require smarter integration by means of miniaturization (More Moore) as well as diversification (More than Moore). In recent years, mechanically flexible electronics as one of the approaches for diversification has caused a disruptive technology evolution and gained prominent market attractiveness for new user-friendly applications such as wearable devices, roll-screen displays and intelligent papers. This allows people as well as environment to interact with the complex information, which are typically processed by the high-performance miniaturized devices, in a more efficient and natural way.

A thin silicon chip is a possible solution, one that takes advantage of the crystalline silicon structure and leads to many high-speed flexible applications [32–35]. Nevertheless, an alternative solution is thin-film circuits that are based on a completely different class of materials such as organic semiconductors. Nowadays, organic materials are forming the basis of a new low-cost microelectronic technology that can be fabricated on large-area and flexible substrates such as polymer foils, papers or even fabrics. This is mainly owed to their low-temperature manufacturability, also to their (thermo) mechanical properties that makes them compatible with such kind of unconventional substrates [9]. One can envisage processing of this kind of materials by printing methods, which enables low-cost, high-volume and high-throughput production [9]. The trend, however, for this large-area and flexible technology is to reduce the *cost per unit area*, instead of increasing the number of *functions per unit area* that is being followed by the crystalline silicon technology [29]. By combining both technologies in a so called hybrid system-in-foil (SiF), one could actually take advantage of both worlds [33].

In fact, the first studies on the electrical activity of organic materials can be traced back to the early twentieth century [36]. Anthracene was the first organic compound in which photoconductivity was observed by Pochettino in 1906 [37] and Volmer in 1913 [38]. Later in the 1950s and 1960s, the potential use of organic materials as photoreceptors in imaging systems was recognized [36, 39]. During the same time, electroluminescence in organic compounds was observed by Bernanose et al. in 1955 by applying an alternating current (AC) in air to compounds such as brilliant acridine orange E [40] and by Pope et al. in 1963 by applying a direct current (DC) in vacuum to anthracene [41]. In spite of these handful preliminary reports and principal demonstrations, the technological use of organic semiconductors was still very limited due to several drawbacks. First, the reproducibility and carrier mobility[1] in organic semiconductors were very low. Second, the demonstrated devices were operating at extremely high voltages (e.g. 400 V) as a consequence of the crystal thickness (in the micrometer to millimeter range) and the difficulties to prepare stable, injection-efficient contacts to the compounds [42]. Third, the poor control of material purity and structure ordering were also obstacles. Finally, the materials used so far did achieve neither sufficient efficiency nor satisfying stability [42]. However, the research and interest in this field were flourished in 1977 by the successful synthesis of electrically conducting organic polymers through controlled halogen doping [43]. This discovery by Alan G. MacDiarmid, Alan J. Heeger, Hideki Shirakawa and co-workers was considered a major breakthrough, opened many new and exciting applications and honored with the 2000 Nobel Prize in Chemistry.

The first available organic materials were intractable, immobile, or even insoluble [44]. Nevertheless, the rapid advancement of the materials and processing has enabled the development of soluble organic compounds. Solubility is a key prominent feature, one that opened the possibility for cheap and high-volume production of printed electronics. Henceforth, the utilization of organic materials by various electronic components has given them, incontrovertibly, a place in the development of this theme [44]. Three components that can be considered as the foundation of organic electronics are organic photovoltaic cells (OPVCs), organic light-emitting diodes (OLEDs) and organic thin-film transistors (OTFTs):

First, the use of conjugated polymers, such as poly(sulphur nitride) and polyacetylene, for the realization of OPVCs were firstly investigated in the 1980s; their power conversion efficiencies, however, were well below 0.1 % [36]. A major breakthrough came in 1986 when American physical chemist Ching W. Tang discovered that a two-layer OPVC by bringing a donor and an acceptor in one cell could dramatically

---

[1] The carrier mobility is a measure of how fast an electric charge is transmitted through a material under an applied electric field and is mostly represented in units of cm^2/Vs.

improve the efficiency to 1 % [45]. Subsequent developments of OPVCs achieved in early-2013 efficiency as high as 12 % according to recent announcements from the German company Heliatek [46].

Second, OLEDs in the form available today were firstly presented in 1987 by Ching W. Tang and Steven Van Slyke using a double layer structure of organic thin films (8-hydroxylquinoline aluminum Alq_3 and aromatic diamine) [47]. Later, in 1990, the research on polymer electroluminescence culminated in the first successful demonstration of green-yellow polymer-based OLED using 100 nm thick film of poly(p-phenylene vinylene) as an active layer [48]. The improved efficiencies combined with increased shelf and operating lifetimes, also superior material properties and manufacturing techniques, have pushed OLEDs already to the market place in applications like OLED-based lighting and displays. For example, the South-Korean company Samsung has just recently launched in August 2013 the first 55 in. full high-definition[2] (Full HD) OLED television with a curved panel (S9C series) [49].

Last, the debut of the field effect in organic semiconductors date back to 1970 [50–53], yet the potential use of the OTFT (at that time mostly referred to as metal-insulator-semiconductor field-effect transistor MISFET) as an electronic device was only identified in 1983 when Ebisawa et al. reported the first attempt to fabricate an OTFT that utilizes polyacetylene as an active semiconducting layer [54]. From this point forward, several studies were devoted to realize successful TFTs based on organic semiconductors such as polyacetylene [54, 55], polythiophenes [56, 57] and metallophthalocyanines [58, 59]. However, their carrier mobilities were very low in the range of 10^{-4} to 10^{-5} cm^2/Vs. It was not until nearly 7 years later, in 1990, that the carrier mobility in organic semiconductors approached and even reached that in amorphous silicon when Garnier et al. reported a carrier mobility as high as 4.3×10^{-1} cm^2/Vs for TFTs that used evaporated hexathiophene as an active material [60]. For comparison, the carrier mobilities in conventional a-Si:H TFTs are in the range of 10^{-1} to 1 cm^2/Vs. The performance and stability of OTFTs have continuously improved since then. Hence, some OTFTs now compete with a-Si:H TFTs to enable revolutionary design possibilities in new large-area and mechanically-flexible applications. The most compelling application of OTFTs is backplanes for flexible active-matrix displays; accordingly, the German company Plastic Logic is currently manufacturing ultra-thin and lightweight plastic displays that are able to bend, twist and even roll-up like a piece of paper [61].

The field of organic electronics has been well-profiled and recognized by several international awards bestowed upon great scholars and scientists working on this subject. As mentioned above, the 2000 Novel Prize in Chemistry was awarded jointly to the Americans Alan G. MacDiarmid and Alan J. Heeger, and the Japanese Hideki Shirakawa for their discovery and development of conductive polymers [43].

[2]The full high-definition (Full HD) is implying a resolution of 1920 × 1080 (2.1 megapixel) in a 16:9 aspect ratio.

Furthermore, the 2010 Millennium Technology Prize[3] was awarded to the British Sir Richard Friend (as one of the three laureates in that year) whose team discovered electroluminescent diodes based on polymers (PLEDs) and greatly participated in the development of OTFTs and OPVCs [48, 55]. In addition, the 2011 Deutscher Zukunftpreis[4] (German Future Prize) was awarded to Karl Leo, Jan Blochwitz-Nimoth and Martin Pfeiffer for their major contribution in the advancement of organic functional materials and manufacturing techniques, especially for applications in lighting and photovoltaics [64–67]. Finally, it is worth mentioning that despite all these advances, the field of organic electronics is still in its infancy and there is still much room for improvement and much to be learned and investigated.

2.2 Materials

Organic electronics have been promising on account of their low-cost, low-temperature and fast manufacturability in addition to their compatibility with various kinds of substrates that are thin, large in area, transparent or mechanically flexible. In principle, organic electronics rely on electrically active materials that are based on conjugated organic compounds whose molecules contain carbon and hydrogen elements. A basic device, such as an organic transistor, is generally comprised of a stack of conducting, semiconducting and insulating thin-film layers. There are many attempts to realize all these layers solely from organic materials [68, 69]; however, they are mostly combined with special inorganic thin-films in order to optimize the device performance [70]. The structures and applications of the different organic-based devices are given in Sect. 2.3, where the different fabrication processes used to deposit and pattern the thin-film layers are discussed in Sect. 4.1. In this section, an overview of the different organic as well as inorganic materials is presented.

The materials are classified in this section as the following: (i) semiconductors, (ii) conductors, (iii) dielectrics, (iv) passivation, and (v) substrates. Each material in every class has its advantages and limitations, where often the process conditions as well as the interplay of the material with other layers have a large influence on the device performance [71]. Therefore, the selection of the materials has to be carefully done to meet application and technology parameters such as thermal, mechanical and optical properties. For example, transparency is very important for the realization of solar panels that are going to be mounted on building facades, but not for displays that are designed for e-book readers. A summary of the key application and technology parameters is listed below [71]:

---

[3]The Millennium Technology Prize is the world's largest technology award. It is awarded every 2 years by Technology Academy Finland, an independent foundation established by Finnish industry and the Finnish state in partnership [62].

[4]The Deutscher Zukunftpreis (German Future Prize) is one of the most prestigious awards conferred for science and innovation within Germany. It is awarded annually by the Federal President of Germany [63].

- Electrical Performance—The performance (operation frequency, current driving capability) of the devices depends on the carrier mobility in the semiconductor, conductivity of the conductor and the dielectrical behavior of the dielectric material.
- Resolution and Registration—The reliability and performance of the devices depend on the lateral distance of the electrodes (pitch, or resolution) within the devices and the overlay accuracy (Registration) between different patterned layers. In addition, scalability is necessary to have a sustainable technology development.
- Environmental Stability—For the sustainability and proper lifetime of the devices, the sensitivity of the materials to oxygen and moisture has to be well considered. This depends also on the barrier properties of the protective sealing layers (substrate and encapsulation).
- Mechanics and Optics—The mechanical and optical properties include thin form factors, flexibility, bending radius, conformability, weight, transparency, color and appearance. Accordingly, the material, design and process have to be carefully chosen.
- Process Parameters—The process parameters include throughput, temperature and ambient conditions. For a reliable production, it is important to adjust the process parameters for the different employed materials.
- Cost and Yield—High volume production is only possible when the processes allow fabrication at an acceptable yield. This includes adjusted materials, circuit designs as well as in-line quality control for low-cost and low-requirement (e.g. disposable sensors) to high-cost and high-performance (e.g. flexible OLED display) products.

Semiconductors

Organic semiconductors are traditionally classified as small molecules[5] or polymers[6] [9]. Polymers often have excellent solubility, which makes them amenable to mass printing processes such as flexographic and gravure printing [70]. On the other hand, small molecules are usually deposited by vacuum sublimation; nevertheless, recent advancements enabled some of the semiconducting small-molecules to be processed in solution or dispersion [71], which makes them no longer restricted to evaporation/sublimation processes.

The carrier mobility is commonly used as a figure of merit to characterize the performance of materials, devices or fabrication methods. It is found that the carrier mobility in organic semiconductors varies greatly depending on the choice of material, its chemical purity and its microstructure, also on the process conditions and the interface to other layers in the device [9, 71]. For example, Fig. 2.2a illustrates the difference in the carrier mobility depending on the substrate temperature during deposition of the different semiconducting small-molecules. Furthermore, amorphous films

[5] An organic small molecule is a compound containing carbon atoms that are bonded into stable individual molecular unit.

[6] An organic polymer is a compound with a molecular structure formed from many identical organic small molecules bonded together.

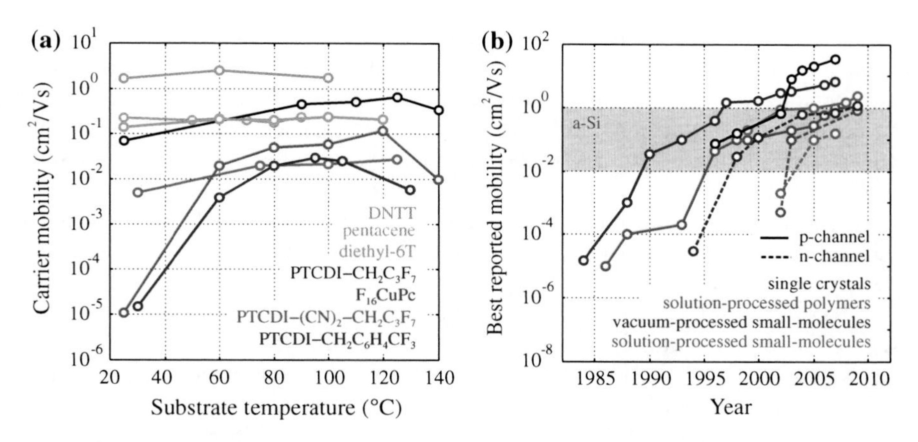

Fig. 2.2 **a** Relationship between the carrier mobility in the transistor channel and the substrate temperature during the deposition of the organic semiconductor layer for five different small molecules: DNTT [74], pentacene [75], diethyl-sexithiophene [76], PTCDI–$CH_2C_3F_7$ [77], $F_{16}CuPc$ [78], PTCDI–$(CN)_2$–$CH_2C_3F_7$ [79] and PTCDI–$CH_2C_6H_4CF_3$ [80]. **b** Development of the carrier mobility in organic transistors based on small molecules, polymers and single crystals. The data is categorized according to the deposition process, material class and type of injected carriers. The seven categories are the following (with reference to the publications from which the highest carrier mobilities are extracted): single crystals (p-channel [81]), vacuum-processed small molecules (p-channel [73] and n-channel [77]), solution-processed small molecules (p-channel [82] and n-channel [83, 84]) and solution-processed polymers (p-channel [72] and n-channel [85]). Adopted from [70] with permission from The Royal Society of Chemistry

of solution-processed semiconducting polymers usually have mobilities in the range of 10^{-6} to 10^{-3} cm^2/Vs [70]. However, the mobilities of certain semiconducting polymers can be increased to about 1 cm^2/Vs through molecular engineering and also by inducing semicrystalline order through better control of the film formation [72]. Small-molecule organic semiconductors, on the other hand, are often forming polycrystalline films when deposited by vacuum sublimation, which results in carrier mobilities as large as about 6 cm^2/Vs [73].

Figure 2.2b depicts the development of the best reported field-effect mobility of p- and n-channel OTFTs based on small-molecule and polymeric semiconductors since 1984. The carrier mobility in organic semiconductors, though still underperform that of the crystalline silicon, has improved dramatically until it approached and even surpassed that of the amorphous silicon (a-Si). Unlike inorganic semiconductors, organic semiconductors are not atomic solids but they are π-conjugated materials for which the charge transport mechanism is based on hopping between the individual conjugated molecules [44]. In this case, the mobility is mainly limited by trapping of charges in localized states [70, 86]. As for inorganic media, in a different manner, defaults such as traps, along with molecular and macromolecular structural irregularities, have a crucial impact on the charge transport [44].

It is very important to note that organic semiconductors are usually undoped (intrinsic semiconductors) and the notions of p- and n-channel OTFTs do not imply the same meaning as for inorganic semiconductors. An n-channel OTFT is one in which electrons are more easily injected than holes. This has something to do with the matching of energy levels of the metal contacts and the semiconductor employed by the transistor as further clarified in the following chapter. The n-channel OTFTs are actually of special concern as they suffer from more than tenfold lower carrier mobility than their p-channel contenders (if in the same organic technology). In addition, they are highly sensitive to ambient conditions, especially to oxygen and moisture [87]. As a result, most organic-based circuits today make use of p-channel designs only [88, 89].

Several research efforts are currently devoted to realize stable n-channel OTFTs with relatively high carrier mobility as this enables the use of complementary circuit topologies, which offer higher-robustness, lower power consumption and larger noise-margin compared to unipolar circuits [90]. There are also other attempts to integrate the p-channel OTFTs with n-channel TFTs that are based on metal-oxide semiconductors (e.g. amorphous indium-gallium-zinc-oxide a-IGZO), as they can achieve electron mobilities larger than 10 cm^2/Vs [91–94]. In general, the ongoing development of organic semiconductors is not limited only to the performance measures, but also extended to other essential issues such as lifetime in real-world environmental conditions, matching over large areas, reproducibility, production yield and operation voltage.

Conductors

The need for conductive traces in all electronic products is indispensable. As each conducting material has its own properties, the choice of the material strongly depends on the application. Conductive inks, which are typically consisting of micron-seized conducting flake particles, organic resins, solvents and rheology modifiers, are offering promising properties [71]. These compositions are compatible with a wide variety of substrates and are suitable for low-cost and high-speed manufacturing techniques such as screen and flexographic printing. For applications that demand highly conductive features, silver inks that can have electrical conductivity as large as 10^4 S/cm is a preferable choice [71]. Examples of mechanically flexible applications that utilize printed silver inks are membrane touch switches, keyboards and on-chip antennas. Another favorable choice for less demanding applications is conductive carbon inks. In addition, some special compositions of carbon inks can also be used as resistors or positive temperature coefficient (PTC) heaters [71].

Moreover, some devices like OLEDs and OPVCs require not only mechanical flexibility and good conductivity, but also translucency or high transparency for their metal electrodes. In this case, inorganic conductors like indium tin oxide (ITO) or polymeric conductors like poly(3, 4-ethylenedioxythiophene):poly(styrenesulfonate) (PEDOT:PSS) are possible solutions [70, 71]. Meanwhile, these materials are used already for applications such as touch screens and electrochromic displays. An alternative for ITO and PEDOT:PSS is a mesh of very thin (20–80 nm) and narrow (15 μm) metal layers (e.g. silver or copper). Such a pattern with a spacing of about 200 μm can achieve a transparency of about 65 % over the entire wavelength range from 400 to 900 nm. For comparison, the transparency of an ITO film is ranging from about 60 to 85 % for wavelengths from 400 to 900 nm, respectively [95].

For the OTFTs, material properties of the conducting layers, especially for the source and drain contacts, are very critical as they affect significantly the devices performance. The choice of the material in this case depends on the architecture employed by the OTFT, i.e., the order of which the device layers are deposited. The typical used materials are aluminum (Al) or chromium (Cr) for the gate electrode, and gold (Au) for the source and drain contacts [70]. For the design of an all-polymer OTFT, conductive polymers such as polyaniline (PANI) or PEDOT:PSS are also suitable for the gate, source and drain electrodes.

Dielectrics

Dielectrics are used in both active and passive devices such as OTFTs and capacitors, respectively. The majority of OTFTs to date have used inorganic dielectrics, mostly silicon oxide. In fact, the performance of the device depends strongly on the quality, physical properties and chemical nature of the insulator-semiconductor interface [9]. For instance, trapping states at the mentioned interface immobilize the carrier charges in the channel and correspondingly limit the performance [70, 87]. Significant improvements can be achieved as demonstrated in literature just by inserting few nanometers of organic single layer between the insulator and the semiconductor [96]. In order to take advantage of the complementary design features while not increasing the production cost, the challenge is to ensure that the dielectric material functions well with both p- and n-channel OTFTs.

Passivation

Passivation materials (encapsulation) are used to protect the devices against environmental influences such as scratches and degradation due to the presence of the water, oxygen or light [71]. In some applications, the use of encapsulation is highly necessary to ensure an adequate lifetime for the devices. As an example, OTFTs that are developed for medical applications can be encapsulated with poly(chloro-para-xylylene) (parylene) and gold layers to protect them against water [97].

Substrates

Finally, the substrate is the base material onto which the devices are manufactured. Key material parameters for choosing the substrate material are: optical transmittance, dimensional stability, surface smoothness, durability, barrier capability, temperature tolerance and mechanical properties (bending radius, deformation and hysteresis behavior) [71]. Nowadays, the majority of applications are using glass (also thin and flexible glass) or stainless steel substrates, also polymer substrates such as poly(ethylene terephthalate) (PET) or poly(ethylene 2,6-naphthalate) (PEN). In addition, paper (cellulose) or textile substrates are sometimes used.

2.3 Devices and Applications

Given the advances in chemicals and materials by international leading firms like DuPont in the USA and Merck in Germany, organic electronics promise real growth opportunities for developers in new innovative products, some of which are already translated into commercial reality. The flexible and large-area from factors as well as the potential low production costs of the organic technology are key benefits over their bulk, or rigid, silicon and other inorganic counterparts. The technology is versatile enough to be used in a wide range of applications as discussed herein. The organic materials can be combined to a number of active electronic components such as transistors, light-emitting diodes, photovoltaic cells, various types of sensors, memories or batteries, also passive devices such as conductive traces, antennas, resistors, capacitors or inductors [71].

2.3.1 Organic Light-Emitting Diodes

The most established and largest sector within the organic electronics industry, even by some margin, is OLEDs. Besides the display market, lighting applications have emerged recently as more than a niche market for OLEDs [98]. The basic device structure of an OLED is shown in Fig. 2.3a. The structure comprises two organic semiconducting layers, which are sandwiched by anode and cathode electrodes laying on a transparent substrate (e.g. glass). Depending on the transparency of the anode and cathode electrodes, the light is transmitted either from top, bottom or both directions of the device. This basic device structure is called heterostructure OLED, or sometimes referred to as bilayer structure, which resembles a pn-junction and is similar to the one used in the very first demonstration of an efficient electroluminescence (EL) OLED based on organic thin-films that was built in 1987 by Ching W. Tang and Steven Van Slyke of Eastman Kodak Co. [47]. In this structure, the two organic semiconductors function as a hole-transporting and light-emitting layers. To gain more insight about the exact role of the organic semiconducting layers, the operation is explained in the following.

The LEDs, regardless whether organic or inorganic, are principally operated by applying an external voltage across the pn-junction to accelerate charge carriers of opposite polarities, namely electrons and holes, from the cathode and anode contacts, respectively [99]. The carriers are driven towards the so called recombination region, which is located at the space charge region of the pn-junction and there the carriers form a neutral bound state, or exciton. It is called a recombination region because this is where the electrons recombine with the holes by falling into a lower energy level and realising energy in the form of a photon. The wavelength (color) of the emitted light depends on the bandgap energy of the materials forming the pn-junction. The recombination region, where the luminescent molecular excited states are generated, is typically very small in the single heterostructure LED and it is located at

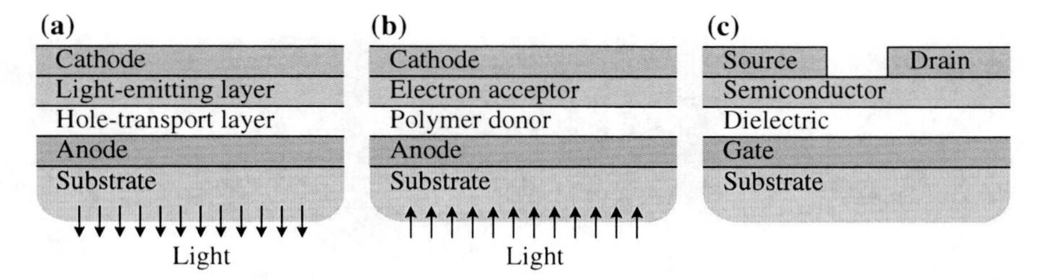

Fig. 2.3 Schematic cross section of basic organic-based devices. **a** Organic Light-Emitting Diode (OLED). **b** Organic Photovoltaic Cell (OPVC). **c** Organic Thin-Film Transistor (OTFT)

the boundary between the two semiconductors. Therefore, to increase the probability of electron-hole recombination and improve the internal quantum efficiency[7] of the device, an additional third semiconducting layer is exploited in a double heterostructure (O)LED. In this case, the three (organic) semiconductors function as electron-transporting, light-emitting and hole-transporting layers.

The organic semiconductors can be made of small-molecules or polymers. Depending on the materials used, the devices differ mainly in three criteria, namely fabrication technique and process controllability, operating voltage and efficiency [99]. Small-molecule thin organic layers are mostly deposited by vacuum evaporation or sublimation, while polymer layers are usually processed in the liquid-state by spinning and solidification by heating. Control of the thickness of the organic thin-films in a spin-on technique is relatively harder than in vapor deposition. However, polymer-based OLEDs can usually operate at lower power than that of small-molecule-based OLEDs. This is owed to the high conductivity of organic polymers. The operation supply voltage of polymer-based OLEDs is in the range of 2–5 V, which is about 1–2 V less than that of small-molecule-based OLEDs. Furthermore, the efficiency of polymer-based OLEDs is typically higher than that of the small-molecule-based OLEDs. Nevertheless, focusing now on one single process or method would not be favorable, as the technologies are not mature enough to determine the optimal method. Display as well as lighting industries are currently focusing on different technical approaches for both solution- and vacuum-processable organic materials to develop cost-effective OLEDs.

For many years, the LCDs has been the norm for the display industries [98]. However, the growing number of laptops, mobile phones, televisions and many other applications increase the demand for higher quality products. In contrast to LCDs, the superior virtues of OLED displays are the thinnest-ever form factor, deeper black levels when individual pixels switch off, and high contrast ratio. The main problem that was setting back the OLEDs for many years was the lifetime, yet it has continued to improve every year reaching now a sufficient level to compete with LCDs.

[7] The internal quantum efficiency is the ratio of the number of emitted photons to the number of injected carriers.

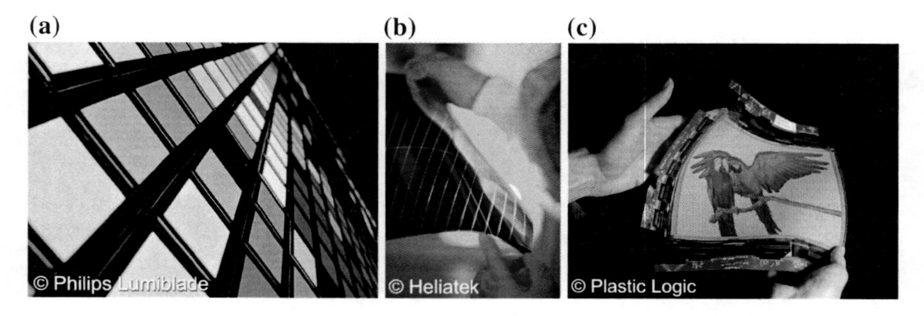

Fig. 2.4 Organic-based applications. **a** Philips Lumiblade OLED lighting. **b** Heliatek OPV film. **c** Plastic Logic e-paper with OTFT backplane. Photography courtesy of Philips Lumiblade, Heliatek and Plastic Logic, respectively

Meanwhile, Asian companies such as Samsung and LG dominate the manufacturing of both LCD and OLED flat panel displays.

Another core competence of OLEDs is the lighting industry. In general, solid-state lightings (SSLs), including EL, LED and OLED lighting, are soon replacing the conventional lighting techniques such as incandescent combustion (candles and incandescent lamps) and gas discharge (fluorescent and induction lamps). SSLs have been promising on account their superior energy efficiency, absence of hazardous metals, flexible form factors, durability and their surface emission for design features. Exclusively, OLED lighting offer prospects for a vast number of unique features; this includes mechanical flexibility, large-area illumination, thinness of light-source, high efficacy and variable colours (including translucent colors) [71]. European companies like Philips and Osram are currently at the forefront in the development and production of OLED lighting. Figure 2.4a shows an example of an OLED lighting demonstration by Philips.

2.3.2 *Organic Photovoltaic Cells*

The other key component in organic electronics is photovoltaic cells. Today, the majority of commercialized solar cell modules are made of inorganic materials such as silicon [98]. However, the interest in organic photovoltaics is growing owing to the inherent capabilities offered by the materials; this includes the compatibility with low-cost reel-to-reel manufacturing, possibility to be fabricated on mechanically flexible substrates, less energy resources needed for production and reduced installation costs. The efficiency and durability of printed or vacuum-deposited OPVCs on flexible plastic or metal foils have been improving in recent years [98]. Rigid silicon will still dominate the fixed grid, large-area applications for many years to come; this is mainly because of the established and commercially proven production of silicon. However, more cost-effective photovoltaic technologies open the possibility to integrate renewable solar power generation in everyday structures and items. In addition, the mechanical flexibility enables to use OPVCs in clothing, bags and awnings to power wearable and portable electronic devices such as music players, mobile phones and tablets [98]. Moreover, the ability to fabricate OPVCs with customized transmission factors (transparency) offer prospects to integrate solar films as energy harvesting components in windows or building facades.

In principle, the progress and development of photovoltaic technology is divided into three generations [98]. The first generation is confined to the bulk, or rigid, silicon and other inorganic PVCs. This kind of PVCs typically comprise thin wafers of single-crystal or polycrystalline silicon, patterned with metallic electrodes and sandwiched between glass plates; these cells have today a maximum efficiency greater than 20 %, where the theoretical maximum efficiency is about 30 % [100]. The second generation is thin-film PVCs, which are made by depositing thin layers of silicon (Si), cadmium telluride (CdTe) or other materials on glass substrates using vacuum-coating techniques; this kind of PVCs are potentially cheaper than bulk silicon PVCs and have cell efficiencies in the range of 12 % to 20 %. Finally, the third generation is the OPVCs, which are mainly targeted to be incorporated into consumer electronic devices and they are expected to be even cheaper than thin-film PVCs. The record efficiency of OPVCs, however, is 12 % as recently reported in January 2013 by Heliatek [46]. There are two types of PVCs that belong to the class of organic photovoltaics, namely dye-sensitized solar cells (DSSCs) and organic photovoltaic cells (OPVCs). In fact, the DSSCs mainly do not comprise organic materials with the exception of a thin film of organic molecular dye. Nevertheless, the DSSCs production and applications are similar to that of OPVCs. There is a mirror analogy between DSSCs, OPVCs and double-, single-heterostructure OLEDs, respectively.

First, the DSSC is composed of the following five layers (from top to bottom) [101]: (i) transparent top conductor with low series resistance, through which the light can be penetrated; (ii) porous semiconducting layer of titanium dioxide nanoparticles, which is used solely for charge transport; (iii) molecular dye that absorbs the light rays such as chlorophyll or porphyrin organic materials; (iv) electrolyte solution based on iodide/triiodide redox system, from which the electrons are recovered by the dye before decomposition; and (v) base conducting substrate that serves also as a counter electrode. The operation principle in this case is called artificial photosynthesis; the light enters the cell through the transparent top conductor, striking the dye at the surface of the semiconductor, where the charge separation takes place. Only photons with enough energy, larger than the bandgap of the dye material, are absorbed and are able to make the electrons in their excited state. Excited electrons are then injected directly into the conduction band of the semiconducting layer, through which they diffuse to the top-electrode (anode) to power the load. Before the dye molecules are decomposed, the dye recovers the lost electrons from the iodide in the electrolyte, oxidising it into triiodide. This reaction takes place before the injected electrons recombine to prevent short circuiting. Finally, the missing electrons needed to recover the trioxide then diffuses from the counter electrode (cathode), from which they are provided by the load.

Second, the structure and operation of OPVCs are similar to inorganic solar cells, where the incoming light creates electron-hole exciton, at the interface between the donor and acceptor. If a polymer donor is used, the electron-hole pair will have a binding energy of 0.1–1.4 eV, which is in fact larger than the few milli electron-volts provided in the case of inorganic materials [102]. By applying an electric field, ideally all excitons are separated into electrons and holes that migrate to the corresponding electrodes, i.e., anode and cathode contacts. Referring to Fig. 2.3b, the simplest configuration that supports this kind of operation principal is the planar heterojunction (PHJ) solar cells; in this case, the donor and acceptor are sandwiched between a transparent anode (such as ITO) and a reflecting cathode (such as Al). However, the efficiency is limited here by the exciton diffusion length,[8] which is about 3–10 nm in most organic semiconductors. Therefore, for a normal organic thin-film thickness of about 100 nm, most of the excited electrons recombine before reaching the electrode and the active region is only limited to a very thin region around the donor/acceptor interface, which is not enough to capture a reasonable amount of incoming light. For this reason, nano-structuring in bulk and ordered heterojunction (BHJ and OHJ) solar cells is investigated by researchers to alleviate this limitation. A photograph of an OPV film produced at Heliatek is shown in Fig. 2.4b.

The most promising organic-based PVC design, however, appear to be the tandem cells, which stack several DSSCs or OPVCs [98]. The advantage of this stacked configuration is to increase the power efficiency by means of the following [102]: (i) reduction of the thermalization losses of high-energy photos by stacking different absorbers, (ii) absorption of the entire spectrum by optimising each subcell to one part of the solar light spectrum, and (iii) compensation of the aforementioned problem of PHJ, i.e., the thin active region, by stacking a large number of subcells.

2.3.3 *Organic Thin-Film Transistors*

Organic thin-film transistors (OTFTs) offer prospects for a vast number of unique circuit applications in mechanically flexible, inexpensive, large-area and biomedical electronics. In principal, OTFTs are metal-insulator-semiconductor (MIS) field-effect transistors (FETs) in which the semiconductor is a conjugated organic material [70]. Figure 2.3c shows a basic structure of an OTFT; it is a three terminal device that consists of four thin-film layers deposited on an insulating substrate, namely gate electrode, gate dielectric, organic semiconductor and source/drain contacts. When a voltage is applied between the gate and the semiconductor, a thin sheet of mobile charges is created in the semiconductor in close vicinity of the semiconductor/dielectric interface [70], i.e., a channel is created between the source and drain contacts. This charge layer balances the charges of opposite polarity located on the gate electrode. By tuning the applied gate voltage, the charge density in the

[8]The diffusion length is the distance over which the excitons travel from the donor to the acceptor before undergoing recombination.

semiconductor channel is varied. Thus, the electric conductance of this channel is accordingly modulated.

There are four different possible (O)TFT planar configurations that are categorized according to the location of the gate and the source/drain contacts relative to the semiconductor layer, namely coplanar, inverted coplanar, staggered, inverted staggered structure [9, 14]. The transistor shown in Fig. 2.3c employs the inverted staggered (bottom-gate, top-contact) device structure. More details about the different device configuration are given in the following chapter. There are many different fields and applications in which the advantages of OTFTs can be exploited, but at this point, it is too early to designate one as a killer application. It is worthwhile to mention that there is a lot of movement and activity in this field through established and new start-up companies to develop novel and previously unexpected applications. Some of the major application areas of OTFTs are reviewed in the following.

One of the most important applications of OTFTs is radio-frequency identification (RFID) tags. They are frequency-coupled portable devices that are used to identify, track, sort and detect items or persons; they can be used in card authentication, ticketing, gaming and surveillance. Together with integrated sensors (e.g. temperature sensor [103]), low-cost RFID labels can be realized, which are able to monitor supply chains, ensuring for example that temperature-sensitive medicines, vaccines and food products are kept cool while being transported along extensive distribution network [98]. In an RFID, the communication takes place between a reader and a transponder (tag). The tag comprises a small chip and an antenna, where the antenna is solely used to couple RF energy from the reader into the tag; there are different coupling techniques, including backscatter, inductive and capacitive coupling. Organic RFIDs are expected to be deployed at the item-level as a smart replacement to barcodes. Using these tags for example in a supermarket would save much time by allowing customers to walk a cart that is full of goods past a scanner and the scanner would read all the items in the cart, create a list of the merchandise and finally calculates the total amount of charge in an instant; they could also be used to collect all the information about the products in the store and generate an inventory within a few seconds.

There are basically two types of RFID tags, namely passive and active tags. The passive tags do not employ an energy supply and they are merely activated by the reader, while the active tags comprise a battery (to enable larger reading distance) or sensors (to monitor data such as temperature). Although, the cost of silicon-based RFIDs has dropped to just few cents, the cost of assembly and packaging with the antenna and products is still high. As a result, the use of silicon-based RFIDs is limited to small-scale volumes. However, the potential ultra-low-cost production of printable OTFT-based RFIDs together with the antenna on thin plastic or paper substrates opens the possibility for individual item-level development. Companies like PolyIC in Germany and OrganicID in the USA are working on printed RFIDs based on OTFTs, while others, such as Kovio in the USA, are developing printed RFIDs based on silicon inks. At this time, the performance and capabilities of printed organic-based RFID tags are lower than that of the silicon-based ones in many aspects like memory size, data rate and reading distances; however, the research is focusing now on the development of these aspects in addition to minimising production costs.

Besides being used in RFID tags, printed memories are an attractive application for OTFTs, which gained also market attention. They can be used in various other devices that require information storage (e.g. games, sound, video and smart cards). Since most printed applications nowadays do not employ an integrated battery, non-volatile memories are generally used. Depending on the application, either ROM (Read Only Memory), WORM (Write Once Read Many memory) or NV-RAM (Non-Volatile Random Access Memory) are used. The company Thin Film Electronics ASA in Norway is an international front runner in the development and commercialization of printed memory devices based on polymeric materials [104].

Another key application of OTFTs is backplanes for mechanically flexible displays. Many flat panel displays are consisting of a matrix of pixels, formed at the intersection of rows and columns, where each pixel is an LED that is capable to emit light by being switched on or off [105]. Colored displays are realized by positioning matrices of red, green and blue pixels very close together. There are two ways to control the pixels and thus forming the desired image, i.e., either passive or active matrix driver methods are used. In a passive matrix, each row and each column of the display has its own driver. In this case, each pixel should maintain its state until it can be refreshed again. Passive drivers tend to be used only for simple and cheap modules because the realization of such kind of drivers gets more difficult when the current required to brighten a pixel increases and also when the area of the display increases. Consequently, active-matrix displays are used in such cases, where each pixel is efficiently addressed by incorporating a transistor (typically TFTs are comprised) in series that provide control over the current and hence the brightness of individual pixels. Therefore, smaller currents can flow in this case through the control wires and the wires can be finer as a result. This means that higher resolution is realizable. Furthermore, the transistor is also capable of holding the current setting, keeping the pixel at the required brightness until it receives another control signal.

This class of display drivers are currently being built by inorganic transistors (mostly using a-Si:H TFTs) on glass substrate, which forbid the capability of being mechanically flexible. Semiconductor materials such as conjugated organic compounds and metal oxides are being explored to enable the flexibility. Sony has demonstrated, in 2008, a bendable OLED display that is driven by OTFTs [106]; they have also proposed conceptual prototypes such as a laptop with flexible OLED screen and a wrist wearable music player at the 2009 CEATEC exhibition in Japan. Moreover, sony has announced in 2010 the first successful demonstration of an OLED panel that is capable of reproducing moving images while being rolled-up around a cylinder with a radius of only 4 mm [107]. These demonstrations are achieved by the integration of OLEDs and OTFTs on top of plastic films; these displays are thin, light-weighted, bendable, foldable and they can be rolled-up or dropped without being broken.

Besides backlit-LCD and LED flat panel displays that emit light, electr-onic-paper (e-paper), also referred to as e-reader, is another technology that imitates the appearance of ordinary ink on paper. The display elements used by e-papers feature a memory function that enable continuous display of the same image even if no current is driven. Furthermore, images can be changed using very low currents. An example of such display elements is Electrophoretic cells.[9] These features open the possibility to develop the backplane drivers for such kind of displays using OTFTs, which offer virtues in manufacturing flexible, thin, light-weighted and large-area e-papers at low production costs. The German company Plastic Logic is commercializing a range of high-quality plastic e-papers, which are shatterproof daylight readable displays and extremely flexible with proven lifetimes. Figure 2.4c shows one of their demonstrated displays that is driven by OTFTs. Just recently, in September 2013, a start-up company popSLATE in the USA had joined forces with Plastic Logic to create an always-on secondary screen display for Apple iPhones [108]. Owing to the unique and novel features of these displays, they can be easily embedded into accessory cases of smartphones or similar devices.

Likewise an OTFT-based active-matrix strategy can be used in large-area, low-end and conformable array of sensors and/or actuators. Electronic-skin (e-skin) that mimic the sensitivity of real skin is a good example [109, 110]. Considering the vast amount of pain receptors in a person's skin and the large surface area of the human's body, active-matrix driver holds a great advantage for such a sensor array. To gain more insight about the complexity, the human skin has more than two million receptors, which is equivalent to the number of pixels found in a typical high-definition television; moreover, an average adult's skin exceeds two square meters of surface area, which is nearly as big as the largest flat-screen LCD television.

Researchers at the University of Tokyo demonstrated recently, in July 2013, the world's lightest and thinnest flexible tactile sensor. This OTFT-driven prototype has a thickness of 2 μm, which is about one-fifth the thickness of plastic kitchen wrap, and comprises an array of 144 pixels over an area of $48 \times 48\,\text{mm}^2$ [111]. In addition, this sensor could stand repeatable bending, be crumpled like a paper and accommodate stretching of up to 230 %. It can also work at high temperatures and aqueous environments (e.g. in saline solutions), which enables the functionality of this sensor even inside the human's body. This is considered as a major breakthrough to successfully mimic the sensitivity of a biological skin and opens up a wide range of new applications in fields ranging from healthcare and biomedicine to welfare [112]. For example, the same research group has used this ultrathin OTFT technology to realize a 64 channel surface electromyogram (EMG) measurement sheet for prosthetic hand control [113].

Other four examples of large-area sensors or actuators that could be driven by OTFTs are the following. First, smart textile carpets or mats, ones that are useful for safety or security purposes. For instance, they can be deployed in patient rooms in hospitals to indicate a dangerous fall, heart attack, stroke, or similar complications and accordingly warn the staff. Second, flexible sheet image sensors or photodetectors as demonstrated by ISORG (a start-up company in France) and Plastic Logic at the 2013 Printed Electronics USA. They have developed the first fully-organic image

[9]Electrophoretic cells are transparent microcapsules incorporating charged pigment particles. A matrix of these cells, sandwiched between two layers (top transparent electrode and bottom patterned line-electrodes), is forming an electrophoretic display. Depending on the individually applied electric field on each cell, or pixel, a visible image is displayed.

sensor with an area of $40 \times 40\,\text{mm}^2$ and 8930 pixels, combining ISORG's organic photodetectors with Plastic Logic's OTFT backplanes [114]. Third, large-scale stress sensors that could be placed on a randomly shaped body such as a surface of a plane, allowing to detect micro-cracks soon enough during the flight. Finally, a sheet of braille actuators, which can be felt and read by blind persons. The same research group from the University of Tokyo has successfully developed in 2005 an integrated circuit card (ICC) that features a mechanically flexible braille display built in an OTFT technology [115].

Given the wide variety and novel features of organic semiconductors, OTFTs have been also explored for chemical and biological sensors [116]. Unlike the silicon-based technology (Chemically Sensing Gate Field-Effect Transistor ChemFET), OTFT-based sensors are promising on account of their upside-down device structure, which makes the semiconductor in direct contact with the analytes [117]. Examples of such parameters are the bulk conductivity, channel conductivity, threshold voltage and field-effect mobility. On the other hand, sensing occurs in the silicon-based technology at the gate or the gate/insulator interface and by indirect modulation of the drain current through capacitive coupling. In principal, organic conjugated compounds can perform weak chemical reactions with a variety of analytes, including both vapor- and water-based analytes as demonstrated in [118] and [119], respectively. Furthermore, it was shown in [118] that the response of different organic semiconducting materials to different analytes is distinctive. This facilitates the possibility to construct an electronic-nose (e-nose) with which an analyte (odor) can be uniquely identified from a coded reception of different organic semiconductors.

In addition, OTFT-based chemical and biological detectors are ideal for cheap, mass-produced, flexible sensors that can be used in a range of new disposable health-care products. Using for example point-of-care (PoC) devices, people could diagnose or monitor illness quite easily in private, much like pregnancy tests and hygiene tests for life sciences and food handling applications. Companies and research facilities, such as Molecular Vision in the UK, Acreo in Sweden and VTT Technical Research Center in Finland, do have the vision to develop lab-on-a-chip technologies using organic semiconductors. This is to promote a more proactive approach to healthcare, with the onus on prevention of diseases and early self-diagnosis [98]. However, such a flexible system would require not only sensors but also circuits and other electronic components; this is commonly known as a system-in-foil (SiF) as discussed in the following.

2.3.4 Systems-in-Foil

Systems-in-Foil (SiF) are an emerging new class of flexible electronic products in which complete systems are integrated in thin polymeric foils [32]. This is achieved by combining sensors, displays, power sources (batteries or photovoltaics), circuits, discrete components and/or interconnect traces all in a thin polymeric foil (or in some cases on the foil) to give a flexible end product. Being very thin and easy to glue to other substrates or even integrated into textiles, these systems can easily make arbitrary objects have smart interactive surfaces. In addition, they can also be fitted to existing equipment as retrofit sensors; for example, the power distribution of an equipment can be monitored using an attached temperature sensor foil [120].

As mentioned above, disposable biosensor systems are one example of SiF that could be used in different application areas such as point-of-care, food safety, environmental monitoring and agriculture [121]. Acreo is currently working on the realization of an all-printed biosensor system that includes battery, sensor, display and electronics on a single foil. Another example of SiF could be a contactless user interface, combining optical sensors, LEDs and flexible interconnects, as being developed by ISORG. Such a platform can allow hand proximity detection, gesture recognition or linear control. Furthermore, trading cards and board games could integrate RFIDs, displays or buttons in a form of SiF to add more functionalities and user engagement as well as authenticity when being susceptible to counterfeits [98].

Packaging is also a viable market for printed and organic electronics. The main role of packaging is to protect, advertise and provide information of its content, all at a cost commensurate with what is essentially a disposable item [98]. Currently, packages use low-cost holograms, watermarks, barcodes and safety labels as tools to indicate whether a product is real or fake. However, the idea of using low-end, mass-produced, printed RFIDs, sensors and optical feedback in a SiF could potentially provide new ways for simultaneously branding, promoting and protecting the products. This can only be feasible once the cost of printing such functional components becomes comparable to that of printing text and graphics on everyday packages, but this will take several years to achieve. Consequently, several technology start-ups and established suppliers of printing equipment are investing into reel-to-reel, high-throughput manufacturing of organic semiconductors. Until this technology comes to commercial reality, packaging companies, such as Karl Knauer in Germany, are developing and producing for example simple luminescent packaging [122].

Since lots of packaging and labels are made from paper-based materials (cellulose), there is also much research activity in paper-based electronics. For instance, a research group at the University of Cincinnati has been working on building circuits, displays and solar cells onto paper substrates [123]. Others, at Harvard University and the University of Washington, are trying to develop paper-based microfluidic devices as sensing elements. In addition, some of the paper manufacturers, such as Felix Schoeller in Germany, are active in this field; they are developing high-end coating techniques to produce high-quality, smooth paper substrates for printed electronics.

2.4 Summary

The electrical activity of organic materials has been explored since the twentieth century. However, these materials remained with no significant technological use for many years mainly due to their very low reproducibility and carrier mobility, poor control of material purity and structural ordering, and process difficulties to prepare stable, injection-efficient contacts to the compounds. This was only until 1977 when the research and interest in this field were flourished by the successful synthesis of electrically conducting organic polymers through controlled halogen doping. Owing to the excellent solubility of many organic polymers, they are amenable to mass printing processes, which have been well established in the graphic art printing industry. This groundbreaking discovery by MacDiarmid, Heeger and Shirakawa was honored with the 2000 Nobel Prize in Chemistry. Henceforth, the utilization of organic materials by various electronic components, including transistors, light-emitting diodes and photovoltaic cells, has given them incontrovertibly a place in the development of this theme.

Organic electronics pose nowadays a strategic challenge to the market and they have shown lately a rapid progressive development towards higher level of integration and better performance. The carrier mobility in organic semiconductors, though still underperform that in the crystalline silicon and is likely to retain this position in the foreseeable future, has improved dramatically until it approached and even surpassed that in the amorphous silicon. In principle, the organic materials have to be carefully exploited since process conditions as well as the interplay with other layers (can be organic or inorganic thin films) have a large influence on the device performance.

The organic technology is versatile enough to be used in a wide range of new applications in fields ranging from healthcare to welfare, especially when flexibility, transparency, large area and/or low cost are paramount. Recent advances in organic materials have permitted a disruptive evolution in nearly all electronic products such as displays, lighting, power sources, sensors and integrated smart systems-in-foil. The following chapters are focusing on organic thin-film transistor (OTFT) fabrication, characterization, modeling and circuits design.

References

1. J.N. Burghartz (ed.), *Guide to State-of-the-Art Electron Devices* (Wiley-IEEE Press, Chichester, 2013)
2. S.M. Sze, *Semiconductor Devices Physics and Technology*, 2nd edn. (Wiley, Hoboken, 2002)
3. K. Eichhorn, *Heinrich Geissler (1814–1879)—Leben, Werk und Wirkung* (Rat der Stadt, Steinach, 1989). (in German)
4. J.J. Thomson, Cathode rays. Philos. Mag. ser. 5 **44**(269), 293–316 (1897)
5. A. Hart-Davis (ed.), *Science* (Dorling and Kindersley, New York, 2009)
6. F. Braun, Ueber ein Verfahren zur Demonstration und zum Studium des zeitlichen Verlaufes variabler Ströme. Ann. Phys. **296**(3), 552–559 (1897). (in German)
7. J.A. Fleming, Instrument for converting alternating electric currents into continuous currents, U.S. Patent 803 684, 7 Nov 1905
8. L. de Forest, Device for amplifying feeble electrical currents, U.S. Patent 841 387, 15 Jan 1907
9. H. Klauk (ed.), *Organic Electronics: Materials, Manufacturing and Applications*, 2nd edn. (Wiley-VCH, Weinheim, 2008)
10. S. Levy, (2013) The brief history of the ENIAC computer. http://www.smithsonianmag.com/history-archaeology/The-Brief-History-of-the-ENIAC-Computer-228879421.html

11. J.E. Lilienfeld, Method and apparatus for controlling electric currents, U.S. Patent 1 745 175, 28 Jan 1930
12. J.E. Lilienfeld, Device for controlling electric current, U.S. Patent 1 900 018, 7 Mar 1933
13. O. Heil, Improvements in or relating to electrical amplifiers and other control arrangements and devices, U.K. Patent 439 457, 6 Dec 1935
14. C.R. Kagan, P. Andry (eds.), *Thin-Film Transistors* (Marcel Dekker, New York, 2003)
15. T.P. Brody, The thin film transistors–A late flowering bloom. IEEE Trans. Electron Devices **31**(11), 1614–1628 (1984)
16. J. Bardeen, W.H. Brattain, The transistor, a semi-conductor triode. Phys. Rev. **74**, 230–231 (1948)
17. J. Bardeen, W.H. Brattain, Three-electrode circuit element utilizing semiconductive materials, U.S. Patent 2 524 035, 3 Oct 1950
18. W.H. Brattain, Bell Telephone Laboratories, Notebook no. 18194 (case no. 38138–7), pp. 192–194, 16 Dec 1947
19. W.H. Brattain, Bell Telephone Laboratories, Notebook no. 21780 (case no. 38138–7), pp. 7–8, 24 Dec 1947
20. W. Shockley, Bell Telephone Laboratories, Notebook no. 20455, pp. 128–132, 24 Jan 1948
21. W. Shockley, Circuit element utilizing semiconductive material, U.S. Patent 2 569 347, 25 Sep 1951
22. M.M. Atalla, E. Tannenbaum, E.J. Scheibner, Stabilization of silicon surface by thermally grown oxides. Bell Syst. Tech. J. **38**(3), 749–783 (1959)
23. D. Kahng, Electric field controlled semiconductor device, U.S. Patent 3 102 230, 27 Aug 1963
24. M.M. Atalla, Insulated-gate field effect transistor with electrostatic protection means, U.S. Patent 3 413 497, 26 Nov 1968
25. P.K. Weimer, An evaporated thin-film triode. IRE Trans. Electron Devices **8**(5), 421 (1961)
26. P.K. Weimer, Evaporated circuits incorporating a thin-film transistor, in IEEE International Solid-State Circuits Conference Technical Digest, pp. 32–33 (1962)
27. P.K. Weimer, The TFT–A new thin-film transistor. Proc. of the IRE **50**(6), 1462–1469 (1962)
28. R.A. Street (ed.), *Technology and Applications of Amorphous Silicon* (Springer, Berlin, 2000)
29. B. Redman-White, B. Murmann, S. Kosonocky, S. Rusu, R. Thewes, K. Zhang, A. Cathelin, H.-J. Yoo, D. Su, D. Friedman, ISSCC 2013 technology trends, in IEEE International Solid-State Circuits Conference, 2013. http://isscc.org/trends/
30. G.E. Moore, Cramming more components onto integrated circuits. Electronics **38**(8), 114–117 (1965)
31. G.E. Moore, Progress in digital integrated electronics, in IEEE International Electron Devices Meeting Technical Digest, 1975, pp. 11–13
32. J.N. Burghartz (ed.), *Ultra-Thin Chip Technology and Applications* (Springer, Berlin, 2011)
33. J.N. Burghartz, You can't be too thin or too flexible. IEEE Spectr. **50**(3), 38–61 (2013)
34. J.N. Burghartz, W. Appel, H.D. Rempp, M. Zimmermann, A new fabrication and assembly process for ultrathin chips. IEEE Trans. Electron Devices **56**(2), 321–327 (2009)
35. K. De Munck, T. Chiarella, P. De Moor, B. Swinnen, C. Van Hoof, Influence of extreme thinning on 130-nm standard CMOS devices for 3-D integration. IEEE Electron Device Lett. **29**(4), 322–324 (2008)
36. H. Spanggaard, F.C. Krebs, A brief history of the development of organic and polymeric photovoltaics. Sol. Energy Mater. Sol. Cells **83**(2–3), 125–146 (2004)
37. A. Pochettino, A. Sella, Photoelectric behavior of anthracene. Acad. Lincei Rendus **15**, 355–363 (1906)
38. M. Volmer, Die verschiedenen lichtelektrischen Erscheinungen am Anthracen, ihre Beziehungen zueinander, zur Fluoreszenz und Dianthracenbildung. Ann. Phys. **345**(4), 775–796 (1913). (in German)
39. P.M. Borsenberger, D.S. Weiss, *Organic Photoreceptors for Imaging Systems* (Marcel Dekker, New York, 1993)
40. A. Bernanose, Electroluminescence in organic compounds. Br. J. Appl. Phys. **6**(S4), S54–S56 (1955)

41. M. Pope, H.P. Kallmann, P. Magnante, Electroluminescence in organic crystals. J. Chem. Phys. **38**(8), 2042–2043 (1963)
42. W. Brütting, Encyclopedia of Physics, ed. by R.G. Lerner, G.L. Trigg, 3rd edn, vol. 2 (Wiley-VCH, Weinheim, BW, Germany, 2005), pp. 1866–1876 (ch. Organic Semiconductors)
43. H. Shirakawa, E.J. Louis, A.G. MacDiarmid, C.K. Chiang, A.J. heeger, Synthesis of electrically conducting organic polymers: halogen derivatives of polyacetylene, $(CH)_x$. J. Chem. Soc. Chem. Commun. 24, 578–580 (1977)
44. A. Moliton, R.C. Hiorns, The origin and development of (plastic) organic electronics. Polym. Int. **61**(3), 337–341 (2012)
45. C.W. Tang, Two-layer organic photovoltaic cell. Appl. Phys. Lett. **48**(2), 183–185 (1986)
46. "Heliatek consolidates its technology leadership by establishing a new world record for organic solar technology with a cell efficiency of 12%," Press Release, Heliatek. http://www.heliatek.com/newscenter/?lang=en. Accessed 16 Jan 2013
47. C.W. Tang, S.A. VanSlyke, Organic electroluminescent diodes. Appl. Phys. Lett. **51**(12), 913–915 (1987)
48. J.H. Burroughes, D.D.C. Bradley, A.R. Brown, R.N. Marks, K. Mackay, R.H. Friend, R.L. Burns, A.B. Holmes, Light-emitting diodes based on conjugated polymers. Nature **347**, 539–541 (1990)
49. "UHD TV S9C OLED," Spec Sheet, Samsung, 2013. [Online]. Available: http://www.samsung.com/us/video/tvs/KN55S9CAFXZA
50. D.F. Barbe, C.R. Westgate, Surface state parameters of metal-free phthalocyanine single crystals. J. Phys. Chem. Solids **31**(12), 2679–2687 (1970)
51. M.L. Petrova, L.D. Rozenshtein, Field effect in the organic semiconductor chloranil. Fiz. Tverd. Tela (Soviet Phys. Solid State) 12, 961–962 (1970)
52. G. Horowitz, Organic field-effect transistors. Adv. Mater. **10**(5), 365–377 (1998)
53. G. Horowitz, Organic field-effect transistors: From theory to real devices. J. Mater. Res. **19**(7), 1946–1962 (2004)
54. F. Ebisawa, T. Kurokawa, S. Nara, Electrical properties of polyacetylene/polysiloxane interface. J. Appl. Phys. **54**(6), 3255–3259(1983)
55. J.H. Burroughes, C.A. Jones, R.H. Friend, New semiconductor device physics in polymer diodes and transistors. Nature **335**, 137–141 (1988)
56. A. Tsumura, H. Koezuka, T. Ando, Polythiophene field-effect transistor: Its characteristics and operation mechanism. Synth. Met. **25**(1), 11–23 (1988)
57. A. Assadi, C. Svensson, M. Willander, O. Inganäs, Field-effect mobility of poly(3-hexylthiophene). Appl. Phys. Lett. **53**(3), 195–197 (1988)
58. R. Mardru, G. Guillaud, M. Al, Sadoun, M. Maitrot, J.-J. André, J. Simon, R. Even, A well-behaved field effect transistor based on an intrinsic molecular semiconductor. Chem. Phys. Lett. **145**(4), 343–346 (1988)
59. C. Clarisse, M.T. Riou, M. Gauneau, M. Le Contellect, Field-effect transistor with diphthalocyanine thin film. Electron. Lett. **24**(11), 674–675 (1988)
60. F. Garnier, G. Horowitz, X. Peng, D. Fichou, An all-organic "soft" thin film transistor with very high carrier mobility. Adv. Mater. **2**(12), 592–594 (1990)
61. Plastic Logic Technology. http://www.plasticlogic.com/technology/
62. Technology Academy Finland. http://www.technologyacademy.fi
63. Deutscher Zukunftpreis. http://www.deutscher-zukunftspreis.de
64. M. Riede, C. Uhrich, J. Widmer, R. Timmreck, D. Wynands, G. Schwartz, W.-M. Gnehr, D. Hildebrandt, A. Weiss, J. Hwang, S. Sundarraj, P. Erk, M. Pfeiffer, K. Leo, Efficient organic tandem solar cell based on small molecules. Adv. Funct. Mater. **21**(16), 3019–3028 (2011)
65. J. Meiss, T. Menke, K. Leo, C. Uhrich, W.-M. Gnehr, S. Sonntag, M. Pfeiffer, M. Riede, Highly effiecient semitransparent tandem organic solar cells with complementary absorber materials. Appl. Phys. Lett. **99**(4), 043301-1–043301-3 (2011)
66. M. Pfeiffer, K. Leo, X. Zhou, J.S. Huang, M. Hofmann, A. Werner, J. Blochwitz-Nimoth, Doped organic semiconductors: Physics and application in light emitting diodes. Org. Electron. **4**(2–3), 89–103 (2003)

67. X. Zhou, J. Blochwitz-Nimoth, M. Pfeiffer, B. Maennig, J. Drechsel, A. Werner, K. Leo, High-efficiency low-voltage stable inverted transparent electrophosphorescent organic light-emitting diodes: Combining electrically doped carrier transport layers and iridium-complex doped emissive layer. Synth. Met. **137**(1–3), 1063–1064 (2003)
68. P. Cosseddu, A. Bonfiglio, Soft lithography fabrication of all-organic bottom-contact and top-contact field effect transistors. Appl. Phys. Lett. **88**(2), 023506-1–023506-3 (2006)
69. M. Halik, H. Klauk, U. Zschieschang, T. Kriem, G. Schmid, W. Radlik, K. Wussow, Fully patterned all-organic thin film transistors. Appl. Phys. Lett. **81**(2), 289–291 (2002)
70. H. Klauk, Organic thin-film transistors. Chem. Soc. Rev. **39**(7), 2643–2666 (2010)
71. W. Clemens, D. Lupo, K. Hecker, S. Breitung (eds.), OE-A roadmap for organic and printed electronics. White Paper, 4th edn, Organic Electronics Association (2011). http://www.oe-a.org/en_GB/
72. I. McCulloch, M. Heeney, M.L. Chabinyc, D. DeLongchamp, R.J. Kline, M. Cölle, W. Duffy, D. Fischer, D. Gundlach, B. Hamadani, R. Hamiliton, L. Richter, A. Salleo, M. Shkunov, D. Sparrowe, S. Tierney, W. Zhang, Semiconducting thienothiophene copolymers: Design, synthesis, morphology, and performance in thin-film organic transistors. Advanced Materials **21**(10–11), 1091–1109 (2009)
73. C.G. Choi, B.-S. Bae, Effects of hydroxyl groups in gate dielectrics on the hysteresis of organic thin film transistors. Electrochemical and Solid-State Letters **10**(11), H347–H350 (2007)
74. T. Yamamoto, K. Takimiya, Facile synthesis of highly π-extended heteroarenes, dinaphtho[2,3-b:2',3'-f]chalcogenopheno[3,2-b]chalcogenophenes, and their application to field-effect transistors. Journal of The American Chemical Society **129**(8), 2224–2225 (2007)
75. C. Kim, A. Facchetti, T.J. Marks, Probing the surface glass transition temperature of polymer films via organic semiconductor growth mode, microstructure, and thin-film transistor response. Journal of The American Chemical Society **131**(25), 9122–9132 (2009)
76. Y. Ishii, R. Ye, M. Baba, K. Ohta, Correlation between morphology and electrical properties of α,α'-diethyl-sexithiophene thin films. Japanese Journal of Applied Physics **47**(2), 1256–1258 (2008)
77. R. Schmidt, J.H. Oh, Y.-S. Sun, M. Deppisch, A.-M. Krause, K. Radacki, H. Braunschweig, M. Könemann, P. Erk, Z. Bao, F. Würthner, High-performance air-stable n-channel organic thin film transistors based on halogenated perylene bisimide semiconductors. Journal of The American Chemical Society **131**(17), 6215–6228 (2009)
78. Z. Bao, A.J. Lovinger, J. Brown, New air-stable n-channel organic thin-film transistors. Journal of The American Chemical Society **120**(1), 207–208 (1998)
79. R.T. Weitz, K. Amsharov, U. Zschieschang, E.B. Villas, D.K. Goswami, M. Burghard, H. Dosch, M. Jansen, K. Kern, H. Klauk, Organic n-channel transistors based on core-cyanated perylene carboxylic diimide derivatives. Journal of The American Chemical Society **130**(14), 4637–4645 (2008)
80. Y. Hosoi, D. Tsunami, H. Ishii, Y. Furukawa, Air-stable n-channel organic field-effect transistors based on N, N'-bis(4-trifluoromethylbenzyl)perylene-3,4,9,10-tetracarboxylic diimide. Chemical Physics Letters **436**(1–3), 139–143 (2007)
81. O.D. Jurchescu, M. Popinciuc, B.J. van Wees, T.T.M. Palstra, Interface-controlled, high-mobility organic transistors. Advanced Materials **19**(5), 688–692 (2007)
82. R. Hamilton, J. Smith, S. Ogier, M. Heeney, J.E. Anthony, I. McCulloch, J. Veres, D.D.C. Bradley, T.D. Anthopoulos, High-performance polymer-small molecule blend organic transistors. Advanced Materials **21**, 1166–1171 (2009)
83. S. Handa, E. Miyazaki, K. Takimiya, Y. Kunugi, Solution-processable n-channel organic field-effect transistors based on dicyanomethylene-substituted terthienoquinoid derivative. Journal of The American Chemical Society **129**(38), 11684–11685 (2007)
84. H. Usta, A. Facchetti, T.J. Marks, Air-stable, solution-processable n-channel and ambipolar semiconductors for thin-film transistors based on the indenofluorenebis(dicyanovinylene) core. Journal of The American Chemical Society **130**(27), 8580–8581 (2008)
85. H. Yan, Z. Chen, Y.Z.C. Newman, J.R. Quinn, F. Dötz, M. Kastler, A. Facchetti, A high-mobility electron-transporting polymer for printed transistors. Nature **457**, 679–686 (2009)

86. F. Ante, "Contact effects in organic transistors," Ph.D. dissertation, Swiss Federal Institute of Technology in Lausanne, Lausanne, Switzerland, Dec. 2011
87. P. Heremans, W. Dehaene, M. Steyaert, K. Myny, H. Marien, J. Genoe, G. Gelinck, E. van Veenendaal, "Circuit design in organic semiconductor technologies," in Proceedings of the European Solid-State Device Research Conference, Sep. 2011, pp. 5–12
88. T. Zaki, F. Ante, U. Zschieschang, J. Butschke, F. Letzkus, H. Richter, H. Klauk, J.N. Burghartz, A 3.3 V 6-Bit 100 kS/s current-steering digital-to-analog converter using organic p-type thin-film transistors on glass. IEEE Journal of Solid-State Circuits **47**(1), 292–300 (2012)
89. H. Marien, M.S.J. Steyaert, E. van Veenendaal, P. Heremans, A fully integrated $\Delta\Sigma$ ADC in organic thin-film transistor technology on flexible plastic foil. IEEE Journal of Solid-State Circuits **46**(1), 276–284 (2011)
90. R. Rödel, F. Letzkus, T. Zaki, J.N. Burghartz, U. Kraft, U. Zschieschang, K. Kern, H. Klauk, Contact properties of high-mobility, air-stable, low-voltage organic n-channel thin-film transistors based on a naphthalene tetracarboxylic diimide. Applied Physics Letters **102**(233303), 2333031–2333035 (2013)
91. J.H. Na, M. Kitamura, Y. Arakawa, High field-effect mobility amorphous InGaZnO transistors with aluminum electrodes. Applied Physics Letters **93**(6), 0635011–0635013 (2008)
92. G. Oike, T. Yajima, T. Nishimura, K. Nagashio, A. Toriumi, "High electron mobility (>16 $cm^2/Vsec$) FETs with high on/off ratio (>10^6) and highly conductive films (σ>10^2 S/cm) by chemical doping in very thin (~20 nm) TiO_2 films on thermally grown SiO_2," in IEEE International Electron Devices Meeting Technical Digest, Dec. 2013, pp. 11.5.1-11.5.4
93. M. Rockelé, D.-V. Pham, A. Hoppe, J. Steiger, S. Botnaras, M. Nag, S. Steudel, K. Myny, S. Schols, R. Müller, B. van der Putten, J. Genoe, P. Hermans, Low-temperature and scalable complementary thin-film technology based on solution-processed metal oxide n-TFTs and pentacene p-TFTs. Organic Electronics **12**(11), 1909–1913 (2011)
94. S.R. Thomas, P. Pattanasattayavong, T.D. Anthopoulos, Solution-processable metal oxide semiconductors for thin-film transistor applications. Chemical Society Reviews **42**(16), 6910–6923 (2013)
95. "PolyTC transparent conductive films," Spec Sheet, PolyIC, 2013. [Online]. Available: http://www.polyic.com/products/polytcr.html
96. U. Zschieschang, F. Ante, D. Kälblein, T. Yamamoto, K. Takimiya, H. Kuwabara, M. Ikeda, T. Sekitani, T. Someya, J. Blochwitz-Nimoth, H. Klauk, Dinaphtho[2,3-b:2',3'-f]thieno[3,2-b]thiophene (DNTT) thin-film transistors with improved performance and stability. Organic Electronics **12**(8), 1370–1375 (2011)
97. K. Kuribara, H. Wang, N. Uchiyama, K. Fukuda, Tomoyuki, U. Zschieschang, C. Jaye, D. Fischer, H. Klauk, T. Yamamoto, K. Takimiya, M. Ikeda, H. Kuwabara, T. Sekitani, Y.-L. Loo, T. Someya, "Organic transistors with high thermal stability for medical applications," Nature. Communications **3**, 7231–7237 (2012)
98. J. Thompson, Ed., +Plastic Electronics, vol. 2, no. 3, Plastic Electronics Foundation, 2009
99. S. Forrest, P. Burrows, M. Thompson, The dawn of organic electronics. IEEE Spectrum **37**(8), 29–34 (2000)
100. T. Tiedje, E. Yablonovitch, G.D. Cody, B.G. Brooks, Limiting efficiency of silicon solar cells. IEEE Transactions on Electron Devices **31**(5), 711–716 (1984)
101. B. O'Regan, M. Grätzel, A low-cost, high-efficiency solar cell based on dye-sensitized colloidal TiO_2 films. Nature **353**, 737–740 (1991)
102. A.C. Mayer, S.R. Scully, B.E. Hardin, M.W. Rowell, M.D. McGehee, Polymer-based solar cells. Materials **10**(11), 741–748 (2007)
103. "Thinfilm demonstrates first integrated printed electronic system with rewritable memory," Press Release, Thin Film Electronics ASA, Dec. 20, 2012. [Online]. Available: http://www.thinfilm.no/news/
104. Thinfilm MemoryTM. http://www.thinfilm.no/products/memory/
105. J. Thompson, Ed., +Plastic Electronics, vol. 1, no. 3, Plastic Electronics Foundation, 2008

106. I. Yagi, N. Hirari, Y. Miyamoto, M. Noda, A. Imaoka, N. Yoneya, K. Nomoto, J. Kasahara, A. Yumoto, T. Urabe, A flexible full-color AMOLED display driven by OTFTs. Journal of the Society for Information Display **16**(1), 15–20 (2008)
107. M. Noda, N. Kobayashi, M. Katsuhara, A. Yumoto, S.-I. Ushikura, R.-I. Yasuda, N. Hirai, G. Yukawa, I. Yagi, K. Nomoto, T. Urabe, A rollable AM-OLED display driven by OTFTs. Society for Information Display Symposium Technical Digest **41**(1), 710–713 (2010)
108. "popSLATE partners with Plastic Logic to create the always-on second screen display for their Apple iPhone 5 & 5S accessory," Press Release, Plastic Logic, Sep. 13, 2013. [Online]. Available: http://www.plasticlogic.com/news/
109. T. Someya, Building bionic skin. IEEE Spectrum **50**(9), 50–56 (2013)
110. T. Someya, Y. Kato, T. Sekitani, S. Iba, Y. Noguchi, Y. Murase, H. Kawaguchi, T. Sakurai, Conformable, flexible, large-area networks of pressure and thermal sensors with organic transistor active matrixes. Proceedings of the National Academy of Sciences of the United States of America **102**(35), 12321–12325 (2005)
111. M. Kaltenbrunner, T. Sekitani, J. Reeder, T. Yokota, K. Kuribara, T. Tokuhara, M. Drack, R. Schwödiauer, I. Graz, S. Bauer-Gogonea, S. Bauer, T. Someya, An ultra-lightweight design for imperceptible plastic electronics. Nature **499**, 458–463 (2013)
112. T. Someya, T. Sekitani, M. Kaltenbrunner, T. Yokota, H. Fuketa, M. Takamiya, T. Sakurai, "Ultraflexible organic devices for biomedical applications," in IEEE International Electron Devices Meeting Technical Digest, Dec. 2013, pp. 8.5.1-8.5.4
113. H. Fuketa, K. Yoshioka, Y. Shinozuka, K. Ishida, T. Yokota, N. Matsuhisa, Y. Inoue, M. Sekino, T. Sekitani, M. Takamiya, T. Someya, T. Sakurai, "1μm-thickness 64-channel surface electromyogram measurement sheet with 2V organic transistors for prosthetic hand control," in IEEE International Solid-State Circuits Conference Technical Digest, Feb. 2013, pp. 104–105
114. "Plastic Logic and ISORG demo thin film plastic image sensor at IDTechEx," Press Release, Plastic Logic and ISORG, Nov. 19, 2013. http://www.plasticlogic.com/news/
115. Y. Kato, T. Sekitani, M. Takamiya, M. Doi, K. Asaka, T. Sakurai, T. Someya, Sheet-type braille displays by integrating organic field-effect transistors and polymeric actuators. IEEE Transactions on Electron Devices **54**(2), 202–209 (2007)
116. A. Dodabalapur, Organic and polymer transistors for electronics. Materials Today **9**(4), 24–30 (2006)
117. R. Shinar, J. Shinar (eds.), *Organic Electronics in Sensors and Biotechnology* (McGraw-Hill, New York, NY, USA, 2009)
118. B. Crone, A. Dodabalapur, A. Gelperin, L. Torsi, H.E. Katz, A.J. Lovinger, Z. Bao, Electronic sensing of vapors with organic transistors. Applied Physics Letters **78**(15), 2229–2231 (2001)
119. T. Someya, A. Dodabalapur, A. Gelperin, H.E. Katz, Z. Bao, Integration and response of organic electronics with aqueous microfluidics. Langmuir **18**(13), 5229–5302 (2002)
120. L. Jamet, "Printed electronics opens up large flexible sensor design opportunities," EETimes Europe, pp. 34–35, 2013
121. A.P.F. Turner, Biosensors: Sense and sensibility. Chemical Society Reviews **42**(8), 3175–3648 (2013)
122. Illuminated Packaging for Bombay Sapphire. http://www.karlknauer.com
123. A.J. Steckl, Circuits on cellulose. IEEE Spectrum **50**(2), 48–61 (2013)

Chapter 3
Organic Thin-Film Transistors

Organic thin-film transistors (OTFTs) are providing impetus in flexible, low-cost and large-area integrated circuit applications. The choice of the OTFT architecture has a considerable impact on the electrical performance, because of differences in the charge injection mechanism, but often limitations exist due to restrictions in processing methods or incompatibility of the materials employed [1]. These issues are discussed herein, together with simplified modeling equations and characteristic parameters to gain deeper insight of the operation principle of the device. Furthermore, stability and performance measures of the OTFTs used in this work are demonstrated.

3.1 Architecture and Materials

OTFTs are three-terminal devices in which the semiconductor is a thin layer of conjugated organic molecules. An OTFT is basically consisting of four layers, namely gate electrode, gate dielectric, semiconductor and source/drain contacts, all deposited on an insulating substrate. The organic semiconductor can either be small molecule or polymer. In this work, only small-molecule organic semiconductors are considered. The other three TFT layers can be either made of organic or inorganic materials [2]. For instance, the source/drain contacts are usually made from a noble metal (e.g. gold or palladium), but they can also be made from a conducting conjugated polymer (e.g. PEDOT:PSS). Depending on the sequence in which the layers are deposited, four different TFT configurations can be distinguished as shown in Fig. 3.1 [3]. The configurations are designated as coplanar (top-gate, top-contact), inverted-coplanar (bottom-gate, bottom-contacts), staggered (top-gate, bottom-contacts) and inverted-staggered (bottom-gate, top-contact) TFTs. Besides these planar TFTs, there are also vertical TFTs in which the critical dimension is a film thickness rather than a lateral distance [4–6]. However, we focus here only on the planar configurations.

It has been shown in literature by experiments [7] and simulations [8] that coplanar OTFTs often have a larger contact resistance in comparison to their staggered

T. Zaki, *Short-Channel Organic Thin-Film Transistors*, Springer Theses,
DOI 10.1007/978-3-319-18896-6_3

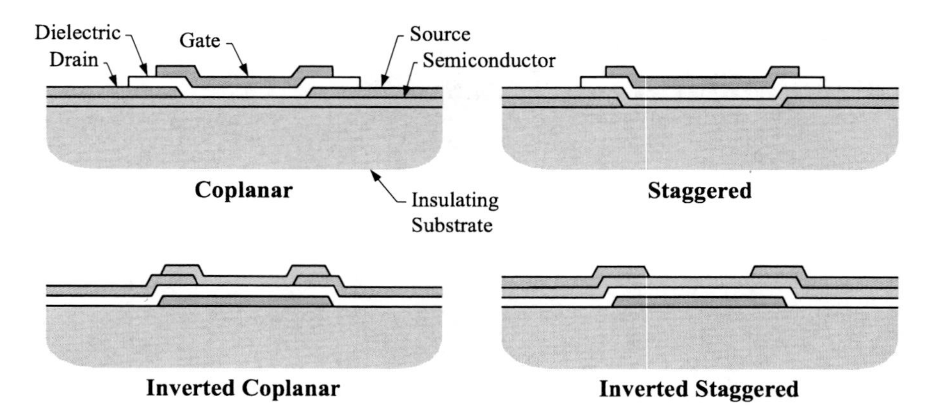

Fig. 3.1 Schematic cross section of the four TFT configurations, namely coplanar (*top*-gate, *top*-contact), inverted-coplanar (*bottom*-gate, *bottom*-contact), staggered (*top*-gate, *bottom*-contact) and inverted-staggered (*bottom*-gate, *top*-contact) TFTs

counterparts (note: this is similar for the inverted configurations as well). This is because the gate fields are partially shielded in the coplanar OTFTs by the source and drain contacts, and thus, a channel is induced only in the region between the contacts. As a result, the efficient charge transfer between the contacts and the semiconductor is limited in this case only to the narrow regions where the gate-induced channel touches the contact edges [9]. In staggered OTFTs, however, the source and drain contacts are located opposite to the gate dielectric; therefore, the channel extends beyond the contact edges along the entire gate-overlap lengths. Consequently, the area available for the charge transfer in staggered OTFTs is larger than in coplanar OTFTs, which explains why the staggered OTFTs often have a smaller contact resistance than the coplanar OTFTs.

In addition, the top-gate configurations (coplanar and staggered OTFTs) are typically useful for polymer-based OTFTs but not for small-molecule-based OTFTs. This is because thin films of small molecules often exhibit a rough three-dimensional growth (in contrast to the smooth amorphous films of spin-coated polymers), so that the dielectric/semiconductor interface would be also rough, and thus, the carrier mobility would be low [2].

The inverted-staggered (bottom-gate, top-contact) configuration, which benefits from a smooth dielectric/semiconductor interface and typically has a small contact resistance, is therefore chosen for our OTFTs. This kind of OTFTs can be fabricated by evaporation through shadow masks [10, 11], by soft lamination [12, 13], by self-aligned printing [14, 15], or by sub-femtoliter inkjet printing [16]. In this work, high-resolution silicon shadow (stencil) masks are employed to fabricate OTFTs with submicrometer accuracy and superb device uniformity [9, 17, 18]. Moreover, the use of shadow masks allow to pattern the source/drain contacts without the need of solvents or elevated temperatures that might cause undesirable phase transition for the high-mobility small-molecule organic semiconductors [17, 19, 20].

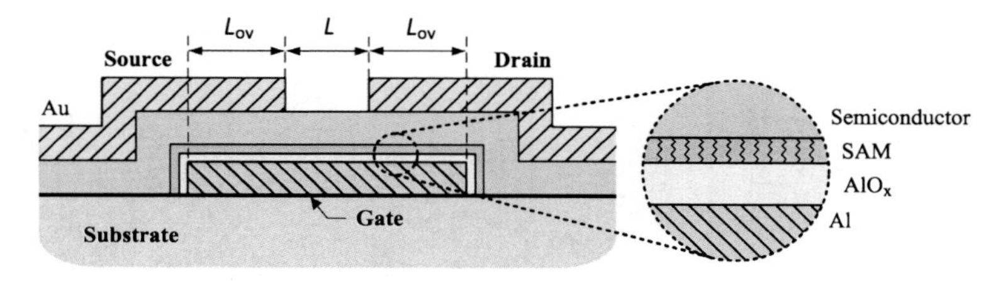

Fig. 3.2 Schematic cross section of the fabricated inverted-staggered (*bottom*-gate, *top*-contact) OTFT, where L is the channel length and L_{ov} is the gate-overlap length [23]

The small-molecule organic semiconductors are deposited in vacuum by sublimation through the silicon shadow masks to form well-ordered polycrystalline films [2]. In principle, the exact morphology and nucleation density on the surface depends mainly on the substrate temperature, substrate surface energy, surface roughness and deposition rate [2, 21]. As a result, an elevated substrate temperature of about 60–100 °C is utilized during the deposition to provide additional energy to the molecules, so that the diffusion length on the surface increases and the number of nucleation centres reduces. Furthermore, glass substrates Eagle^{2000} from Corning are used because of their very smooth surface roughness of about 0.2 nm (root mean square) [22], noting that this fabrication process is also compatible with mechanically flexible substrates. The silicon shadow masks are manufactured at IMS CHIPS, while the OTFTs are prepared at MPI-SSR. The fabrication process is explained in detail in Chap. 4.

Figure 3.2 depicts the schematic cross section of the fabricated inverted-staggered OTFT. The OTFT layers, namely 30 nm-thick aluminium gate, 11 nm-thick organic semiconductor and 25 nm-thick gold source/drain contacts, are deposited in vacuum through shadow masks. Prior to the semiconductor deposition, a hybrid dielectric, which consists of an oxygen-plasma-grown AlO_x layer (3.6 nm thick) and a solution-processed self-assembled monolayer (SAM) of *n*-tetradecylphoshonic acid (1.7 nm thick), is formed. This hybrid dielectric features a high capacitance of the order of 1 μF/cm^2, which allows the OTFTs to operate at low supply voltages ($\leq$3 V) [24, 25]. The organic semiconductor dinaphtho[2,3-b:2',3'-f]thieno[3,2-b]thiophene ($C_{22}H_{12}S_2$; DNTT) is used for our p-channel OTFTs as it features high intrinsic mobility, promising shelf-life and bias-stress stability, as well as little hysteresis behavior [25, 26]. As for the n-channel OTFTs, the air-stable organic semiconductors hexadecafluorocopperphthalocyanine ($F_{16}CuPc$) or N,N'-bis-(heptafluorobutyl)-2,6-dichloro-1,4,5,8-naphthalene tetracarboxylic diimide [NTCDI-Cl_2-$(CH_2C_3F_7)_2$] are employed [17, 27]. The molecular structure of the used organic materials are shown in Fig. 3.3.

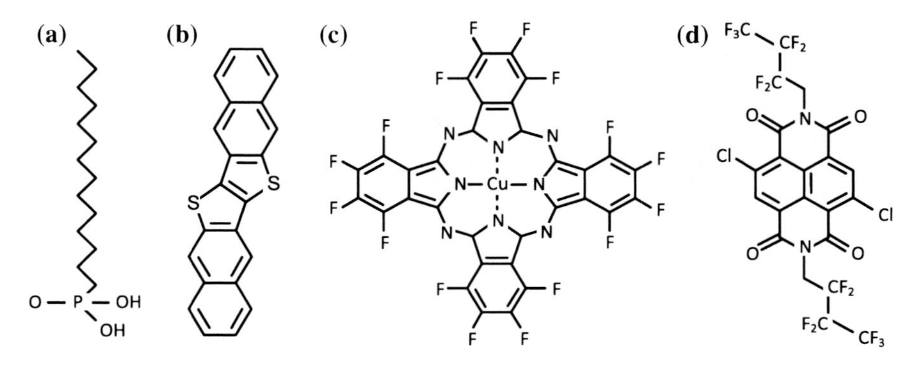

Fig. 3.3 Molecular structure of the used organic materials. **a** The tetradecylphosphonic acid with alkyl chain length of 14 carbon atoms [$C_{14}H_{29}PO(OH)_2$] self-assembled monolayer (C14-SAM), which is the second layer of the hybrid gate dielectric. **b** The dinaphtho[2, 3-b:2', 3'-f]thieno[3, 2-b]thiophene (DNTT), which is the organic semiconductor used for p-channel OTFTs. **c** The hexadecafluorocopperphthalocyanine ($F_{16}CuPc$), which is the organic semiconductor used for n-channel OTFTs. **d** The N,N'-bis-(heptafluorobutyl)-2,6-dichloro-1,4,5,8-naphthalene tetracarboxylic diimide [$NTCDI\text{-}Cl_2\text{-}(CH_2C_3F_7)_2$], which is the organic semiconductor used for n-channel OTFTs as an alternative to $F_{16}CuPc$ for high-performance applications. All these materials are processed at MPI-SSR

3.2 Operation Principle

The OTFT is an insulated-gate field-effect transistor (IGFET) in which the current flowing between the source and drain contacts—under the influence of an applied drain-source voltage—is modulated by means of a variable voltage applied between the gate and source electrodes. In other words, the resistance of the semiconductor between the source and drain contacts is controlled by the applied gate-source bias, noting that the gate electrode is capacitively separated from the semiconductor by the gate dielectric [2]. Referring to Fig. 3.2, if no potential difference is applied between the gate and source nodes, the charge carriers will be homogenously distributed within the semiconducting layer. Therefore, the conductance between the source and drain contacts will be very small due to the low density of carriers. If, however, a negative gate-source potential is applied (assuming that the DNTT organic semiconductor is used to realize a p-channel OTFT), excess positive carriers (holes) will be induced and accumulated at the semiconductor-insulator interface. This implies that a conducting channel is created. As a result, the conducting channel, through which a large current can flow, connects the source and drain contacts. Varying the gate-source potential in this case modulates the conductance of the channel. To gain more insight of the operation, Fig. 3.4 shows measured current–voltage (I–V) characteristics of a p-channel OTFT with a channel length (L) of 200 μm, a channel width (W) of 400 μm and a gate-overlap length (L_{ov}) of 10 μm. The different operation modes of the OTFT are explained hereinafter. Throughout the following discussion, the source serves as the reference (grounded) electrode.

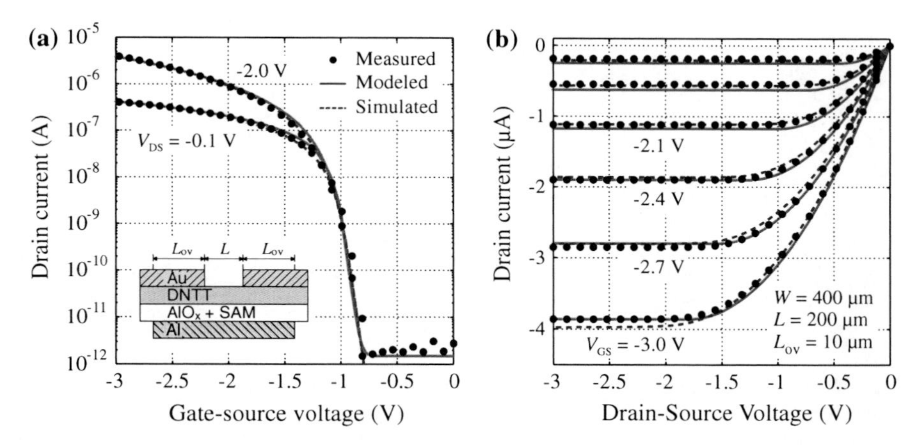

Fig. 3.4 Measured versus simulated (on Sentaurus Device Simulator) and modeled (on SPICE) static I–V characteristics of an OTFT with $W = 400$ μm, $L = 200$ μm and $L_{ov} = 10$ μm. **a** Linear and saturation transfer characteristics. **b** Output characteristics. The *inset* shows the simplified OTFT schematic representation used for the simulation

Assuming a p-channel OTFT, Fig. 3.5 shows simplified energy band diagrams of a metal-insulator-semiconductor (MIS) diode, which corresponds to the intrinsic part of the OTFT and a metal-semiconductor contact interface as part of the extrinsic part of the OTFT. The energy scheme indicates the respective positions of the Fermi level of aluminum and gold as well as the frontier orbitals, namely highest occupied molecular orbital (HOMO) and lowest unoccupied molecular orbital (LUMO), of DNTT. There is roughly an analogy between the HOMO and the LUMO of an organic semiconductor to the valence band maximum (E_V) and the conduction band minimum (E_C) of an inorganic semiconductors, respectively. In particular for the DNTT, the HOMO is about -5.19 eV and the LUMO is about -1.81 eV [23, 28]. This introduces a sufficiently large HOMO-LUMO energy gap of 3.38 eV that is adequate for a transistor operation.

Only intrinsic organic semiconductors are used in this work, meaning that they are not intentionally doped with impurities. In fact, there are always some dopants in the form of unintentional impurities, which are very hard to eliminate during processing; however, this kind of impurities are neglected here for simplicity. This implies that the Fermi energy[1] level (E_F) is located near the middle of the HOMO-LUMO gap as shown in Figs. 3.5a, b. Therefore, the work function of the DNTT ($q\phi_s$) is 3.5 eV. Given that the work function of aluminum gate electrode ($q\phi_{Al}$) is 4.1 eV, a small work function difference ($q\phi_{Al\text{-}s} \equiv q\phi_{Al} - q\phi_s$) of 0.6 eV is resulted. In addition to the work function difference, the MIS diode is affected by the concentration of charges in the insulator (Q_i) and of trapped charges at the insulator/semiconductor

[1]The Fermi energy is the energy at which the probability of occupation by an electron is exactly one-half.

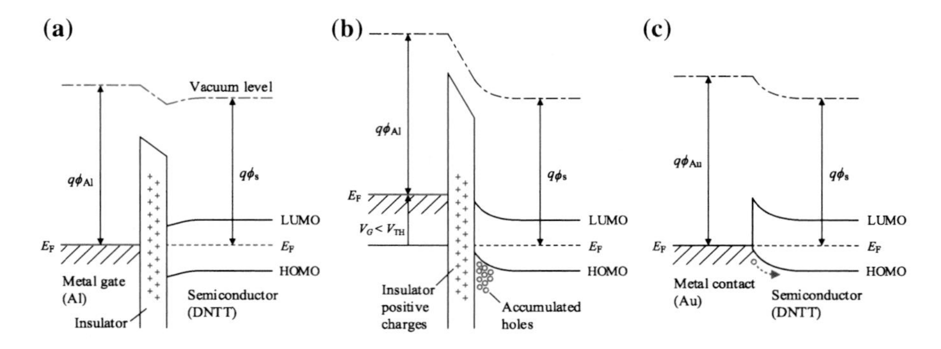

Fig. 3.5 Simplified energy band diagrams of a metal-insulator-semiconductor (MIS) diode and a metal-semiconductor contact interface to explain the operation principle of a p-channel OTFT (gate: Al; contacts: Au; organic semiconductor: DNTT), assuming no unintentional doping exists and neglecting changes in the work functions of the metals that results from the comprised SAM insulator layer. **a** Energy band diagram of the MIS diode at thermal equilibrium with positive insulator charges. There is a small work function difference ($q\phi_{\text{Al-s}}$) of 0.6 eV between the aluminum and the DNTT. The interface-trapped charges are neglected here. **b** Energy band diagram of the MIS diode when an excess negative gate bias is applied ($V_G < V_{TH} < 0$). Accordingly, a gate-induced accumulation channel is created in close vicinity of the semiconductor/insulator interface. **c** Energy band diagram of the gold-DNTT contact interface. The Fermi level (E_F) of gold is very close to the HOMO level of DNTT. Therefore, holes are more easily injected than electrons from the contact metal to the semiconductor

interface (Q_{if}). To achieve the flat-band condition, a negative voltage (V_{FB}) has to be applied to the metal gate:

$$V_{FB} = \phi_{\text{Al-s}} - \frac{(Q_i + Q_{if})}{C_I}. \tag{3.1}$$

To construct the energy band diagram of the MIS diode at thermal equilibrium, the Fermi level must be constant and the vacuum level must be continuous [29]. To accommodate the work function difference and positive insulator charges, assuming negligible interface-trapped charges, the semiconductor bands bend downward as shown in Fig. 3.5a.

In single-crystalline silicon MOSFETs, the threshold voltage is defined as the gate-source voltage at the onset of inversion in the transistor channel [2]. Organic transistors, however, do not operate in inversion; they are rather operating in the accumulation mode. Instead, the threshold voltage in OTFTs is often considered as the gate-source voltage necessary to achieve the flat-band condition ($V_{TH} = V_{FB}$). When an excess negative voltage ($V_G < V_{TH}$) is applied to the gate, the energy bands are bent upwards as shown in Fig. 3.5b, which in turn gives rise to an enhanced concentration, and an accumulation of holes near the insulator/semiconductor interface is created.

Considering now the metal-semiconductor contact interface (Fig. 3.5c), when a positive voltage is applied to the gate electrode, negative charges are accordingly induced at the source contact. However, the LUMO level of DNTT is quite far away from the Fermi level of gold (with $q\phi_{Au} = 5.0$ eV), so there is a substantial energy barrier that the electrons need to overcome. Thus, electron injection is very unlikely and nearly no current passes through the semiconducting layer [1]. In contrast, when a negative gate voltage is applied, holes can be easily injected to the semiconductor because the Fermi level of gold is very close to the HOMO level of DNTT and the barrier height is relatively low. Given that a gate-induced accumulation channel is created at the insulator/semiconductor interface, charge carriers can now be driven from the source to the drain by applying a second, independent, bias to the drain. Because holes are more easily injected than electrons, the transistors based on DNTT are said to be p-channel OTFTs. Similarly, the transistors are said to be n-channel OTFTs, when the LUMO of the organic semiconductor (such as $F_{16}CuPc$) is closer to the Fermi level of the gold contact than the HOMO.

This concept differs from that of doping in conventional inorganic semiconductors [1]. In particular, the effect of intentional doping in organic semiconductors only appears for large densities of dopants, thus making the materials act as conductors rather than semiconductors. Only interstitial doping can be exploited in the organic materials, where the dopants "flee" in the organic lattice and do not change the composite [30]. The charge transfer between alkali dopants and organic materials lead to n-doping, while the charge transfer between halogen dopants and organic materials lead to p-doping. However, the usual substitutional doping, which is used for inorganic materials, cannot be performed with organics because this process requires the breaking of the chemical bonds that inevitably lead to the loss of the chemical characteristics of the organic material [30]. As noted above, only intrinsic (undoped) organic semiconductors are used in this work.

Furthermore, the small (ohmic) contact barrier shown in Fig. 3.5c for the transfer of holes from the metal into the organic semiconductor contributes to the contact resistance of the OTFT. Besides surface modification techniques [2], this barrier is typically reduced by choosing a source/drain contact metal with a proper work function that matches the LUMO or HOMO energy levels of the organic semiconductor for n-channel or p-channel OTFTs, respectively. By choosing the proper materials, the contact resistance can be accordingly minimized.

Borkan and Weimer have presented in [31] a simplified model for the evaluation of the drain current I_D of TFTs based on gradual channel approximation. In the following, the model is going to be derived for n-channel TFTs; for p-channel TFTs, the polarities of the voltages and currents are just inverted. Referring to Fig. 3.2, the charge at an arbitrary distance x from the source is given by $C_I(V_G - V_S - V_x)$, where C_I is the capacitance per unit area of the insulator layer of thickness d_i ($C_I = \varepsilon_0\varepsilon_i/d_i$), V_G is the gate voltage, V_S is the source voltage and V_x is the potential in the semiconducting film at an arbitrary distance x from the source node ($0 \leq x \leq L$, where L is the channel length). The potential V_x is varying continuously along the channel from V_S at the source node to V_D at the drain node, whereas V_G is constant along the channel. For an initial number of free carriers in the semiconductor per unit area of n_o, the total conducting charge per unit area can be expressed as $n_o q + C_I(V_{GS} - V_x)$, where q is the absolute electron charge. Hence, the drain current I_D can be expressed as

$$I_D = \mu W[n_o q + C_I(V_{GS} - V_x)]\frac{dV_x}{dx}, \tag{3.2}$$

where μ is the charge carrier mobility and dV_x is the potential difference across an incremental section of the channel. By defining $V_{TH} = -n_o q/C_I$, we obtain

$$I_D \int_0^L dx = \mu C_I W \int_{V_S}^{V_D} [(V_{GS} - V_{TH}) - V_x] dV_x. \tag{3.3}$$

After integration, the drain current I_D can be written as

$$I_D = \mu C_I \frac{W}{L}\left[(V_{GS} - V_{TH})V_{DS} - \frac{V_{DS}^2}{2}\right], \tag{3.4}$$

where V_{TH} is the gate threshold voltage on which the conducting channel is created. For $V_{TH} > 0$, the transistor is said to be in the enhancement mode (normally OFF), while for $V_{TH} < 0$, the transistor is said to be in the depletion mode (normally ON). Equation (3.4) is only valid at $V_{GS} > V_{TH}$; the drain current is very close to zero if the gate-to-source voltage is less than the threshold voltage. If a small drain-source voltage is applied, electrons will flow from the source to the drain (the corresponding current will flow from the drain to the source) through the conducting channel [29]. Thus, the channel acts as a resistor and I_D is proportional to the drain voltage. This is the *linear* region, as indicated by the constant-resistor lines (at $V_{DS} < V_{GS} - V_{TH}$) in the measured output characteristics shown in Fig. 3.4b.

When the drain voltage increases, eventually it reaches a maximum of $V_{DS} = V_{DS,sat} = V_{GS} - V_{TH}$ (pinch-off point), at which the drain current I_D starts to remain essentially the same as depicted in Fig. 3.4b. Beyond the pinch-off point (at $V_{DS} \geq V_{GS} - V_{TH}$) is the *saturation* region, since I_D is constant regardless of an increase in the drain voltage. Substituting $V_{DS} = V_{DS,sat}$ in (3.4) yields the following expression for the saturated drain current $I_{D,sat}$:

$$I_{D,sat} = \mu C_I \frac{W}{2L}(V_{GS} - V_{TH})^2. \tag{3.5}$$

Figure 3.4 shows the measured versus simulated data for the I–V output and transfer characteristics of the OTFT with $L = 200\ \mu m$. Here we distinguish between two different simulations. The first is a 2-D device simulation developed by our partners at the University of Ilmenau on Sentaurus Device simulator from Synopsys (designated as simulated in Fig. 3.4), while the second is a simple compact DC model adopted from the equations presented above and implemented on a SPICE simulator (designated as modeled in Fig. 3.4).

Table 3.1 Simulation and material parameters of the p-channel OTFTs [23]

Parameter	Notation	Value	Unit
Al-gate work function	$q\phi_{Al}$	4.1	eV
Au-contacts work function	$q\phi_{Au}$	5.0	eV
DNTT-semiconductor thickness	d	11	nm
AlO_x-dielectric thickness	d_{ox}	3.6	nm
SAM-dielectric thickness	d_{SAM}	1.7	nm
AlO_x/SAM effective permittivity	ε_i	3.37	–
Dielectric capacitance per unit area	C_I	560	nF/cm^2
DNTT highest occupied molecular orbital	HOMO	−5.19	eV
DNTT lowest unoccupied molecular orbital	LUMO	−1.81	eV
DNTT band gap	E_G	3.38	eV
Intrinsic charge carrier mobility	μ_o	2.0	cm^2/Vs
Doping concentration in the DNTT	N_A	1×10^{16}	cm^{-3}
Fixed interface charges concentration	N_{if}	5×10^{11}	cm^{-2}

For the 2-D device simulation, the TFT design is slightly simplified as illustrated in the inset of Fig. 3.4. However, comparative simulations for both structures showed negligible error. The used simulation parameters are summarized in Table 3.1. As for the compact DC model, the following model parameters are used: $\mu = 2.0$ cm^2/Vs, $V_{TH} = -1.0$ V and $C_I = 560$ nF/cm^2. The extraction of the simulation/model parameters is described in the following section. As shown in Fig. 3.4, an excellent agreement between measured, simulated and modeled data using the same device parameters is achieved.

3.3 Transistor Parameters

There are eleven important parameters that are used to characterize and benchmark the (O)TFTs, namely on/off current ratio (I_{on}/I_{off}), subthreshold slope (S_{sth}), turn-on voltage (V_{ON}), transconductance (g_m), channel conductance (g_{ch}), sheet resistance (R_{sheet}), threshold voltage (V_{TH}), insulator capacitance per unit area (C_I), charge carrier mobility (μ), contact resistance (R_C) and current-gain cutoff frequency (f_T). The parameters of the OTFT, of which the measured characteristics are shown in Fig. 3.4, are summarized in Table 3.2. The description and extraction of these parameters are presented in detail herein.

Table 3.2 Measured and extracted characteristic parameters of the p-channel OTFT with $W = 400$ µm, $L = 200$ µm and $L_{ov} = 10$ µm, operating at ≤ 3 V

Parameter	Notation	Value	Unit
Maximum on/off current ratio	I_{on}/I_{off}	10^6	–
Subthreshold slope	S_{sth}	62.3	mV/decade
Transconductance[a]	g_m	4.5	µA/V
Channel conductance[a]	g_{ch}	4	µA/V
Turn-on voltage	V_{ON}	−0.8	V
Threshold voltage	V_{TH}	−1.0	V
Insulator capacitance per unit area	C_I	560	nF/cm^2
Effective charge carrier mobility	μ	2.0	cm^2/Vs
Contact resistance[a]	R_C	2.5	kΩ
Sheet resistance[a]	R_{sheet}	420	kΩ/□
Current-gain cutoff frequency[a]	f_T	1.35	kHz

[a]These transistor parameters are extracted at a gate-source voltage of −3 V.

On/off Current Ratio

The on/off current ratio (I_{on}/I_{off}) is a measure of how efficient the current can be modulated by the gate-source voltage; it is simply the ratio between the on-state drain current at maximum gate-source voltage ($|V_{GS}| = 3$ V) and the off-state drain current at minimum gate-source voltage ($|V_{GS}| = 0$ V). From the measured input transfer characteristics shown in Fig. 3.4a, the on/off current ratios can be easily extracted; the OTFT has I_{on}/I_{off} of about 10^6 and 10^5 in the saturation and linear regimes, respectively.

Subthreshold Slope and Turn-On Voltage

The subthreshold slope ($S_{s\text{-}th}$) is a measure of how fast the transition between the off- and on-states takes place. The subthreshold region is for gate-source voltages below the threshold voltage (V_{TH}) and above the turn-on voltage (V_{ON}), which is the voltage on which the off-state current starts to increase dramatically. In principle, a steep subthreshold slope can be achieved by maximizing the gate dielectric capacitance and/or by utilizing a gate dielectric that forms less trap states at the dielectric/semiconductor interface [2]. Again from the measured input transfer characteristics shown in Fig. 3.4a, the turn-on voltage and subthreshold slope can be extracted; the OTFT has V_{ON} of −0.8 V and $S_{s\text{-}th}$ of 62.3 mV/decade.

Transconductance

The transconductance (g_{m}) describes the response of the drain current to changes of the gate-source voltage at a constant drain-source voltage:

$$g_{\mathrm{m}} = \left.\frac{\partial I_{\mathrm{D}}}{\partial V_{\mathrm{GS}}}\right|_{V_{\mathrm{DS}}}. \tag{3.6}$$

By substituting for I_{D} using (3.4) and (3.5), the transconductance can be calculated in the linear and saturation regimes, respectively:

$$g_{\mathrm{m}} = \begin{cases} \mu C_{\mathrm{I}} \frac{W}{L} V_{\mathrm{DS}} & \text{for } V_{\mathrm{DS}} < V_{\mathrm{GS}} - V_{\mathrm{TH}}, \\ \\ \mu C_{\mathrm{I}} \frac{W}{L} (V_{\mathrm{GS}} - V_{TH}) & \text{for } V_{\mathrm{DS}} \geq V_{\mathrm{GS}} - V_{\mathrm{TH}}. \end{cases} \tag{3.7}$$

Thus, the transconductance is proportional to the gate-overdrive voltage ($V_{\mathrm{GS}} - V_{\mathrm{TH}}$) in the saturation regime, but it is constant in the linear regime.

By substituting for I_{D} using the measured transfer characteristics (Fig. 3.4a) in (3.6), the transconductance (g_{m}) is extracted. The transconductance in the saturation regime (at $V_{\mathrm{DS}} = -2$ V) is found to be proportional to the gate-overdrive voltage ($V_{\mathrm{GS}} - V_{\mathrm{TH}}$) with a maximum of $g_{\mathrm{m,sat}} = 4.5$ μA/V (at $V_{\mathrm{GS}} = -3$ V) and a slope of about 2.4 μA/V^2 (slope $= \mu C_{\mathrm{I}} W/L$). Accordingly, the threshold voltage (V_{TH}) can be deduced; the threshold voltage is found to be $V_{\mathrm{TH}} = -1.0$ V. On the other hand, the transconductance in the linear regime (at $V_{\mathrm{DS}} = -0.1$ V) is found to be fairly constant with a value of $g_{\mathrm{m,lin}} = 0.23 \pm 0.04$ μA/V. All the results correspond closely to what is expected from (3.7).

Channel Conductance and Sheet Resistance

The channel conductance (g_{ch}), which is also referred to as the output conductance, describes the response of the drain current to changes of the drain-source voltage at a constant gate-source voltage:

$$g_{\mathrm{ch}} = \left.\frac{\partial I_{\mathrm{D}}}{\partial V_{\mathrm{DS}}}\right|_{V_{\mathrm{GS}}}. \tag{3.8}$$

Again by substituting for I_{D} using (3.4) and (3.5), the channel conductance can be calculated in the linear and saturation regimes, respectively:

$$g_{\mathrm{ch}} = \begin{cases} \mu C_{\mathrm{I}} \frac{W}{L} (V_{\mathrm{GS}} - V_{\mathrm{TH}} - V_{\mathrm{DS}}) & \text{for } V_{\mathrm{DS}} < V_{\mathrm{GS}} - V_{\mathrm{TH}}, \\ \\ 0 & \text{for } V_{\mathrm{DS}} \geq V_{\mathrm{GS}} - V_{\mathrm{TH}}. \end{cases} \tag{3.9}$$

Note that at very small drain-source voltages ($V_{DS} \ll V_{GS} - V_{TH}$), the expression of g_{ch} reduces to $\mu C_I(W/L)(V_{GS} - V_{TH})$. Ideally, the channel conductance is equal to zero in the saturation regime; in practice, however, it has a finite value, which corresponds to an effect called *channel length modulation* as explained in Chap. 5.

Furthermore, at sufficiently small drain-source voltage ($V_{DS} \ll V_{GS} - V_{TH}$), the gate-induced accumulation channel is assumed to have a nominally uniform thickness. Accordingly, the sheet resistance (R_{sheet}), which is a measure of the resistance of the channel, can be calculated from the channel conductance (g_{ch}) as follows:

$$\frac{1}{R_{sheet}} = \frac{g_{ch} L}{W} = \mu C_I (V_{GS} - V_{TH}) \quad \text{for } V_{DS} \ll V_{GS} - V_{TH}. \tag{3.10}$$

By substituting for I_D using the measured output characteristics (Fig. 3.4a) in (3.8), the channel conductance (g_{ch}) is extracted. In the linear regime, the channel conductance is found to be ranging from 0.7 to 4 μA/V (this corresponds to R_{sheet} of 2.9 MΩ/□ and 500 kΩ/□, respectively) at gate-source voltages ranging from −1.5 to −3 V. In the saturation regime, however, the channel conductance is found to be very close to zero. These values conform closely to what is expected from (3.9).

Threshold Voltage

As mentioned before, the threshold voltage (V_{TH}) is the gate-source voltage on which the conducting channel is created and it is often considered in literature to be the gate-source voltage necessary to achieve the flat-band condition [2]. Using (3.4) and (3.5), the threshold voltage can be estimated by extrapolating the measured I_D (or $\sqrt{I_D}$) in the linear (or saturation) transfer characteristics to the intersection with the x-axis. An alternative approach is to use (3.7), i.e., the transconductance, to extract the threshold voltage in a similar way. All methods yield to approximately the same value of $V_{TH} = -1.08 \pm 0.02$ V.

Insulator Capacitance

To extract the insulator capacitance per unit area (C_I), dedicated metal-insulator-metal (MIM) devices are measured using an LCR (Inductance-Capacitance-Resistance) meter as described in Chap. 6. In principle, the MIM is a two-terminal pendant of the OTFT sharing the same layer structure except for the semiconductor. The average value measured for MIMs with different dimensions (200 × 200, 100 × 100, 50 × 50 and 10 × 10 μm^2) is found to be $C_I = 560$ nF/cm^2. With the hybrid insulator thickness of about $d_i = 5.3$ nm, an effective value of $\varepsilon_i = 3.37$ resulted for the relative dielectric constant. Given that $\mu C_I W/L = 2.4$ μA/V^2, which is found during the extraction of the tansconductance, a charge carrier mobility (μ) of 2.1 cm^2/Vs can be calculated.

Carrier Mobility and Contact Resistance

As mentioned in the previous chapter, the charge carrier mobility is a measure of how fast an electric charge is transmitted through a material under an applied electric field. However, one should distinguish between the intrinsic (μ_o) and the effective (μ) charge carrier mobilities. The intrinsic charge carrier mobility (μ_o) is a property of the material and thus it is a constant. On the other hand, the effective charge carrier mobility (μ) includes other effects such as the contact resistance (R_C). The expression of R_C is derived in Appendix C.2, assuming an ohmic distributed resistance network for our inverted-staggered (bottom-gate, top-contact) OTFTs. Furthermore, the effective charge carrier mobility (μ) decreases as the channel length reduces. This is owed to the increase in the relative contribution of the contact resistance to the total device resistance [9]. The relation between μ, μ_o and R_C can be derived as follows.

Assuming an ohmic contact resistance (R_C) for a symmetrical OTFT, i.e., both the source and drain resistances are equal to the contact resistance ($R_{S,D} = R_C$), the drain current I_D given by (3.4) can be rewritten as the following (V_{DS} is replaced by $V_{DS} - 2I_D R_C$, V_{GS} is replaced by $V_{GS} - I_D R_C$ and μ is replaced by μ_o):

$$\begin{aligned} I_D &= \mu_o C_I \frac{W}{L} \cdot (V_{DS} - 2I_D R_C) \cdot \left(V_{GS} - I_D R_C - V_{TH} - \frac{V_{DS}}{2} + I_D R_C \right) \\ &= \mu_o C_I \frac{W}{L} \cdot (V_{DS} - 2I_D R_C) \cdot \left(V_{GS} - V_{TH} - \frac{V_{DS}}{2} \right) \\ &= \frac{\mu_o L}{L + 2\mu_o W C_I R_C (V_{GS} - V_{TH} - V_{DS}/2)} \cdot C_I \frac{W}{L} \cdot \left[(V_{GS} - V_{TH}) V_{DS} - \frac{V_{DS}^2}{2} \right]. \end{aligned}$$

Hence, the effective charge carrier mobility (μ) in the linear and saturation regimes can be expressed as

$$\mu = \begin{cases} \dfrac{\mu_o L}{L + 2\mu_o W C_I R_C (V_{GS} - V_{TH} - V_{DS}/2)} & \text{for } V_{DS} < V_{GS} - V_{TH}, \\ \dfrac{\mu_o L}{L + \mu_o W C_I R_C (V_{GS} - V_{TH})} & \text{for } V_{DS} \geq V_{GS} - V_{TH}. \end{cases} \tag{3.11}$$

Similar to the extraction of the threshold voltage, the effective charge carrier mobility (μ) can be estimated from the measured linear or saturation transfer characteristics using (3.4) or (3.5), respectively, or from the transconductance using (3.7). By applying any of these approaches to the measured OTFT characteristics (Fig. 3.4), an effective charge carrier mobility of $\mu = 2.1$ cm^2/Vs is found.

As for the intrinsic mobility (μ_o), along with other parameters, such as V_{TH}, R_{sheet} and R_C, they are typically extracted from a general approach called the transmission line method (TLM), which considers the total resistance of transistors with different channel lengths measured directly at a fixed gate-overdrive voltage [27, 32]. Using TLM analysis on several OTFTs at $V_{DS} = -0.1$ V, a set of parameters are extracted; $\mu_o = 2.1$ cm^2/Vs and $V_{TH} = -1.04$ V are obtained, and at $V_{GS} = -3$ V,

$R_{\text{sheet}} = 420\ \text{k}\Omega/\square$ and $R_C \cdot W = 0.1\ \text{k}\Omega\,\text{cm}$ are found (Table 3.2). Detailed description of the TLM is given in Chap. 5. Since the OTFT has a relatively long channel length ($L = 200\ \mu\text{m} \gg L_{\text{ov}} = 10\ \mu\text{m}$), contact effects are insignificant; accordingly, the effective charge carrier mobility in both the saturation and linear operation regimes are very close to the intrinsic mobility.

Current-Gain Cutoff Frequency

The dynamic performance of an OTFT is typically quantified by the current-gain cutoff frequency (f_T), also referred to as transition or relaxation frequency, which is defined as the frequency at which the current gain is unity. In this case, the current gain is the ratio between drain and gate currents. For an (O)TFT, the drain and gate currents are given by $i_D = g_m v_{GS}$ and $i_G = j2\pi f C_G v_{GS}$, respectively, where C_G is the total gate capacitance. The frequency-independent contribution of the gate leakage current is insignificant ($\sim 10^{-12}$–10^{-11} A), and thus, it is neglected.

One of the main objectives of this work is to characterize the dynamic performance of the OTFTs, i.e., the current-gain cutoff frequency (f_T). Unlike other user-configured methods that directly measure gate and drain modulation currents [33–37], the use of self-contained methods in this work by measuring the device admittance or S-parameters are superior, since they only require either an LCR meter or a vector network analyzer (VNA), respectively, to characterize the entire AC electrical properties of the OTFT [23, 38]. These characterization techniques, which are used to extract f_T, are discussed in detail in Chaps. 6 and 7.

Regardless of the device structure (staggered or coplanar) and the fabrication process (printing or vacuum evaporation), the usual lack of self-alignment OTFT process typically necessitates the design of non-negligible overlaps between the source/drain contacts and the gate electrode [14]. Therefore, the total gate capacitance of an OTFT is considered as the sum of the channel capacitance (C_{ch}) and the parasitic gate-overlap capacitances (C_{ov}), i.e., $C_G = C_{ch} + 2C_{ov} \cong C_I W(L + 2L_{ov})$. Consequently the current-gain cutoff frequency can be estimated in the saturation regime by the following expression:

$$f_T = \frac{g_m}{2\pi C_G} \cong \frac{\mu(V_{GS} - V_{TH})}{2\pi L(L + 2L_{ov})}. \tag{3.12}$$

This is often designated in literature as the gain-bandwidth product. Using the extracted parameters for the measured OTFT with $L = 200\ \mu\text{m}$ and $L_{ov} = 10\ \mu\text{m}$ (Fig. 3.4): $\mu = 2.1\ \text{cm}^2/\text{Vs}$ and $V_{TH} = -1.08$ V, a current-gain cutoff frequency (f_T) of 1.46 kHz at $V_{GS} = -3$ V is calculated. This value conforms closely to $f_T = 1.35$ kHz that is extracted by means of admittance measurements as shown in Chap. 6. Finally, equation (3.12) shows that the current-gain cutoff frequency can be increased by using an (organic) semiconductor with a higher carrier mobility or by reducing the (O)TFT lateral dimensions, i.e., the channel or gate-overlap lengths [2]. Using an OTFT technology based on high-resolution silicon stencil masks, we demonstrate in Sect. 7.3 a record cutoff frequency of 3.7 MHz for air-stable, high-mobility p-channel OTFTs with $L = 0.6$ μm and $L_{ov} = 5$ μm operating at low supply voltages ($\leq$3 V).

3.4 Stability and Circuit Performance Measures

The organic semiconductor dinaphtho-thieno-thiophene (DNTT), which was introduced by Yamamoto and Takimiya in 2007 [28, 39], is mainly used in this work as it offers high charge carrier mobility. Haas and co-workers reported a hole mobility as high as 8.3 cm^2/Vs and we have measured a hole mobility of 3 cm^2/Vs for single-crystal and polycrystalline DNTT-based TFTs, respectively [2, 40]. This makes DNTT a good candidate for high-performance applications. The potential use of our OTFTs in the marketplace, however, depends critically on the uniformity of the transistors and the ability of the DNTT to sustain its electrical performance without significant degradation. In general, the electrical stability and performance of OTFTs depend on how they are fabricated and how the chemical compositions of the used materials change over time. The transistors in this work are fabricated by a new OTFT technology based on high-resolution silicon stencil masks, offering a dramatic performance boost to the circuits through sub-micrometer channel lengths and far improved device matching.

This section is organized as follows. In the first part, the device uniformity of our OTFTs is demonstrated. In the second part, the shelf-life stability of the DNTT during exposure to ambient conditions is explored. In the third part, the dynamic performance of p- and n-channel OTFTs by measuring the propagation delay of ring oscillators are presented; in addition, a comparison to the results reported in literature is given. All the measurements are performed at MPI-SSR in ambient air and at room temperature.

3.4.1 Device Matching

As presented in the previous section, The OTFTs are modeled and characterized through parametric testing. The resulting (mostly DC) device parameters either consist of directly measurable observables (currents or voltages under specific biasing and environmental conditions), or of derived model parameters such as resistances, capacitances, threshold voltages, current gains, etc. [41]. These parameters and their spreads determine the performance specs of the OTFTs, including their limits and their performance variations. The functionality of many analog and mixed-signal integrated circuit blocks, such as data converters, comparators, operational amplifiers, depend substantially on the availability of pairs or larger groups of identical transistors (matched transistors). These transistors are typically placed in a close proximity and in an identical layout environment. Furthermore, they experience the

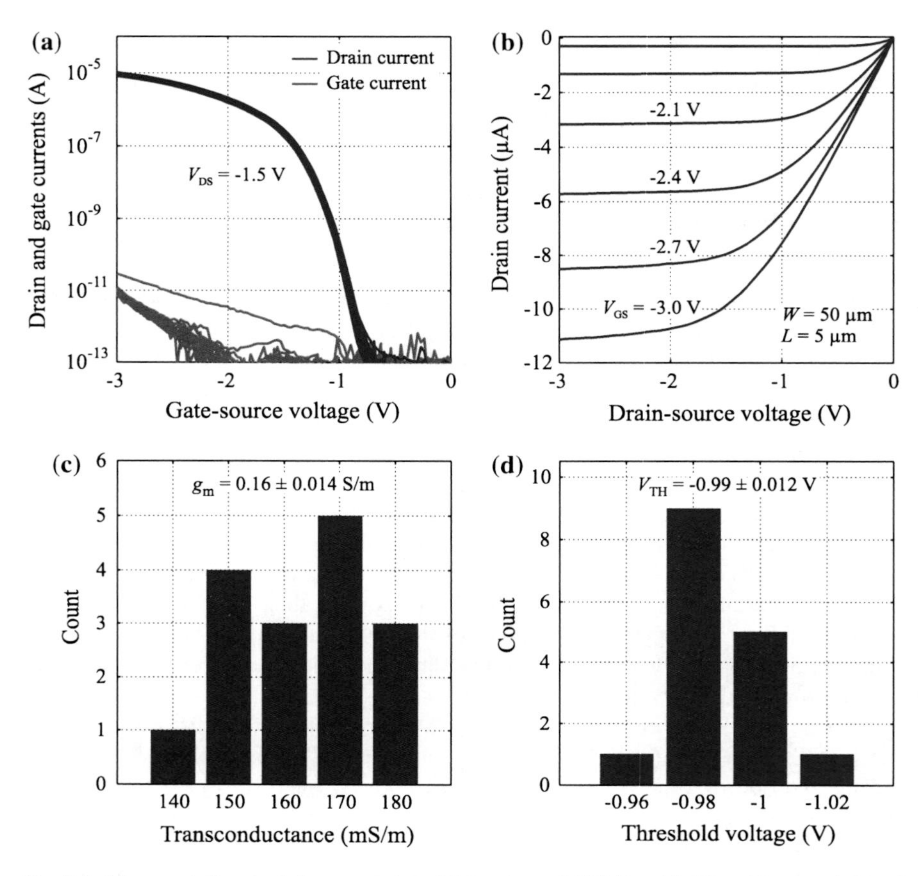

Fig. 3.6 Measured electrical characteristics of 16 p-channel OTFTs with $W = 50$ μm and $L = 5$ μm. **a** Transfer characteristics and gate leakage currents. **b** Output characteristics. **c** Extracted transconductance (g_m). **d** Extracted threshold voltage (V_{TH}) [42]

same technological treatment, and thus, they are generally more identical than devices made on different dies at different times in the process life-cycle. Nevertheless, parametric differences are generally observed even between such matched devices. These parametric mismatches for our OTFTs are discussed herein.

Figure 3.6 depicts the electrical characteristics of 16 DNTT-based p-channel OTFTs with channel lengths of 5 μm fabricated using high-resolution silicon stencil masks on the same substrate. The drain current in the output characteristics shows a linear increase for small drain-source voltages ($V_{DS} < V_{GS} - V_{TH}$) and good current saturation for large drain-source voltages ($V_{DS} \geq V_{GS} - V_{TH}$). Moreover, the transistors show excellent matching with an average transconductance (normalized to the channel width) of 0.16 ± 0.01 Sm^{-1} and an average threshold voltage (defined as the gate-source voltage for which $I_D = 100$ pA) of -0.99 ± 0.01 V. The threshold voltage variation is less than one percent with respect to the maximum supply

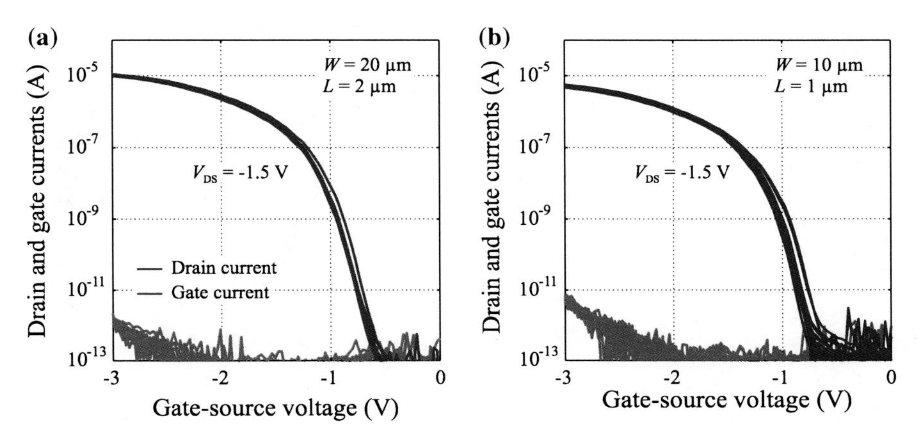

Fig. 3.7 Measured electrical characteristics of matched transistor arrays. **a** 15 p-channel OTFTs with $W = 20$ μm and $L = 2$ μm. **b** 16 p-channel OTFTs with $W = 10$ μm and $L = 1$ μm [9]

voltage of 3 V. From the transfer characteristics, a hole mobility of ~0.6 cm²/Vs, a subthreshold slope of 90 mV/dec and an on/off current ratio of more than 10^8 are extracted.

Furthermore, Fig. 3.7 demonstrates the excellent parameter uniformity of the p-channel OTFTs with channel lengths of 1 and 2 μm. Given that the transistors shown in both Figs. 3.6 and 3.7 have the same aspect ratio (W/L), all the measurements of the 47 p-channel OTFTs show nearly the same characteristics. In contrast to the conventional polyimide shadow mask technology, which provides a minimum of 10 μm for both the channel length and the gate-overlap [43], the high-resolution silicon stencil mask technology allow minimum channel length and gate-overlap down to 0.6 and 2 μm, respectively [9, 18, 38]. Accordingly, a dramatic improvement of the transistors dynamic performance is expected. A comparison between the measured saturation transfer characteristics of 16 identical p-channel OTFTs fabricated using both the polyimide shadow masks and the silicon stencil masks is depicted in Fig. 3.8, noting the noticeable edge roughness in the case of Fig. 3.8a with a channel length of 30 μm in comparison to the very smooth edges in the case of Fig. 3.8b with a channel length of only 2 μm. The figure shows that the 16 OTFTs with $L = 30$ μm and fabricated by the polyimide shadow masks have $\sigma_I/I = 11\,\%$, whereas the 16 OTFTs with $L = 2$ μm and fabricated by the silicon stencil masks have $\sigma_I/I = 4\,\%$.

3.4.2 Air Stability

One of the main challenges in the development of OTFTs is to find a conjugated semiconductor that provides both a large charge carrier mobility and a good stability during operation in ambient air [25]. Taking for example the small-molecule hydrocarbon pentacene, it has been a popular candidate for OTFTs as it can be operated in

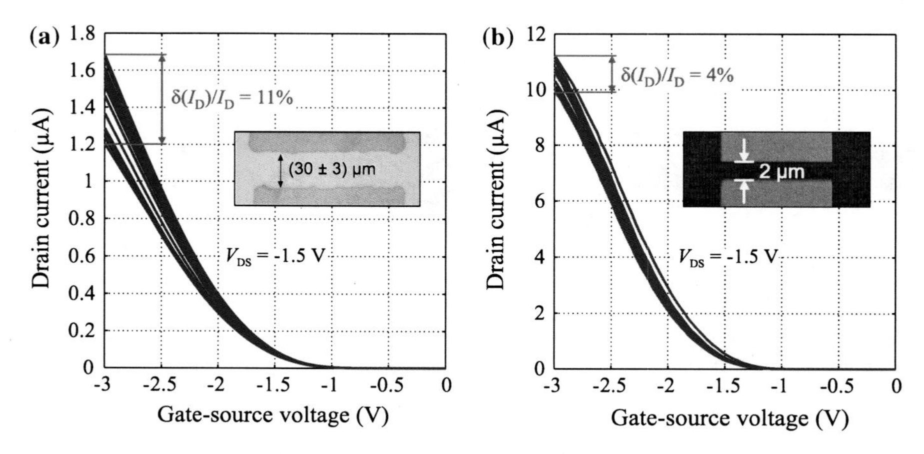

Fig. 3.8 Comparison between the measured saturation transfer characteristics of 16 p-channel OTFTs fabricated by two different technologies. **a** OTFTs with $L = 30$ μm fabricated by conventional polyimide shadow masks. **b** OTFTs with $L = 2$ μm fabricated by high-resolution silicon stencil masks. The insets show the dimensions and edge features of both technologies, exhibiting noticeable edge roughness in case (**a**) and very smooth edges in case (**b**) [42]

air and initially provide a relatively large carrier mobility of about 1 cm^2/Vs [44–46]. However, the main problem of pentacene is that the molecules get easily oxidized in air, and thus, the carrier mobility decreases over time when the devices are exposed to oxygen or water vapor (humidity) [47]. In principle, the mobility degradation can be reduced by encapsulation or storage under an inert atmosphere [48, 49]; nevertheless, air-stable organic semiconductors are very desirable as they greatly simplify the handling of the materials during manufacturing and characterization. Results obtained on DNTT-based OTFTs reveal electrical characteristics with a high carrier mobility of ~3 cm^2/Vs, and more importantly, without significant degradation during exposure to ambient conditions over a period of several months [2, 25, 26]. This is owed to the crystal structure and the thin-film morphology of DNTT, also to its larger ionization potential (5.4 eV [28]) compared to that of pentacene (5.0 eV [50]), which promote higher carrier mobility and better air stability, respectively.

Figure 3.9 shows the evolution of the saturation mobility of DNTT- and pentacene-based OTFTs with $W = 200$ μm and $L = 100$ μm in the course of five months when the substrates are kept in ambient air with a humidity of 40–70 % [26]. The figure depicts that the DNTT-based OTFTs have a substantially better air stability in comparison to the pentacene-based OTFTs fabricated with the same technology, i.e., same type of substrate and same dielectric as well as contact materials. In fact, a closer inspection of the data show that the effective mobility of DNTT-based OTFT initially increases from 2.48 cm^2/Vs for the pristine transistor to a maximum of 3.08 cm^2/Vs after one week, followed by a steady reduction to a minimum of 1.91 cm^2/Vs after five months in air [26]. This initial 24 % improvement in the effective mobility suggest a time-dependent improvement of the contact resistance [2]. Furthermore, the transfer characteristics of the DNTT-based OTFT depicts a threshold voltage

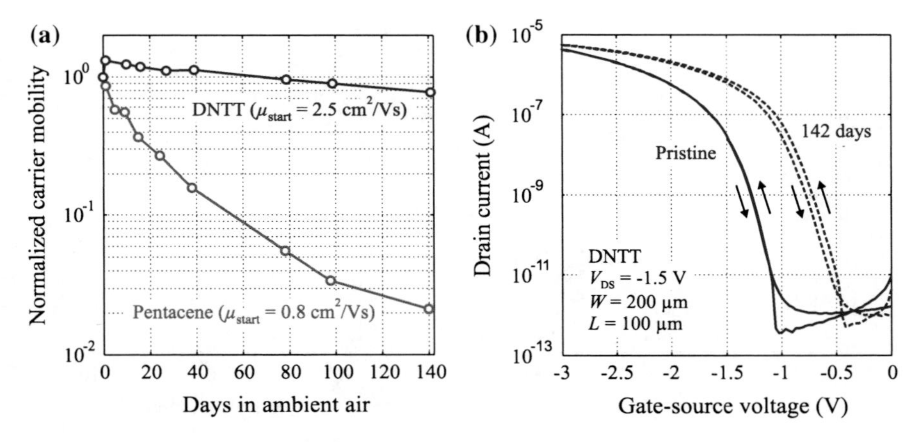

Fig. 3.9 Time-dependent air-induced degradation of the OTFT electrical performance. **a** Evolution of the normalized saturation mobility of DNTT- and pentacene-based OTFTs with $W = 200$ µm and $L = 100$ µm in the course of five months when the substrates are kept in ambient air with a humidity of 40–70 %. **b** Saturation transfer characteristics of the DNTT-based OTFT measured directly after the fabrication (pristine/fresh) and after five months [26]

shift towards positive values by ~0.5 V after five months, but a minor change in the hysteresis behavior, the subthreshold slope ($S_{sth} \cong 90$ mV/dec) and the on/off current ratio ($I_{on}/I_{off} \cong 10^6$).

A general consensus to explain the mechanism for this slow degradation of the carrier mobility in DNTT has not yet been achieved. Although oxidation of the DNTT molecules cannot be completely ruled out, it is unlikely given the molecular structure of DNTT [25]. Another possible explanation is that the carrier transport is hindered when air-borne molecules, such as H_2O, O_2 and O_3, penetrate into the grain boundaries of the polycrystalline DNTT film and then interact with the mobile charge carriers or the DNTT molecules at or near the grain boundaries [51, 52]. Besides the high carrier mobility, promising shelf-life and device matching, the DNTT-based OTFTs feature good bias-stress stability and little hysteresis behavior as demonstrated thoroughly in [25]. In the following, the dynamic performance of our OTFTs is evaluated using ring oscillators; in addition, a comparison to the state-of-the-art results is presented.

3.4.3 Ring Oscillator Speed

The dynamic performance of the OTFTs are typically characterized by measuring the propagation delays of ring oscillators or by extracting the current-gain cutoff frequency from admittance or S-parameter measurements on stand-alone OTFTs. The latter approach is one of the main objectives of this work and is discussed in detail in Chaps. 6 and 7. In this section, however, the frequency limit of both p- and n-channel OTFTs are derived using the former approach, i.e., using ring oscillators.

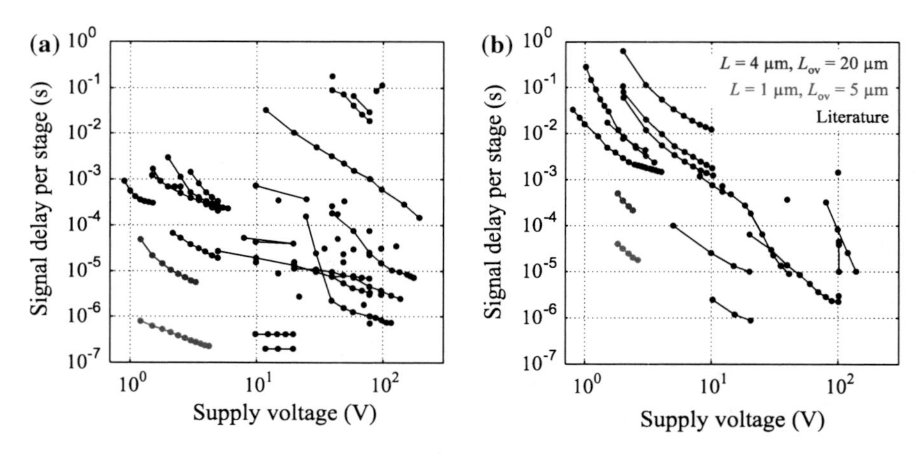

Fig. 3.10 Signal propagation delay per stage of (**a**) unipolar and (**b**) complementary organic ring oscillators. The measured data for the 11-stage ring oscillators are given in blue ($L = 4$ µm, $L_{ov} = 20$ µm, $W_{source}/W_{load} = 110$ µm/24 µm for unipolar, and $W_n/W_p = 80$ µm/40 µm for complementary) and red ($L = 1$ µm, $L_{ov} = 5$ µm, $W_{source}/W_{load} = 72$ µm/24 µm for unipolar, and $W_n/W_p = 80$ µm/40 µm for complementary) colors [9, 27]. The semiconductors employed for the p- and n-channel OTFTs are DNTT and NTCDI-Cl_2-$(CH_2C_3F_7)_2$, respectively. The data given in black color are collected from literature since 1995 [2, 55, 56]. Adopted from [3] with permission from The Royal Society of Chemistry

A ring oscillator is composed of a cascade of odd number of delay stages, namely inverters, connected in a closed loop chain. The inverter computes the logical NOT of its input, and thus, the feedback of the last output to the input results in an oscillation. Each inverter in the loop needs a finite propagation delay τ_d to charge or discharge the parasitic capacitances connected to the output node. Therefore, the oscillation frequency of the ring oscillator depends on the propagation delay τ_d per stage and the number of stages used [53]. To achieve a self-sustained oscillation, the ring must provide a phase shift of 2π and have at least a unity voltage gain. This means that for an n-stage ring oscillator, the oscillating signal must go through each of the n delay stages twice to have a total delay of $2n\tau_d$. Accordingly, the frequency of oscillation (f_{rio}) can be expressed as

$$f_{rio} = \frac{1}{2n\tau_d}. \tag{3.13}$$

Therefore, adding pairs of inverters to the ring oscillator increases the total delay and thereby decreases the oscillation frequency.

Several unipolar and complementary 11-stage ring oscillators are designed and measured. The ring oscillators comprise OTFTs with different channel lengths (L), gate overlaps (L_{ov}) and channel widths for the source (L_s) and the load (L_l) transistors. The design and dimensions are summarized in Appendix A. For the unipolar (p-channel OTFTs) ring oscillator, a biased-load inverter stage is used [54]. The organic semiconductors DNTT and NTCDI-Cl_2-$(CH_2C_3F_7)_2$ are comprised for p- and n-channel OTFTs, respectively. Furthermore, the oscillation frequency (f_{rio})

of the ring oscillators is measured at different supply (and load bias) potentials by an oscilloscope and the signal propagation delay per stage (τ_d) is calculated using (3.13), where $n = 11$. Figure 3.10 shows τ_d as a function of the supply voltage for two complementary and two unipolar ring oscillators along with other data from the literature since 1995 [2, 3, 9, 27, 55, 56]. One should note that the data collected from literature include ring oscillators that are measured not only in air but also under inert or vacuum conditions.

The minimum measured stage delays are 0.2 μs at 4.2 V (the load transistors are biased at -1 V) for the unipolar ring oscillators and 17 μs at 2.6 V for the complementary ring oscillators [9, 27]. For supply voltages below 10 V, the air-stable organic ring oscillators demonstrated in this work are—to our best knowledge—the fastest reported so far.

3.5 Summary

The processing method used for the fabrication of OTFTs in addition to the device architecture and materials play a significant role in realizing the desired electrical performance. Only small-molecule organic semiconductors, namely DNTT for p-channel and $F_{16}CuPc$ or $NTCDI\text{-}Cl_2\text{-}(CH_2C_3F_7)_2$ for n-channel OTFTs are considered. The p-channel OTFTs, in particular, are mostly employed as they are more stable and offer higher carrier mobility than their n-channel contenders. The inverted-staggered (bottom-gate, top-contact) TFT configuration, which benefits from a smooth dielectric/semiconductor interface and provides a small contact resistance, is selected for our OTFTs. The transistors are fabricated by high-resolution silicon stencil masks, allowing to pattern the OTFT layers with sub-micrometer accuracy and without the need of solvents or elevated temperatures. The OTFTs feature excellent device uniformity, promising shelf-life and bias-stress stability, as well as little hysteresis behavior. Organic unipolar and complementary ring oscillators are fabricated and the measured signal delay per stage for both are the lowest reported values at supply voltages below 10 V.

References

1. H. Klauk (ed.), *Organic Electronics: Materials, Manufacturing and Applications*, 2nd edn. (Wiley-VCH, Weinheim, 2008)
2. F. Ante, Contact effects in organic transistors, Ph.D. dissertation, Swiss Federal Institute of Technology in Lausanne, Lausanne, Switzerland, Dec 2011
3. H. Klauk, Organic thin-film transistors. Chem. Soc. Rev. **39**(7), 2643–2666 (2010)
4. Y. Yang, A.J. Heeger, A new architecture for polymer transistors. Nature **372**, 344–346 (1994)
5. M. Yi, X. Xia, T. Yang, Y. Liu, L. Xie, X. Zhou, W. Huang, Vertical n-type organic transistors with tri(8-hydroxyquinoline) aluminum as collector and fullerene as emitter. Appl. Phys. Lett. **98**(7), 073309-1–073309-3 (2011)

6. R. Parashkov, E. Becker, S. Hartmann, G. Ginev, D. Schneider, H. Krautwald, T. Dobbertin, D. Metzdorf, F. Brunetti, C. Schildknecht, A. Kammoun, M. Brandes, T. Riedl, H.-H. Johannes, W. Kowalsky, Vertical channel all-organic thin-film transistors. Appl. Phys. Lett. **82**(25), 4579–4580 (2003)
7. D.J. Gundlach, L. Zhou, J.A. Nichols, T.N. Jackson, P.V. Necliudov, M.S. Shur, An experimental study of contact effects in organic thin film transistors. J. Appl. Phys. **100**(2), 024509-1–024509-13 (2006)
8. C.H. Shim, F. Maruoka, R. Hattori, Structural analysis on organic thin-film transistor with device simulation. IEEE Trans. Electron Devices **57**(1), 195–200 (2010)
9. F. Ante, D. Kälblein, T. Zaki, U. Zschieschang, K. Takimiya, M. Ikeda, T. Sekitani, T. Someya, J.N. Burghartz, K. Kern, H. Klauk, Contact resistance and megahertz operation of aggressively scaled organic transistors. Small **8**(1), 73–79 (2012)
10. S.H. Jin, K.D. Jung, H. Shin, B.G. Park, J.D. Lee, Grain size effects on contact resistance of top-contact pentacene TFTs. Synth. Met. **156**(2–4), 196–201 (2006)
11. K. Sidler, N.V. Cvetkovic, V. Savu, D. Tsamados, A.M. Ionescu, J. Brugger, Organic thin film transistors on flexible polyimide substrates fabricated by full wafer stencil lithography. Sens. Actuators A: Phys. **162**(2), 155–159 (2010)
12. J. Zaumseil, T. Someya, Z. Bao, Y.L. Loo, R. Cirelli, J.A. Rogers, Nanoscale organic transistors that use source/drain electrodes supported by high resolution rubber stamps. Appl. Phys. Lett. **82**(5), 793–795 (2003)
13. J. Zaumseil, K.W. Baldwin, J.A. Rogers, Contact resistance in organic transistors that use source and drain electrodes formed by soft contact lamination. J. Appl. Phys. **93**(10), 6117–6124 (2003)
14. Y.-Y. Noh, N. Zhao, M. Caironi, H. Sirringhaus, Downscaling of self-aligned, all-printed polymer thin-film transistors. Nat. Nanotechnol. **2**, 784–789 (2007)
15. M. Caironi, E. Gili, T. Sakanoue, X. Cheng, H. Sirringhaus, High yield, single droplet electrode arrays for nanoscale printed electronics. ACS Nano **4**(3), 1451–1456 (2010)
16. T. Sekitani, Y. Noguchi, U. Zschieschang, H. Klauk, T. Someya, Organic transistors manufactured using inkjet technology with subfemtoliter accuracy. Proc. Nat. Acad. Sci. USA **105**(13), 4976–4980 (2008)
17. F. Ante, F. Letzkus, J. Butschke, U. Zschieschang, J.N. Burghartz, K. Kern, H. Klauk, Top-contact organic transistors and complementary circuits fabricated using high-resolution silicon stencil masks, in *Device Research Conference*, pp. 175–176 (2010)
18. F. Ante, F. Letzkus, J. Butschke, U. Zschieschang, K. Kern, J.N. Burghartz, H. Klauk, Submicron low-voltage organic transistors and circuits enabled by high-resolution silicon stencil masks, in *IEEE International Solid-State Circuits Conference Technical Digest*, pp. 21.6.1–21.6.4 (2010)
19. T. Ji, S. Jung, V.K. Varadan, On the correlation of postannealing induced phase transition in pentacene with carrier transport. Org. Electron. **9**(5), 895–898 (2008)
20. D.J. Gundlach, T.N. Jackson, D.G. Schlom, S.F. Nelson, Solvent-induced phase transition in thermally evaporated pentacene films. Appl. Phys. Lett. **74**(22), 3302–3304 (1999)
21. A.A. Virkar, S. Mannsfeld, Z. Bao, N. Stingelin, Organic semiconductor growth and morphology considerations for organic thin-film transistors. Adv. Mater. **22**(34), 3857–3875 (2010)
22. H.C. Shah, R.W. Davis Jr, The effect of the roughness of the glass substrate on the roughness of the barrier layer used during fabrication of poly-Si TFTs. Soc. Inf. Disp. Symp. Tech. Dig. **34**(1), 1516–1519 (2003)
23. T. Zaki, S. Scheinert, I. Hörselmann, R. Rödel, F. Letzkus, H. Richter, U. Zschieschang, H. Klauk, J.N. Burghartz, Accurate capacitance modeling and characterization of organic thin-film transistors. IEEE Trans. Electron Devices **61**(1), 98–104 (2014)
24. H. Klauk, U. Zschieschang, J. Pflaum, M. Halik, Ultralow-power organic complementary circuits. Nature **445**, 745–748 (2007)
25. U. Zschieschang, F. Ante, D. Kälblein, T. Yamamoto, K. Takimiya, H. Kuwabara, M. Ikeda, T. Sekitani, T. Someya, J. Blochwitz-Nimoth, H. Klauk, Dinaphtho[2,3-b:2',3'-f]thieno[3,2-b]thiophene (DNTT) thin-film transistors with improved performance and stability. Org. Electron. **12**(8), 1370–1375 (2011)

26. U. Kraft, U. Zschieschang, M.J. Kang, K. Takimiya, T. Zaki, F. Letzkus, J.N. Burghartz, E. Weber, H. Klauk, Evolution of the field-effect mobility and the contact resistance of low-voltage organic thin-film transistors based on air-stable, high-mobility thioacenes, in *Materials Research Society Spring Meeting and Exhibit* (2013) (poster)
27. R. Rödel, F. Letzkus, T. Zaki, J.N. Burghartz, U. Kraft, U. Zschieschang, K. Kern, H. Klauk, Contact properties of high-mobility, air-stable, low-voltage organic n-channel thin-film transistors based on a naphthalene tetracarboxylic diimide. Appl. Phys. Lett. **102**(233303), 233303-1–233303-5 (2013)
28. T. Yamamoto, K. Takimiya, Facile synthesis of highly π-extended heteroarenes, dinaphtho[2,3-b:2',3'-f]chalcogenopheno[3,2-b]chalcogenophenes, and their applications to field-effect transistors. J. Am. Chem. Soc. **129**(8), 2224–2225 (2007)
29. S.M. Sze, *Semiconductor Devices Physics and Technology*, 2nd edn. (Wiley, Hoboken, NJ, USA, 2002)
30. A. Moliton, R.C. Hiorns, The origin and development of (plastic) organic electronics. Polym. Int. **61**(3), 337–341 (2012)
31. H. Borkan, P.K. Weimer, An analysis of the characteristics of insulated-gate thin-film transistors. RCA Rev. **24**, 153–165 (1963)
32. D. Natali, M. Caironi, Charge injection in solution-processed organic field-effect transistors: physics, models and characterization methods. Adv. Mater. **24**(11), 1357–1387 (2012)
33. M. Kitamura, Y. Arakawa, High current-gain cutoff frequencies above 10 MHz in n-channel C_{60} and p-channel pentacene thin-film transistors. Japan. J. Appl. Phys. **50**, 01BC01-1-01BC01-4 (2011)
34. M. Kitamura, Y. Arakawa, Current-gain cutoff frequencies above 10 MHz for organic thin-film transistors with high mobility and low parasitic capacitance. Appl. Phys. Lett. **95**(2), 023503-1–023503-3 (2009)
35. V. Wagner, P. Wöbkenberg, A. Hoppe, J. Seekamp, Megahertz operation of organic field-effect transistors based on poly (3-hexylthiophene). Appl. Phys. Lett. **89**(24), 243515-1–243515-3 (2006)
36. M. Caironi, Y.-Y. Noh, H. Sirringhaus, Frequency operation of low-voltage, solution-processed organic field-effect transistors. Semicond. Sci. Technol. **26**(3), 034006-1–034006-8 (2011)
37. M. Jaiswal, R. Menon, Equivalent circuit for an organic field-effect transistor from impedance measurements under dc bias. Appl. Phys. Lett. **88**(12), 123504-1–123504-3 (2006)
38. T. Zaki, R. Rödel, F. Letzkus, H. Richter, U. Zschieschang, H. Klauk, J.N.Burghartz, S-parameter characterization of submicrometer low-voltage organic thin-film transistors. IEEE Electron Device Lett. **34**(4), 520–522 (2013)
39. T. Yamamoto, K. Takimiya, FET characteristics of dinaphthothienothiophene (DNTT) on Si/SiO_2 substrates with various surface-modifications. J. Photopolym. Sci. Technol. **20**(1), 57–59 (2007)
40. S. Haas, Y. Takahashi, K. Takimiya, T. Hasegawa, High-performance dinaphtho-thieno-thiophene single crystal field-effect transistors. Appl. Phys. Lett. **95**(2), 022111-1–022111-3 (2009)
41. H. Tuinhout, N. Wils, P. Andricciola, Parametric mismatch characterization for mixed-signal technologies. IEEE J. Solid State Circ. **45**(9), 1687–1696 (2010)
42. T. Zaki, F. Ante, U. Zschieschang, J. Butschke, F. Letzkus, H. Richter, H. Klauk, J.N. Burghartz, A 3.3 V 6-Bit 100 kS/s current-steering digital-to-analog converter using organic p-type thin-film transistors on glass. IEEE J. Solid State Circ. **47**(1), 292–300 (2012)
43. H. Klauk, U. Zschieschang, M. Halik, Low-voltage organic thin-film transistors with large transconductance. J. Appl. Phys. **102**(7), 074514-1–074514-7 (2007)
44. M. Kutamura, Y. Arakawa, High-performance pentacene thin-film transistors with high dielectric constant gate insulators. Appl. Phys. Lett. **89**(22), 223525-1–223525-3 (2006)
45. C.Y. Han, W.M. Tang, C.H. Leung, C.M. Che, P.T. Lai, High-performance pentacene thin-film transistors with high-κ HfLaON as gate dielectric. IEEE Electron Device Lett. **34**(11), 1397–1399 (2013)

46. M.W. Alam, Z. Wang, S. Naka, H. Okada, Mobility enhancement of top contact pentacene based organic thin film transistors with bi-layer GeO/Au electrodes. Appl. Phys. Lett. **102**(6), 061105-1–061105-3 (2013)
47. F. De Angelis, M. Gaspari, A. Procopio, G. Cuda, E. Di Fabrizio, Direct mass spectrometry investigation on pentacene thin film oxidation upon exposure to air. Chem. Phys. Lett. **468**(4–6), 193–196 (2009)
48. H.E. Katz, Recent advances in semiconductor performance and printing processes for organic transistor-based electronics. Chem. Mater. **16**(23), 4748–4756 (2004)
49. D. Bode, K. Myny, B. Verreet, B. van der Putten, P. Bakalov, S. Steudel, S. Smout, P. Vicca, J. Genoe, P. Heremans, Organic complementary oscillators with stage-delays below 1 μs. Appl. Phys. Lett. **96**(13), 133307-1–133307-3 (2010)
50. M. Schwoerer, H.C. Wolf, *Organic Molecular Solids* (Wiley, Weinheim, 2007)
51. D. Li, E.-J. Borkent, R. Nortrup, H. Moon, H. Katz, Z. Bao, Humidity effect on electrical performance of organic thin-film transistors. Appl. Phys. Lett. **86**(4), 042105-1–042105-3 (2005)
52. R.T. Weitz, K. Amsharov, U. Zschieschang, M. Burghard, M. Jansen, M. Kelsch, B. Rhamati, P.A. van Aken, K. Kern, H. Klauk, The importance of grain boundaries for the time-dependent mobility degradation in organic thin-film transistors. Chem. Mater. **21**(20), 4949–4954 (2009)
53. M.K. Mandal, B.C. Sarkar, Ring oscillators: characteristics and applications. Indian J. Pure Appl. Phys. **48**, 136–145 (2010)
54. S.-M. Kang, Y. Leblebici, *CMOS Digital Integrated Circuits: Analysis and Design*, 3rd edn. (McGraw-Hill, New York, 2003)
55. M. Ha, Y. Xia, A.A. Green, W. Zhang, M.J. Renn, C.H. Kim, M.C. Hersam, C.D. Frisbie, Printed, sub-3V digital circuits on plastic from aqueous carbon nanotube inks. ACS Nano **4**(8), 4388–4395 (2010)
56. S. Geib, U. Zschieschang, M. Gsänger, M. Stolte, F. Würthner, H. Wadepohl, H. Klauk, L.H. Gade, Core-brominated tetraazaperopyrenes as n-channel semiconductors for organic complementary circuits on flexible substrates. Adv. Funct. Mater. **23**(31), 3866–3874 (2008)

Chapter 4
Fabrication Process

This chapter addresses the fabrication process of the low-voltage submicrometer OTFTs using vacuum evaporation through high-resolution silicon stencil masks. A brief overview of the different deposition and patterning techniques is presented and the reasons of using the stencil lithography, in particular, are explained herein. The stencil masks and the OTFTs are fabricated at IMS CHIPS and MPI-SSR, respectively. Characterization of the stencil masks with respect to refurbishment, defects and uniformity, as part of this work, are discussed in detail. Importance is also given to the design rules of the physical layout. Finally, the design of three mask sets is introduced.

4.1 Deposition and Patterning Techniques

A wide range of fabrication techniques can be used for organic electronics, particularly organic thin-film transistors. The key challenge now is to integrate the organic devices using a cost-effective process at an acceptable yield, resolution and registration. A reasonable classification of the fabrication processes can be made according to the the deposition technique, namely vapor or solution processing [1]. The vapor deposition typically yields smaller dimensions and thus higher thin-film transistor performance. The solution processing, however, is preferable for high-volume, low-cost production. As a rule of thumb, processes with higher resolution often have a lower throughput. Furthermore, solution processing simplifies the manufacturing process compared to vapor deposition, especially for large areas. It is preferable that all layers, and not only the organic semiconductor, are processed from solution. Besides the deposition, solution processing opens up various alternative patterning techniques such as printing, stamping and selective dewetting. Other typical patterning methods used for the fabrication of organic devices are photolithography, stencil lithography and laser ablation. In principle, there is no single standard fabrication process existing today for organic devices and circuitry; however, different

T. Zaki, *Short-Channel Organic Thin-Film Transistors*, Springer Theses,
DOI 10.1007/978-3-319-18896-6_4

deposition and patterning techniques are often combined to optimize for the requirement of each device and/or thin-film layer.

Organic materials are deposited at or near room temperature. Therefore, the fabrication may involve a host number of unconventional large area, flexible or transparent substrates such as glass, paper, textile or plastic films. Each of these substrates has its individual strengths and weaknesses in processability, appearance, yield, lifetime, scalability, sustainability and cost. The appropriate substrate is chosen depending on the target application, functional materials and fabrication technique. Flexible substrates, in particular low-cost, low-temperature plastic substrates such as PET and PEN, limit the process temperature window to achieve the optimum material performance and stability to less than 150 °C [1]. Otherwise, the substrate would exhibit significant distortion when exposed to excess temperatures and in some cases also when exposed to excess humidity variations or mechanical stress. If these distortions occurred during device processing in between the deposition and patterning of two layers, an undesired change in the dimensions would be resulted and thus malfunctioning might occur.

In former times, it was very convenient to take advantage of the well-established inorganic-based fabrication techniques such as oxidation and photolithography. In this context, the OTFTs are built on a highly doped silicon wafer that serves as a global bottom gate. The gate dielectric is thermally grown and then the source/drain contacts are patterned using photolithography. The organic semiconducting layer is in this case deposited lastly on the top to avoid exposure to high temperatures. This results in an inverted coplanar OTFT configuration. However, such process would not make use of the unique properties of the organic semiconductors such as the mechanical flexibility.

The ability to develop materials with well defined structural, chemical, physical or biological functionality that can be processed from solution has enabled the possibility to adapt graphic art printing techniques to manufacture functional devices and circuits [1]. Examples of printing techniques that can be used to process functional materials are flexographic, rotogravure, offset, screen and inkjet printing. However, numerous technological difficulties need first to be overcome before reliable electronic products can be produced by commercial printing techniques. For example, to fabricate all-printed OTFTs, a complete set of solution-processable functional materials must be developed to meet the required processing conditions and desired performance criteria. This includes not only semiconductors, conductors and dielectrics, but also planarization, passivation and isolation materials. Furthermore, the printing equipment and processes must also be improved to meet the rather strict design rules of OTFTs. The typical resolution capability of printing processes is in the range of 20–100 μm.

An alternative to the above printing methods is printing by thermal imaging [1]. It is a digital technique used by the proofing industry and proceeds via the laser-ablative transfer of solid layers. Therefore, the limitations of chemical compatibility in solution-based printing are entirely avoided. In this process, a fully addressable laser beam is directed to a donor film, which is coated with the material to be printed (e.g. conducting layer). A light-to-heat conversion takes place, decomposing an ejecting layer, which is placed in between the donor substrate and the target material, into gaseous state. As a result, a high-temperature bubble is generated that pushes the target film to the substrate with the desired pattern. This fabrication method is compatible with flexible substrates and enables printing of areas up to about 1 m^2 with a resolution down to 5 μm.

For cost-driven applications, such as RFID tags, printing methods are most prominent. At present, silicon RFID tags have relatively high costs (in the range of $0.2). As a result, their use is limited only to applications that can tolerate their price level such as access control and animal tagging. To enable ubiquitous replacement of optical barcodes by RFID tags (maybe integrated with smart sensors), manufacturing solutions for integrated circuits that can be printed directly onto the packaging or an adhesive label at a cost of $0.01 or less must be found [1]. An ultimate goal here is to develop a commercial reel-to-reel solution-based printing technology to manufacture ultra-low cost electronics. However, to allow an early market entry for the organic technology and to investigate its performance limits, industrial companies as well as research facilities tend to use standard, mature processing techniques such as spin coating followed by photolithography (solution-based process) or evaporation through stencil masks (vapor-based process).

On the other hand, high-performance organic devices, especially those that are based on small-molecule organic compounds, are fabricated using vapor-based processes. The pattern definition in this case takes place by stencil lithography using shadow masks. The shadow mask is a thin membrane with a pattern cut out of it, through which the compounds can be deposited on the substrate. Unlike photolithography, this process does not involve any heat or chemical treatment of the substrates.

Conventional vacuum thermal evaporation (VTE) is a commonly used vapor-based deposition technique. It is a high-vacuum evaporation process, where the functional material evaporates from a hot source (e.g. heating filament) and condenses on the target substrate (e.g. glass). The vacuum allows the vapor particles to travel directly to the substrate without collision with the background gas and thus prevents any contamination of parasitic surfaces. VTE is an ideal technique for laboratory usage but it has limitations in scalability and cost effective manufacturing capabilities.

For industrial scale fabrication, however, organic vapor phase deposition (OVPD) is a prominent solution. OVPD employs the gas-phase transport principle, in which the evaporation and the condensation are decoupled. Consequently, the evaporation design can be optimized to any chemical needs without affecting the design for uniform and controlled large-area organic film deposition [1]. In this process, the organic materials are evaporated in individual and decoupled quartz pipes and then carried to the cooled target substrate using an inert carrier gas (e.g. nitrogen). Two or more organic materials (e.g. a host and a dopant or any desired organic composite) can be also simultaneously evaporated, uniformly mixed and finally deposited onto the substrate. In this controlled gas-phase transport technique, no unintended deposition of expensive organic molecules occurs in the main OVPD chamber; consequently, high material utilization is enabled. Furthermore, the use of carrier gas in

OVPD allows the deposition at controlled pressure of 1–10^3 Pa, thus allowing faster deposition because the setup does not have to be pumped down to high-vacuum as in VTE. Finally, the continuous purge of carrier gas in OVPD prevents contamination, which increases the reproducibility of the deposited organic film quality. Using an appropriate industrial OVPD apparatus (e.g. AIXTRON Gen2 OVPD equipment [2]), substrates as large as 400×400 mm^2 can be coated uniformly by more than ten different organic sources.

One of the main objectives in this work is to investigate the maximum performance of organic transistors and circuits. Consequently, top-contact OTFTs based on high-mobility small-molecule conjugated semiconductors with short channel lengths are selected for this study. Given that the operating frequency of a TFT is inversely proportional to its lateral dimension [3], there have been many attempts to fabricate OTFTs with feature sizes below 1 μm using, for example, photolithography [4], sub-femtoliter inkjet printing (SIJ) [5] and self-aligned inkjet printing (SAP) [6]. However, these methods either have relatively small throughput (SIJ, SAP) or involve organic solvents or elevated temperatures (photolithography, SIJ) [3]. Since small-molecule organic compounds often undergo phase transition when exposed to solvents or heat [7, 8], these methods are not suitable to pattern the source/drain contacts on top of such semiconductor. Alternatively, stencil lithography offer a virtue in patterning the top contacts without the need of solvents or elevated temperatures. Moreover, it allows to pattern the organic semiconductor without exposing it to potentially harmful process chemicals (e.g. photoresist), organic solvents or even water. This makes the clean, simple and versatile fabrication technique by evaporation through shadow masks the most favorable for our purpose. In cooperation with the Institute for Microelectronics Stuttgart (IMS CHIPS) and the Max Planck Institute for Solid-State Research (MPI-SSR), we have developed a five-level shadow-mask process that utilizes high-resolution silicon stencil masks to fabricate submicrometer OTFTs with excellent transistor matching [3, 9–11]. This process has led to a considerable improvement in the dynamic performance of OTFTs as demonstrated in this work. In the following sections, the process steps for the fabrication of the shadow masks as well as the OTFTs are presented.

A rough comparison between all the fabrication processes with respect to their throughput and resolution capabilities is given in Table 4.1. The processes are classified to three different technology levels, namely wafer, hybrid and continuous technologies [12]. First, the wafer technologies are those that allow the fabrication of high resolution features but at relatively high production cost; however, they are compatible with mechanically flexible substrates and can be extended to rotary (sheet-to-sheet) processes. Second, the hybrid technologies are those that include large-area processes that make use of flexible substrates and/or produced at medium cost level. Finally, the continuous technologies are the reel-to-reel processes that enable high-volume, low-cost mass production but so far with medium resolution.

Table 4.1 Comparison between the fabrication processes used for organic electronics with respect to resolution and throughput

Technology level	Fabrication process	Feature size (μm)	Throughput (m^2/s)
Wafer	VTE[a]	1–10	$<10^{-4}$
	OVPD	5–20	$<10^{-4}$
Hybrid	Rotary photolithography	3–5	10^{-4}–10^{-1}
	Thermal imaging	5–40	10^{-4}–10^{-3}
	Inkjet printing	15–50	10^{-3}–1
	Screen printing	20–100	10^{-2}–10
Continuous	Offset printing	10–50	1–10^{2}
	Rotogravure printing	20–80	10^{-1}–10^{2}
	Flexographic printing	40–80	1–10^{2}

Data is adopted from [1, 12]
[a] Using high-resolution silicon stencil masks, it is also feasible to fabricate sub-micrometer feature sizes as demonstrated in Chap. 7

4.2 Silicon Stencil Masks

Stencil masks, which are also referred to as shadow masks, are used in stencil lithography to fabricate micro- and nano-meter scale patterns. A stencil mask is a membrane with a pattern cut out of it, through which evaporated or sputtered particles can be deposited marking the same pattern on the substrate. In principle, there are numerous application areas for stencil masks and they vary from medical cell filters to exposure masks for semiconductor lithography. Depending on the used radiation, which can be photons, charged particles, or atoms, different technical implementations of stencil masks are reported in literature. For example, they have been exploited in classical beam shaping in semiconductor applications for silicon wafer resist exposures with single/multi e-beam lithography [13], ion-beam lithography [14], ion-assisted etching and milling [15] and ion-implantation for patterned magnetic media [16].

The quality of the stencil masks is very critical as they define the final pattern properties, including minimum feature size and line edge roughness, thus directly affecting the device performance. Different materials have been explored for the fabrication of stencil masks such as metals, polymers and silicon nitride; however, these materials suffer from patterning and stability problems for membrane areas above 100 mm^2. As a result, monocrystalline silicon stencil masks are selected for this study as they offer virtues with respect to mechanical stability, stiffness and nano-patterning capability [10].

A novel silicon stencil mask process flow is developed at IMS CHIPS for the fabrication of OTFTs. This groundbreaking work in cooperation with the MPI-SSR led to a considerable improvement in the device uniformity and performance [9, 10]. One should note that the stencil mask properties are changing during operation due to an interaction between the used radiation and the mask itself. This produces stress variations within the membrane and changes in the critical dimensions (CD) of the mask pattern due to residues in the openings, which worsen the image of the stencil mask during illumination. During the fabrication of the OTFTs, materials are continuously deposited onto the mask membrane and stencil openings, which results in clogging, stress induced pattern distortions and membrane deformations [10]. Therefore, etching of the deposited materials, cleaning and inspection is mandatory to maintain a good pattern fidelity. The silicon stencil mask process flow is explained in detail in this section. Furthermore, details about minimum feature sizes, refurbishment and defects, which are part of this work, are presented.

4.2.1 Fabrication Process

The fabrication of the stencil masks can be implemented in two techniques, namely membrane flow process (MFP) and wafer flow process (WFP) [17, 18]. In the MFP, a membrane has to be first prepared from a solid wafer, then all the patterning processes are applied on the membrane. In the WFP, however, all the critical patterning steps are carried out on the solid wafer substrate and the membrane etching is applied later. The latter option is favorable for our purpose because it is compatible with the standard process tools and silicon wafer fabrication processes.

The fabrication process of the silicon stencil masks is described in detail in Table 4.2 and a corresponding illustration is depicted in Fig. 4.1. A 150 mm (6 in.) silicon-on-insulator (SOI) wafer is used as a base substrate. The buried silicon oxide layer serves as an etch stop for the stencil etching from the wafer frontside and the membrane etching from the wafer backside. This offers a clear separation between the *microelectrical* patterning steps and the *micromechanical* membrane etching steps. As a result, risk of contamination is significantly reduced.

The fabrication of the fully-patterned top-contact OTFTs requires at least a set of four masks for the following layers: (i) gold metal interconnects, (ii) bottom aluminum gate, (iii) organic semiconductor, and (iv) top gold metal contacts. In the case of a complementary circuit design, which comprises both p- and n-channel OTFTs, an extra mask is employed. The 150 mm (6 in.) SOI wafer is capable of carrying nine individual masks with an area of $30 \times 30\ \text{mm}^2$. Therefore, the complete mask set can be easily fabricated on a single wafer. To guarantee proper yield, the remaining five (or four) slots on the wafer are utilized for mask duplicates, especially for the last mask, i.e., the top gold metal contacts mask, which has the most critical structures with the minimum feature size. The design rules for the physical layout of a particular circuit, which are necessary to ensure a successful fabrication, are discussed in Sect. 4.4. Furthermore, the different mask sets designed for this work are presented in Sect. 4.5.

In the following, emphasis is made on the characterization of the silicon stencil masks, with respect to refurbishment, defect analysis and uniformity of evaporated patterns, to achieve a good pattern fidelity and a cost effective mask technology.

Table 4.2 Fabrication process steps of the silicon stencil masks [10]

No.	Process	Description
1	SOI wafer preparation	A 150 mm (6 in.) wafer with 20-μm-thick Si membrane and 200-nm-thick buried SiO_2 is used
2	Silicon nitride (Si_3N_4) deposition on the wafer backside	This layer acts as a mask for later wet membrane etching
3	Tetraethyl orthosilicate (TEOS) deposition on the wafer frontside	This layer acts as a hardmask for later silicon trench etching. A thickness of 1 μm is carefully chosen to minimize the stress on the wafer
4	Resist coating on the wafer frontside and pattern definition by e-beam lithography	A chemically amplified resist FEP 171 (Fuji) is used and the stencil pattern is written in a 50 kV Leica SB350 MW variable shaped e-beam writer
5	TEOS hardmask etching and resist removal	This is to transfer the resist pattern into the TEOS hardmask
6	Silicon trench etching	Deep reactive ion etching (DRIE) using gas chopping etching technique is implemented
7	Resist coating on the wafer backside and pattern definition by e-beam lithography	This is to define the size of each membrane to be 20×20 mm^2. A total of 9 masks are patterned on the wafer
8	Si_3N_4 etching and resist removal	This is to transfer the resist pattern into the Si_3N_4 masking layer
9	TEOS deposition on the wafer frontside	This layer is to seal the stencil pattern from later wet etching processes
10	Silicon membrane etching from the wafer backside	The 9 membranes are etched in a potassium hydroxide (KOH) solution to a preliminary thickness of 30 μm
11	Wafer sawing and etching of the remaining silicon from the wafer backside	The wafer is dismantled and sawn into 9 single masks with 30×30 mm^2 outer dimensions. The remaining silicon on the backside of each separated mask is etched in tetra methyl ammonium hydroxide (TMAH) solution
12	Removal of all remaining layers, cleaning and inspection of each of the 9 masks	Remaining dielectrics are removed in a HF solution. This leaves 20-μm-thick patterned Si membranes anchored to robust 5-mm-wide frames

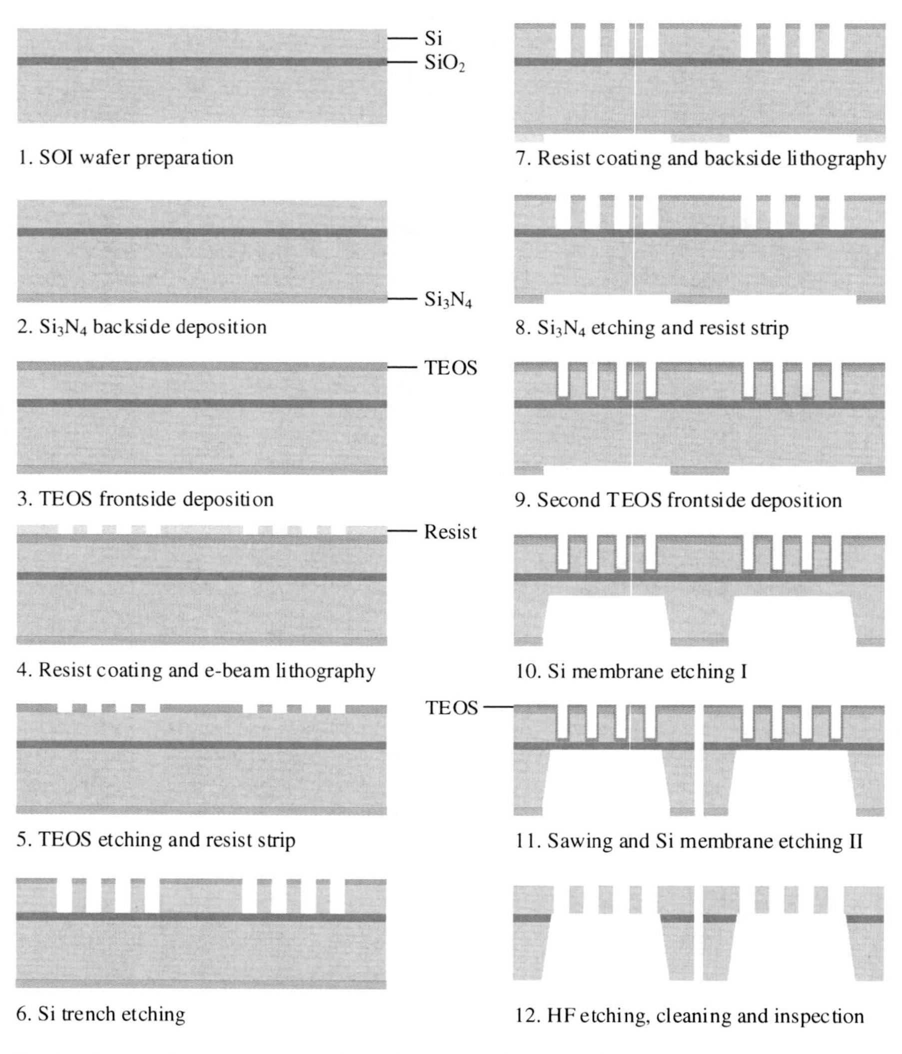

Fig. 4.1 Fabrication process steps of the silicon stencil masks [10]

To carry out this study on the manufacturing process, a special testing mask is designed thereto. Figure 4.2 shows the layout of this mask, containing a matrix of source/drain and gate patterns with various dimensions. The mask comprises a total of 5×7 cells that cover the complete active mask area, where each cell has eight row modules (seven for the source/drain and one for the gate patterns). The source/drain patterns have channel widths and lengths of 100–2 μm and 4–0.3 μm, respectively, whereas the gate patterns have widths of 8–0.3 μm and lengths of 100 μm.

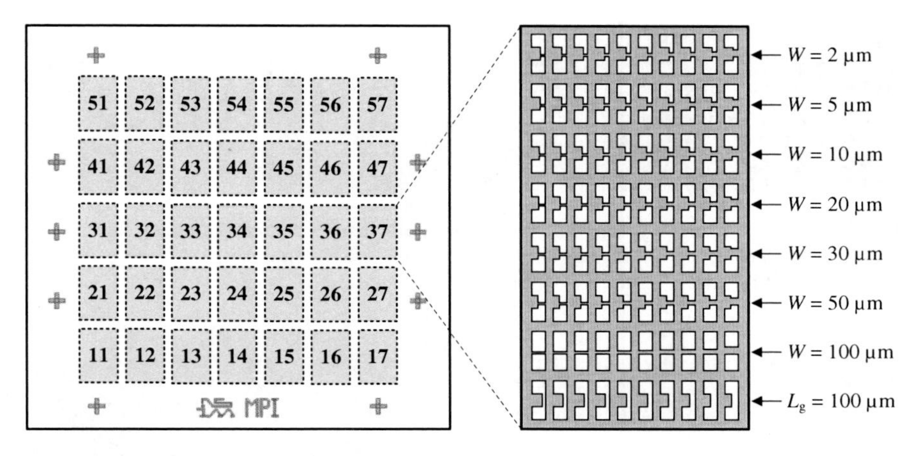

Fig. 4.2 Simplified layout of the test mask, containing a matrix of 5 × 7 cells, where each cell has seven row modules for the source/drain contacts with channel widths (W) of 2–100 μm and channel lengths (L) of 4, 2, 1, 0.9, 0.8, 0.7, 0.6, 0.5, 0.4 and 0.3 μm, and one row module for the gate patterns with lengths (L_g) of 100 μm and widths (W_g) of 8, 6, 4, 2, 1, 0.8, 0.6, 0.5, 0.4 and 0.3 μm

4.2.2 Refurbishment

For the fabrication of the OTFTs, two different metals, i.e., aluminum and gold, are used to define the gate and the source/drain contacts. Therefore, two different wet-chemical metal etching processes are developed at IMS CHIPS for the selective removal of the two metals to the base silicon stencil mask material [10]. This is necessary to ensure a cost effective and multiple usage of the silicon stencil masks in the deposition processes. To evaluate the reliability of the etching processes, initial CD measurements are carried out for two silicon stencil masks, then after aluminum and gold deposition steps, and finally after the metal removal and cleaning.

Figure 4.3 shows scanning electron microscope[1] (SEM) top-down images of a 0.5-μm-wide isolated line (bridge) initially before any processing, after a sequence of gold depositions and finally after gold etching. Furthermore, Fig. 4.4 shows SEM top-down images of a 2-μm-wide isolated space (hole) initially before any processing, after a sequence of aluminum deposition and finally after aluminum etching. The widening of the bridge and the closure of the hole due to the metal deposition are obvious in both figures. Note that the deposition steps are carried out in this experiment on an older mask sample using sputter tool and not on the standard OTFT evaporation tool due to time and availability reasons. The gold layer is removed in a concentrated nitro-hydrochloric acid (HNO_3 + 3 HCL) solution with an etch rate of approximately 7 nm/s. Moreover, the aluminum layer is removed in a hydrofluoric

[1]The scanning electron microscope (SEM) is an electron microscope that produces an electronic image by moving a beam of focused electrons over an object and reading both the electrons scattered by the object and the secondary electrons produced by it.

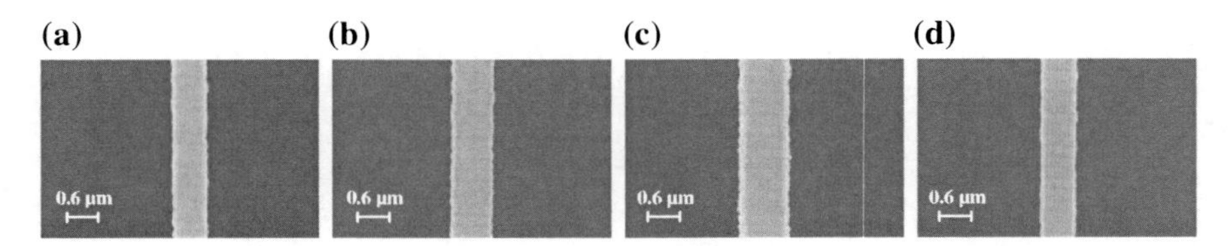

Fig. 4.3 SEM top-down images of a 0.5-μm-wide isolated line (bridge) in the silicon stencil mask. **a** Initially before any processing. **b** After the deposition of 0.2-μm-thick gold layer. **c** After the deposition of 0.5-μm-thick gold layer. **d** After gold etching [10]

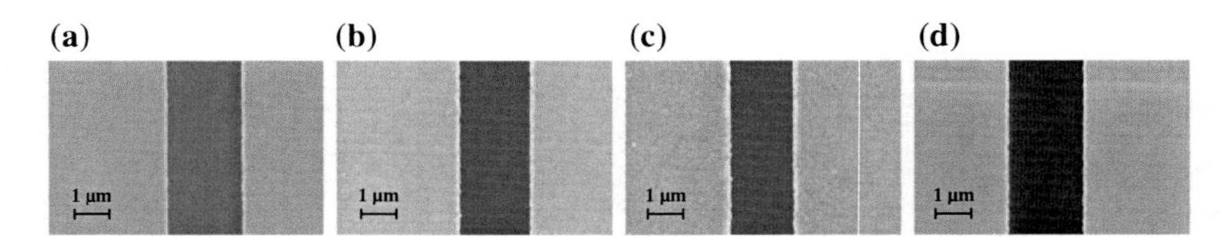

Fig. 4.4 SEM top-down images of a 2-μm-wide isolated space (hole) in the silicon stencil mask. **a** Initially before any processing. **b** After the deposition of 0.1-μm-thick aluminum layer. **c** After the deposition of 0.3-μm-thick aluminum layer. **d** After aluminum etching [10]

acid (15 %wt. HF) solution at an etch rate of approximately 18 nm/s. For both metal etching processes a 100 % over-etch time is applied.

The CD measurement results after the gold and aluminum deposition sequence for 0.5–1-μm-wide bridges and 1–3-μm-wide holes, respectively, are shown in Fig. 4.5. A comparison between the initial and the final CD measurements demonstrates the successful removal of both metal layers for a wide range of dimensions.

4.2.3 Defect Analysis

The evaporated gate patterns are investigated by evaporating a 30-nm-thick gold layer and measuring the corresponding current–voltage (*I*–*V*) characteristics for the different widths. A linear *I*–*V* behavior should be measured indicating no physical separation in the gate lines. Figure 4.6a depicts the measurements for widths of 8, 6, 4, 2 and 1 μm. For the width of 1 μm, no electrical conductance is measured, which results from an insufficient deposition of gold through the 20 μm deep and 1-μm-wide stencil trench opening. Assuming a homogeneous gold thickness, the gradient of the *I*–*V* plot should reflect the electrical conductance of the evaporated gold layer, which should be proportional to the cross-section area of the conductor. This complies for widths of 8, 6 and 4 μm, but not for 2 μm, indicating that the evaporation limit in this process is at an aspect ratio of 1:10.

It is found that most defects in this fabrication process occur during the final membrane etch step in the TMAH solution. This results due to defects in the TEOS

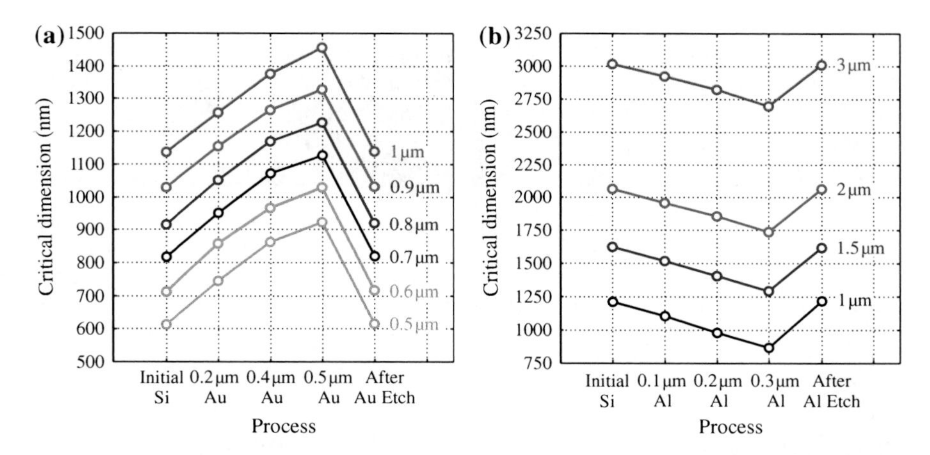

Fig. 4.5 CD measurements of silicon stencil mask patterns initially before any processing, after a sequence of gold or aluminum depositions and finally after metal etching for **a** 0.5–1-μm-wide bridges and **b** 1–3-μm-wide holes. A comparison between the initial and the final CDs for all the given dimensions demonstrates the successful removal of both gold and aluminum layers [10]

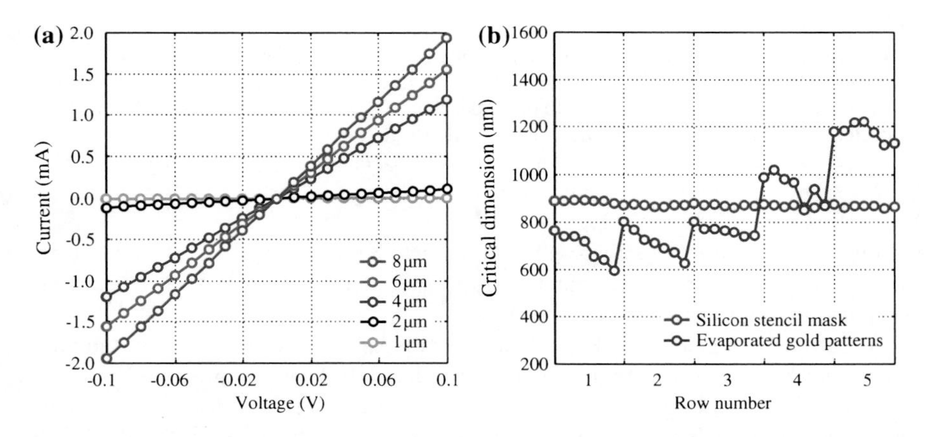

Fig. 4.6 **a** Measured *I*–*V* characteristics of 1–8-μm-wide evaporated gate modules. **b** CD measurements of 0.8-μm-wide and 30-μm-long bridges in the stencil mask and the corresponding evaporated gold source/drain patterns on a test sample; for each row module, there are seven data points, which correspond to the seven columns from the *left* to the *right side* of the sample [10]

protection layer. As shown in Fig. 4.7, this kind of defects occurs at the wafer frontside and is determined by a pyramidal shaped anisotropic silicon etch that is located at the edge of the bridge. In the figure, the defect is located only at one end of the bridge, but sometimes it occurs at both ends. To resolve this issue, a change in the membrane etch strategy is implemented into the process flow. By just reducing the etch time in the TMAH solution by about 50 %, the number of defects are reduced by a factor of five. This is because the exposure time of the wafer frontside to the etch solution

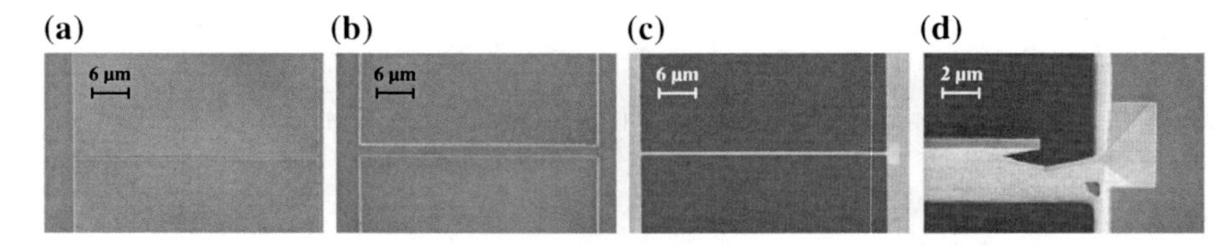

Fig. 4.7 SEM top-down images of the silicon stencil mask pattern that is used to define the source/drain contacts of an OTFT with a channel width of 50 μm and a channel length of 0.5 μm. **a** After silicon trench etching during the process flow of the stencil mask. **b** After TEOS protection layer deposition during the process flow of the stencil mask. **c** After the stencil mask is ready but with a defect on the right edge of the bridge. **d** A close-up image of the defect region showing a pyramidal shaped anisotropic silicon etch [10]

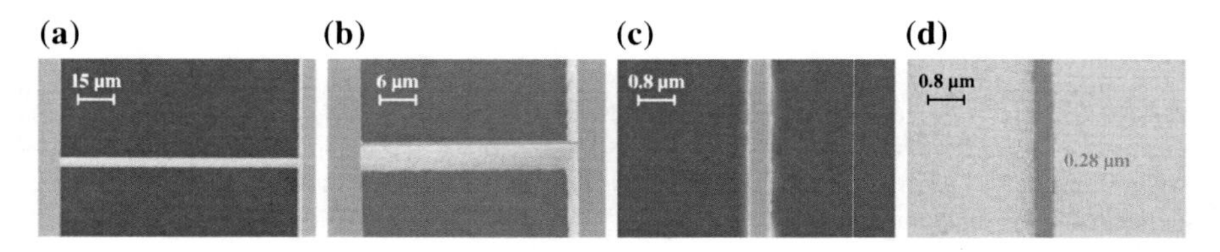

Fig. 4.8 SEM images of the minimum feature size of 0.3 μm in the silicon stencil mask and on a test sample. **a** Source/drain contacts pattern in the stencil mask with a channel width of 100 μm and a channel length of 0.3 μm (10° SEM tilt angle). **b** Source/drain contacts pattern in the stencil mask with a channel width of 30 μm and a channel length of 0.3 μm (10° SEM tilt angle). **c** A close up image of the 0.3-μm-wide stencil mask bridge (top-down SEM). **d** The corresponding 25-nm-thick gold source/drain contacts that are evaporated on a test sample (top-down SEM) [10]

is much less than before. Table 4.3 summarizes the number of defects when using the two etching schemes for channel widths of 100 and 30 μm and various channel lengths ranging from 4 μm down to 0.3 μm.

Figure 4.8 depicts stable stencil features with excellent sharp edges for the minimum feature size of 0.3 μm and channel widths of 100 and 30 μm. By evaporating a 30-nm-thick gold layer onto a test sample, the channel length of 0.3 μm could be successfully realized without electrical shortage as illustrated in Fig. 4.8d. This implies the potential use of the silicon stencil masks to fabrication submicrometer OTFTs.

4.2.4 Uniformity of Evaporated Patterns

The uniformity of the evaporated gold pattern is investigated by measuring the CDs over the complete $20 \times 20\ \text{mm}^2$ test sample and compare them to the CDs on the stencil mask. Figures 4.9 and 4.10 show the SEM top-down images of the 0.8-μm-wide bridges in the stencil mask and the corresponding evaporated gold source/drain patterns on a test sample, respectively. The figures illustrate the variations in the

Table 4.3 Comparison between the number of defects of the stencil mask bridges when using the old and the new etching process strategies

W/L	Number of defects		W/L	Number of defects	
(μm/μm)	Old	New	(μm/μm)	Old	New
100/4	1	0	30/4	0	0
100/2	4	1	30/2	0	0
100/1	7	0	30/1	5	0
100/0.9	11	2	30/0.9	4	0
100/0.8	13	3	30/0.8	5	1
100/0.7	15	2	30/0.7	5	0
100/0.6	15	1	30/0.6	10	3
100/0.5	15	4	30/0.5	14	2
100/0.4	16	3	30/0.4	11	3
100/0.3	24	4	30/0.3	10	3
Total	121	21	Total	64	12

By reducing the etch time in the TMAH solution by about 50 % in the new process, the number of defects are reduced by a factor of five [10]

evaporated patterns compared to the uniformity of the bridge structures on the silicon stencil mask at the center and the four corners of the entire sample.

Figure 4.6b depicts the complete analysis with respect to the position of the structures on the sample. All the bridges with the widths of 0.8 μm and lengths of 30 μm are measured over the complete five rows and seven columns of the sample (Fig. 4.2). The mean values of the CDs are found to be 0.88 ± 0.01 μm for the silicon stencil mask and 0.86 ± 0.19 μm for the evaporated source/drain gold patterns. As shown in the figure, the CD reduces within every single row, which indicates different distances between the mask and the substrate during evaporation. Given also the less sharp edges of the evaporated gold patterns at the top-left and top-right corners (at positions 51 and 57) shown in Fig. 4.10d, e, respectively, the larger CDs found for positions 41–57 indicate a tilt of the stencil mask during evaporation and/or the mask is outside the perpendicular projection of the evaporator source.

In the fabrication process of the OTFTs, the stencil masks are manually mounted on the substrate and the alignment is done under a light microscope with the help of box-in-box and vernier structures. In an industrialized manufacturing process, the mask treatment should be automated, and thus, this kind of non-uniformity would be minimized.

4.3 Organic Transistor Fabrication

After the preparation of the high-resolution silicon stencil masks at IMS CHIPS, they are delivered to the MPI-SSR to fabricate the organic transistors and circuitries. The fabrication of the OTFT-based circuits requires at least a set of four different

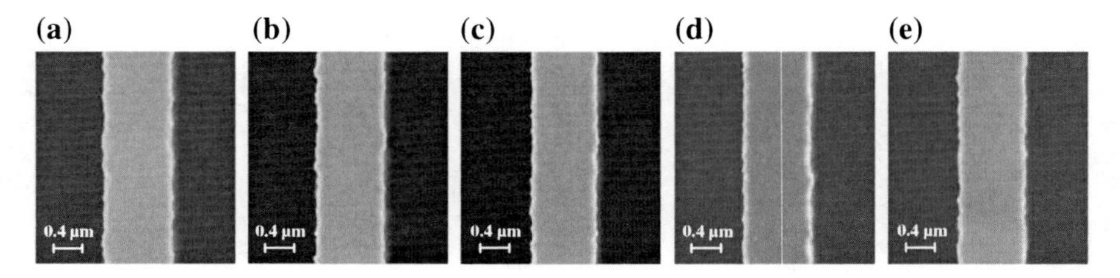

Fig. 4.9 SEM top-down images of 0.8-µm-wide isolated lines (bridges) in a silicon stencil mask at the center and the four corners of the entire pattern: **a** *bottom-left corner*, **b** *bottom-right corner*, **c** center, **d** *top-left* corner, and **e** *top-right* corner [10]

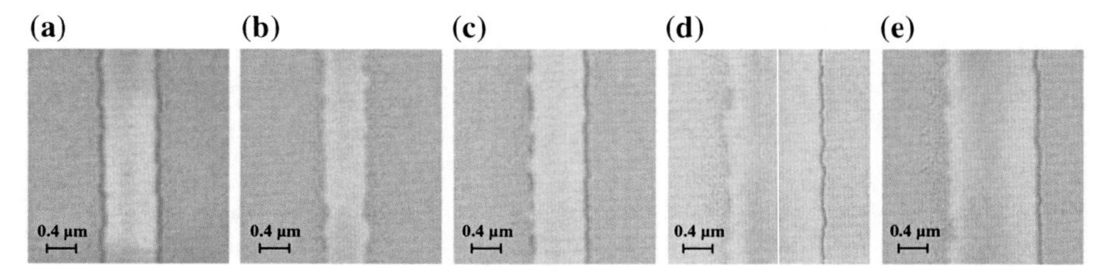

Fig. 4.10 SEM top-down images of 0.8-µm-wide evaporated gold souce/drain patterns on a test sample at the center and the four corners of the entire ample: **a** *bottom-left corner*, **b** *bottom-right corner*, **c** center, **d** *top-left* corner, and **e** *top-right* corner [10]

stencil masks for the following layers: (i) gold metal interconnects, (ii) aluminum gate electrodes, (iii) organic semiconductor, and (iv) gold source/drain contacts. In general, stencil lithography is compatible with a variety of substrates such as rigid silicon wafers, transparent glass or flexible polyethylene naphthalate foils [19]. In this work, glass substrates Eagle^{2000} from Corning are used because of their very smooth surface roughness of about 0.2 nm (root mean square) [20], which makes it easier to fabricate high-performance OTFTs. Atomic force microscopy (AFM) measurements of a 30-nm-thick aluminum on a silicon wafer and on a glass substrate show a surface roughness of about 1 nm for both materials. As a result, no differences in the electrical characteristics of the transistors fabricated on either of the substrates are expected [19, 21]. However, glass is preferable because glass as a non-conducting substrate is better than conducting silicon for the dynamic measurements, which is one of the main subjects of this work. This is because the large parasitic capacitances between the probe pads and the conducting substrate are avoided [19].

The fabrication process of the organic transistors is described in detail in Table 4.4 and the corresponding illustration is depicted in Fig. 4.11. For simplification, the stencil masks are shown in the figure to be suspended from the substrate during evaporation. In practice, however, three steps are necessary with every single mask. At first the mask is aligned and fixed to the substrate, second evaporation of material with the desired thickness takes place, and finally the mask is released carefully from the substrate to avoid any scratches on the sample. The alignment of the stencil masks to the substrate is done under a light microscope with the help of box-in-box and vernier structures with an accuracy of about 2–5 μm. In the final evaporation step, the alignment is very critical due to the definition of the gate-source and gate-drain overlaps, which determine the dominating parasitics of the transistor and thus the dynamic performance [22].

The gate dielectric of the OTFTs is composed of two-stacked layers, namely a 3.6-nm-thick aluminum oxide (AlO_x) layer and a 1.7-nm-thick self-assembled monolayer (SAM). The hybrid gate dielectric has a capacitance of the order of 1 $\mu F/cm^2$, which is sufficiently high to allow the transistors to operate at voltages as low as 3 V [23]. Although, the SAM adds only 1.7 nm to the total dielectric thickness, it reduces the leakage current density by three orders of magnitude from 5×10^{-5} A/cm^{-2} to $(5 \pm 1) \times 10^{-8}$ A/cm^{-2} at 2 V [24]. Furthermore, the gate dielectric has a low surface energy of 20 mN/m [25], which is useful to achieve a high carrier mobility in the small-molecule organic films deposited on top of this gate dielectric [19].

Referring to Fig. 4.12, the design cycle of the organic integrated circuits (ICs) involves several steps. In the beginning, a feasibility study and an IC size estimate is made. Accordingly, one or more ICs along with duplicates, alignment and test structures are planned to be included on the active die area of 20×20 mm^2. Using OTFT models, the circuit(s) are designed and simulated on a SPICE (Simulation Program with Integrated Circuit Emphasis) simulator (e.g. LTspice or Cadence Virtuoso).

Subsequently, the physical layout of the die is implemented (Fig. 4.12), including floorplanning, design for manufacturability (DFM) and design rule checking (DRC). Attention must be given to the interconnect crossings and the positioning of the transistors as they affect significantly the device performance and matching, respectively. For DFM, the design is modified to make it easier and more efficient to produce, also to comply to the design rules, which are set by IMS CHIPS and MPI-SSR. This is achieved for example by improving the stability of the stencil masks by adding support grids (anchors), where possible, to hold sensitive structures on the mask.

Table 4.4 Fabrication process steps of the organic transistors [9, 19, 24]

No.	Process	Description
1	Substrate preparation	An alkali-free glass substrate Eagle^{2000} from Corning is used. The substrate is initially cleaned with deionized water, 2-propanol and acetone. After that, the substrate is covered with 8-nm-thick Al_2O_3 using atomic layer deposition to increase the adhesion to the following processed layers
2	Gold deposition	A 20-nm-thick gold (Au) layer is deposited through the **first shadow mask** in a turbo-pumped vacuum evaporator using a resistively heated tungsten filament with a 1–2 Å/s deposition rate and a 10^{-6} mbar base pressure. This bottom-metal layer defines the interconnects between the transistors

(continued)

Table 4.4 (continued)

No.	Process	Description
3	Aluminum deposition	A 30-nm-thick aluminum (Al) layer is deposited by thermal evaporation through the **second shadow mask** in a similar way as the previous step but with a 10–12 Å/s deposition rate. This layer defines the bottom-gate electrodes of the transistors. A direct contact is created wherever the aluminum gate is deposited over the gold bottom metal
4	Oxygen plasma treatment	The substrate is exposed to an oxygen plasma with a 200 W power and a 0.13 mbar base pressure to increase the thickness of the native aluminum oxide (AlO_x) layer to 3.6 nm and to create a density of hydroxyl groups that is sufficient for molecular adsorption in the following process step
5	Self-assembled monolayer formation	A 1.7-nm-thick self-assembled monolayer (SAM) is formed on top of the aluminum oxide surface by immersing the substrate into a 2-propanol solution of *n*-tetradecylphosphonic acid [$C_nH_{(2n+1)}PO(OH)_2$] for an hour. The substrate is then rinsed in a pure 2-propanol solution, blown by dry nitrogen and briefly backed on a hot plate for 10 minutes at approximately 100°. This completes the formation of a 5.3-nm-thick hybrid gate dielectric with a capacitance of the order of 1 μF/cm^2, which is sufficiently high to allow the transistors to operate at voltages as low as 3 V
6	Organic semiconductor deposition	A 30-nm-thick organic semiconductor is deposited by sublimation through the **third shadow mask** in vacuum using a current-heated molybdenum source with a 0.3 Å/s deposition rate and a 10^{-6} mbar base pressure. The organic semiconductors dinaphtho[2, 3-b:2', 3'-f]thieno[3, 2-b]thiophene (DNTT) for p-channel OTFTs and hexadecafluorocopperphthalocyanine ($F_{16}CuPc$) for n-channel OTFTs are employed. In the case of fabricating a complementary circuit, which comprises both p- and n-channel OTFTs, the deposition of both organic materials are done consecutively through two different shadow masks. The substrate is kept at an elevated temperature of 60° or 90° during the deposition of DNTT or $F_{16}CuPc$, respectively, to fascilitate the formation of well-ordered organic films with a high charge carrier mobility
7	Gold deposition	A 25-nm-thick gold (Au) layer is deposited by thermal evaporation through the **fourth shadow mask** with a 0.3 Å/s deposition rate and a 10^{-6}–10^{-7} mbar base pressure. This layer defines the source and drain top-contacts and it is used as well for interconnect crossings. Owing to the hybrid gate-dielectric layer, no connection is created between the gold top metal and the aluminum gate, and a parasitic capacitance of the order of 1 μF/cm^2 is formed. However, a direct contact is created wherever the gold top metal is deposited over the gold bottom metal

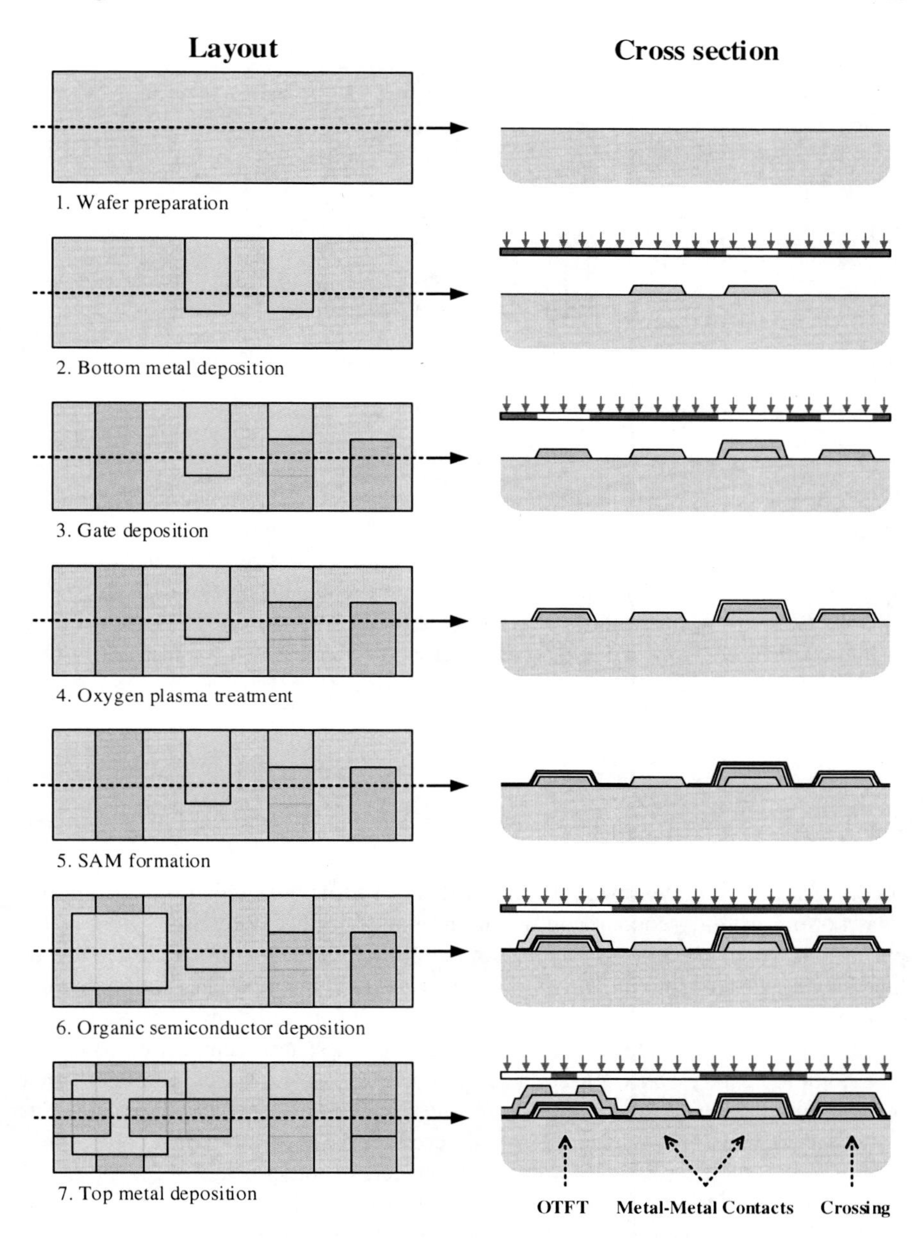

Fig. 4.11 Fabrication process steps of the organic thin-film transistors [26]

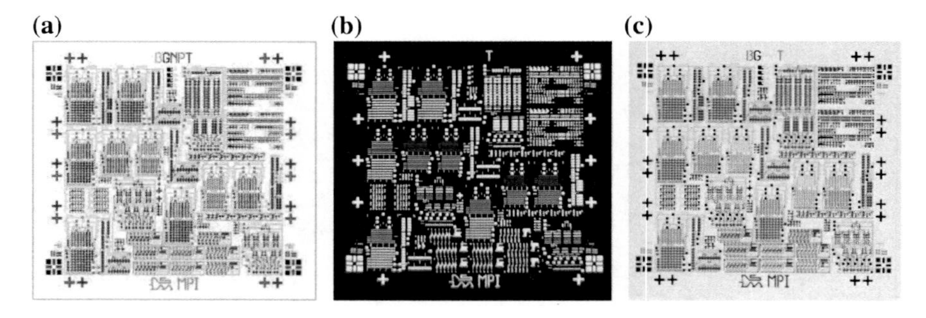

Fig. 4.12 Design flow of the organic ICs. **a** Physical layout of organic ICs designed on an active die area of 20×20 mm^2. **b** Photograph of the silicon stencil mask used to pattern the top-metal contacts. **c** Die photograph of the organic ICs fabricated on a glass substrate

A special DRC code is written to check the design rules of the stencil masks and the organic ICs; these are discussed in detail in the following section.

After that, the layout is sent to IMS CHIPS in the GDSII[2] (Graphic Database System) standard format for tapeout to fabricate the stencil mask set (Fig. 4.12b). The masks are then inspected under the microscope and the critical structures are characterized by CD measurements if necessary. Afterwards, the mask set is sent to MPI-SSR to fabricate organic die samples (Fig. 4.12c). Finally, the samples are brought back to IMS CHIPS for testing.

4.4 Layout Design Rules

As discussed before, the inverted-staggered OTFT structure, which is also referred to as bottom-gate/top-contact configuration, is used in this study as it offers better performance compared to its coplanar counterparts. Table 4.5 summarizes the layout layers and their corresponding materials, mask and GDSII numbers, noting that the dielectric layers (AlO_x/SAM) as well as the substrate are not drawn layers. To gain more insight of a typical OTFT circuit layout design, a sample schematic and layout of a complementary NAND logic gate is shown in Fig. 4.13. The numbers given in the figure designate the layout design rules, which are going to be explained in detail herein. The top-metal layer is mainly used for the source/drain contacts, while the bottom-metal layer is used for the interconnects. Another usage of the top-metal layer is to connect long bottom-metal lines that are cut into shorter fragments for mask stability reasons. As for the crossing, the top-metal layer is separated from the bottom-gate layer by a stack of two dielectric layers, namely a 3.6-nm-thick AlO_x layer and a 1.7-nm-thick SAM. The total thickness of the hybrid gate dielectric is

[2]The GDSII (Graphic Database System) is an industry standard database file format that is used for data exchange of IC layout artwork.

Table 4.5 List of layout layers and their corresponding GDSII numbers

Layer type	Name	Material	Mask	GDSII
Drawn	Bottom metal	Gold	1st	51
	Bottom gate	Aluminium	2nd	46
	Semiconductor n-channel	$F_{16}CuPc$	3rd	40
	Semiconductor p-channel	DNTT	4th	39
	Top metal	Gold	5th	49
Not drawn	Substrate	Glass	–	–
	Substrate coat	Al_2O_3	–	–
	Oxide insulator	AlO_x	–	–
	SAM insulator	$C_nH_{(2n+1)}PO(OH)_2$	–	–

5.3 nm with an equivalent capacitance of the order of 1 μF/cm^2. Exact capacitance measurements are given in Chap. 6.

In general, the Manhattan layout design style is employed, where angles between zero and 90° are ruled out, with the exception for the text labels, which do not have any electrical activity. Furthermore, the coordinates of the layout components, such as wires, contacts, cells and modules, must be placed on grid with a spacing of 0.1 μm. All dimensions drawn are supposed to be equal to the final dimensions on the wafer as well as the stencil masks. The origin (0,0) of the top hierarchy of a design has to be located at the lower left corner.

Referring to Fig. 4.13, the most important layout design rules are summarized in Tables 4.6 and 4.7. The dimensions given in these tables are minimum values. It is recommended, however, to use less critical values whenever significant increase in the area is not resulted. This applies particularly to the width and the spacing of long interconnection lines, which are realized by the bottom-metal layer.

Starting with the OTFT design rules, a minimum channel length (L) of 2 μm defined by the top-metal layer is recommended (Rule 1). Furthermore, the minimum gate-overlap length (L_{ov}), which is defined by the overlap between the top metal and the bottom gate within the transistor active region, is set to 20 μm (Rule 2). However, during the characterization of the performance limits of this OTFT technology through AC measurements, the channel length and the gate overlap are both reduced down to 0.6 and 5 μm, respectively. The enclosure rules, including the enclosure of the semiconductor inside the top metal, enclosure of the top metal inside the semiconductor, enclosure of the gate inside the semiconductor and finally the enclosure of the semiconductor inside the gate, are all set to 20 μm (Rule 3). Nevertheless, if the circuit area is relatively large, it is recommended to increase the enclosures of the gate and the top metal inside the semiconductor. This is because the semiconductor materials are translucent, and thus, they are hard to align when being deposited on the substrate.

The remaining OTFT design rules, including related (Rule 4) and non-related (Rule 5) clearances, and metal layer widths (Rule 6), they are all set to 20 μm.

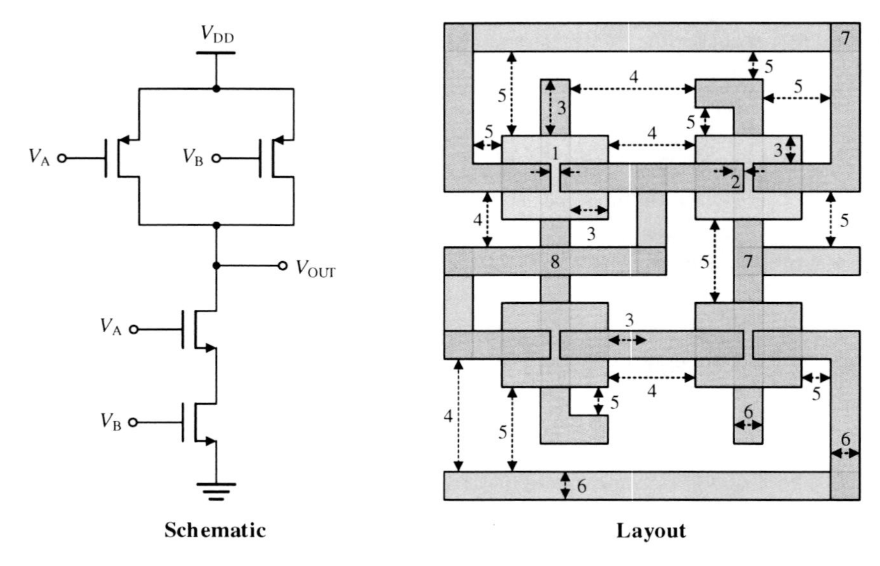

Fig. 4.13 Sample schematic and layout of a complementary NAND logic gate. The designated numbers represent the layout design rules for the DRC

Moreover, areas of direct contacts (Rule 7) and crossings (Rule 8) are set to 20 × 20 μm². However, one should note that crossings should be avoided wherever possible as they significantly reduce the circuit performance. This is because a very high parasitic capacitance of the order of 1 μF/cm² (4 pF for 20 × 20 μm²) would be created by each crossing.

As for the design rules of the silicon stencil masks, which are valid for all layout layers and are necessary for the mechanical stability of the masks, are the following. The remaining silicon area of the mask (average over 1 mm²) should be greater than 50 % (Rule 9). Furthermore, the aspect ratio of long silicon beams, which are attached either at both ends (Rule 10) or at one end only (Rule 11), should be smaller than 50 and 15 μm, respectively. The maximum die area used so far is 20 × 20 mm². All the alignment structures, layer labels and logos are placed at a 4-mm-wide frame. This leaves an active die area of 16 × 16 mm². Finally, donut-shaped isolated features that could fall out during stencil etching should be completely eliminated. It has to be ensured that all features are mechanically and firmly attached by the use of more than one anchor/bridge.

The software suites developed by Cadence Design Systems, namely Cadence Virtuoso and Cadence Assura Physical Verification, are used during the physical layout design of the organic integrated circuits. In this connection, SKILL[3] codes are written to define all the aforementioned design rules that are necessary for the physical verification and also to describe parameterized cells (PCells) that are helpful for automated layout design. Samples of the implemented SKILL codes can be found in Appendix D.1.

[3]SKILL is a scripting language used in many software suites from Cadence Design Systems.

Table 4.6 Layout design rules of the organic integrated circuits

Number	Rule	Layers	Value	Unit
1	Channel length	Top-to-Top	2[a]	μm
2	Gate Overlap	Top-over-Gate	20[a]	μm
3	Enclosures	Top-enclose-Semi[b]	20	μm
		Semi-enclose-Top	20	μm
		Semi-enclose-Gate	20	μm
		Gate-enclose-Semi	20	μm
4	Related clearances	Top-to-Top	20	μm
		Gate-to-Gate	20	μm
		PSemi-to-PSemi	20	μm
		NSemi-to-NSemi	20	μm
		Bottom-to-Bottom	20	μm
5	Non-related spacings	Top-to-Gate	20	μm
		Top-to-PSemi	20	μm
		Top-to-NSemi	20	μm
		Top-to-Bottom	20	μm
		Gate-to-PSemi	20	μm
		Gate-to-NSemi	20	μm
		Gate-to-Bottom	20	μm
		PSemi-to-NSemi	20	μm
		PSemi-to-Bottom	20	μm
		NSemi-to-Bottom	20	μm
6	Metal widths	Top	20	μm
		Gate	20	μm
		Bottom	20	μm
7	Direct contact areas	Top-Bottom-contact	20 × 20	μm^2
		Gate-Bottom-contact	20 × 20	μm^2
8	Crossing area	Top-Gate-crossing	20 × 20	μm^2

[a] In AC measurements, the L and L_{ov} are reduced to 0.6 and 5 μm, respectively
[b] The *Semi* layer is for both p- and n-channel semiconductors

Table 4.7 Layout design rules of the silicon stencil masks (valid for all layers)

Number	Rule	Value	Unit
9	Remaining silicon area (average over 1 mm^2)	≥50 %	–
10	Length-to-width ratio of beams attached at both ends	≤50	μm
11	Length-to-width ratio of beams attached at one end	≤15	μm

4.5 Mask Set Designs

In addition to the test mask presented in Sect. 4.2.1, a total of 5 other mask sets are designed, three of which are relevant to this work. The layout of each is built in a cell-based structure forming a tree-like hierarchical design. Figure 4.14 depicts the top-level hierarchical cell of the three mask sets. The illustrations are not arranged in a chronological order; their order rather depends on the presentation flow of this work. Each layout has several test structures and/or circuits, these are explained herein. The layout of each is designed in a way to maximize the area utilization to have a final compact die that brings together as much circuits and test structure as possible. A minimum distance of 20 μm between the structures is guaranteed and an active area of 16×16 mm^2 on a 20×20 mm^2 die area is always realized. The layout floorplan is a crucial design decision as it affects significantly the matching and the data-related interference. Accordingly, the following criteria are necessary during the design of the layouts:

- The area of each circuit has to be minimized to allow further circuits, or even duplicates of the same circuit, to be fabricated and tested on the same die.
- When device uniformity is paramount for some circuit topologies, certain blocks should be designed with the minimum possible area. This is because the increased distance between the transistors makes them more susceptible to mismatch effects and gradient errors.
- Again when the device uniformity is critical, identical transistors should be placed as close as possible to improve the matching capability. In addition, symmetry should be achieved wherever possible to ensure the same environment for all the transistors.
- Only one metal layer should be used for the design of a complete circuit, noting that the top metal and the bottom metal are both on the same electrical layer. Therefore, any overlap between them is considered as a physical connection.
- Crossings between the top metal and the bottom gate can be fabricated. Nevertheless, they should be avoided wherever possible to minimize cross coupling and improve the circuit performance.

Referring to Fig. 4.14, the three designed mask sets are labeled as *Die 1*, *Die 2* and *Die 3*. The first design, Die 1, includes OTFTs with channel widths of 200 μm, channel lengths of 100–1 μm and gate overlaps of 100–2 μm, which are dedicated for the static characterization. Furthermore, relatively large OTFTs with channel widths of 400 and 1000 μm, channel lengths of 400–5 μm and gate overlaps of 50–5 μm in addition to metal-insulator-metal (MIM) and metal-insulator-semiconductor (MIS) structures with different dimensions (200×200, 100×100, 50×50, 10×10 μm^2) are designed on the same die for the dynamic characterization using admittance measurements.

The second design, Die 2, includes aggressively-scaled OTFTs for S-parameter measurements. For the stability of the stencil masks, the OTFTs are dissected into four parallel transistors sharing the same semiconductor. The OTFTs have channel

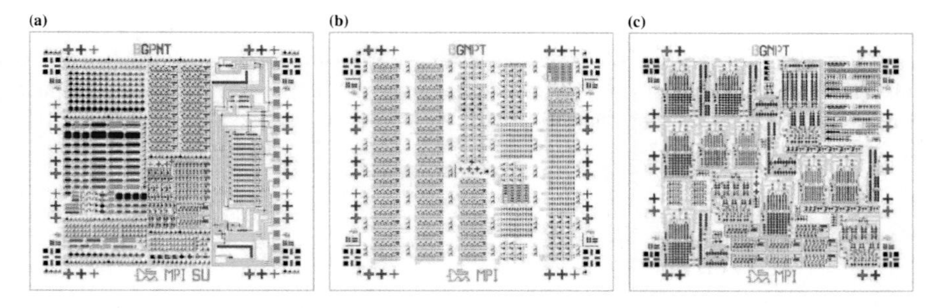

Fig. 4.14 Top-level hierarchical cell layout of the three designed $20 \times 20\ \text{mm}^2$ organic dies. **a** Die 1 that includes OTFTs for static and dynamic characterization. It also includes metal-insulator-metal (MIM) and metal-insulator-semiconductor (MIS) structures for admittance measurements. **b** Die 2 that includes aggressively-scaled OTFTs for scattering-parameter (S-parameter) measurements. **c** Die 3 that includes ring oscillators and digital-to-analog converters

widths of 100 μm (4 × 25 μm), channel lengths of 10–0.4 μm and gate overlaps of 20 and 5 μm. To study the impact of misalignment between the gate-source (L_{gs}) and gate-drain (L_{gd}) overlaps, where $L_{gs} + L_{gd} = 2L_{ov}$, the OTFTs with gate overlaps of 5 μm and channel lengths of 1 and 0.6 μm are duplicated and intentionally misaligned on the mask level in order to realize OTFTs with well-defined $L_{gs} = 1, 2, \ldots, 9$ μm, while keeping $L_{gs} + L_{gd} = 10$ μm. Vernier structures are employed to measure the particular degree of misalignment. The standard calibration structures, i.e., short, open and through, are also designed on the same die to remove the parasitics and to obtain the desired S-parameters at the device terminals.

The third design, Die 3, includes fast and compact organic ICs, namely unipolar (only p-channel OTFTs) and complementary (both p- and n-channel OTFTs) ring oscillators (presented before in Sect. 3.4.3), and 6-bit binary and 3-bit unary current-steering digital-to-analog converters (DACs). Finally, more detailed description of the layout design of the three dies can be found in Appendix A.

4.6 Summary

High-resolution silicon stencil masks are used for the fabrication of OTFTs with submicrometer channel lengths. The masks have a silicon membrane thickness of 20 μm and an area of $20 \times 20\ \text{mm}^2$. A minimum channel length of 0.3 μm for the top-metal evaporation process is realized. Characterization of the masks show that the evaporated gold contacts and aluminum gates can be easily etched with excellent pattern fidelity. The OTFTs employ a hybrid gate dielectric that is composed of a 3.6-nm-thick aluminum oxide and a 1.7-nm-thick organic self-assembled monolayer. The gate dielectric has a capacitance of the order of 1 μF/cm^2, which is sufficiently high to allow the transistors to operate at voltages as low as 3 V. The stencil masks and the organic ICs have recommended layout design rules; accordingly, a Cadence SKILL code is written for the physical verification of the designs. A total of three mask sets are designed during the course of this work to fabricate fast and compact organic transistors and circuits.

References

1. H. Klauk (ed.), *Organic Electronics: Materials, Manufacturing and Applications*, 2nd edn. (Wiley-VCH, Weinheim, 2008)
2. OVPD mass production equipment, Spec Sheet, AIXTRON, 2013. http://www.aixtron.com/en/products/product-portfolio/
3. F. Ante, F. Letzkus, J. Butschke, U. Zschieschang, J.N. Burghartz, K. Kern, H. Klauk, Top-contact organic transistors and complementary circuits fabricated using high-resolution silicon stencil masks, in *Device Research Conference, June 2010*, pp. 175–176
4. G.H. Gelinck, H.E.A. Huitema, E. van Veenendaal, E. Cantatore, L. Schrijnemakers, J.B.P.H. van der Putten, T.C.T. Geuns, M. Beenhakkers, J.B. Giesbers, B.-H. Huisman, E.J. Meijer, E.M. Benito, F.J. Touwslager, A.W. Marsman, B.J.E. van Rens, D.M. de Leeuw, Flexible active-matrix displays and shift registers based on solution-processed organic transistors. Nat. Mater. **3**, 106–110 (2004)
5. T. Sekitani, Y. Noguchi, U. Zschieschang, H. Klauk, T. Someya, Organic transistors manufactured using inkjet technology with subfemtoliter accuracy. Proc. Natl. Acad. Sci. USA **105**(13), 4976–4980 (2008)
6. Y.-Y. Noh, N. Zhao, M. Caironi, H. Sirringhaus, Downscaling of self-aligned, all-printed polymer thin-film transistors. Nat. Nanotechnol. **2**, 784–789 (2007)
7. D.J. Gundlach, T.N. Jackson, D.G. Schlom, S.F. Nelson, Solvent-induced phase transition in thermally evaporated pentacene films. Appl. Phys. Lett. **74**(22), 3302–3304 (1999)
8. T. Ji, S. Jung, V.K. Varadan, On the correlation of postannealing induced phase transition in pentacene with carrier transport. Org. Electron. **9**(5), 895–898 (2008)
9. F. Ante, F. Letzkus, J. Butschke, U. Zschieschang, K. Kern, J.N. Burghartz, H. Klauk, Submicron low-voltage organic transistors and circuits enabled by high-resolution silicon stencil masks, in *IEEE International Solid-State Circuits Conference Technical Digest, Dec. 2010*, pp. 21.6.1–21.6.4
10. F. Letzkus, T. Zaki, F. Ante, J. Butschke, H. Richter, H. Klauk, J.N. Burghartz, Si stencil masks for organic thin film transistor fabrication, in *Proceedings of the SPIE Photomask Technology, Sept 2011*, pp. 81662B-1–81662B-12
11. F. Letzkus, T. Zaki, F. Ante, J. Butschke, H. Richter, H. Klauk, J.N. Burghartz, Si Stencil-Masken für die Herstellung organischer Dünnschichttransistoren, in *MikroSystemTechnik Kongress*, pp. 181–184 (2011) (in German)
12. W. Clemens, D. Lupo, K. Hecker, S. Breitung, (eds.), OE-A roadmap for organic and printed electronics. White Paper, 5th edn. (Organic Electronics Association, 2013). http://www.oe-a.org/en_GB/
13. E. Platzgummer, S. Cernusca, C. Klein, J. Klikovits, S. Kvasnica, H. Loeschner, eMET: 50 keV electron mask exposure tool development based on proven multi-beam projection technology, in *Proceedings of the SPIE Photomask Technology, Sept 2010*, pp. 782308-1–782308-12
14. R. Kaesmaier, A. Ehrmann, H. Löschnerc, Ion projection lithography: status of tool and mask developments. Microelectr. Eng. **57–58**, 145–153 (2001)
15. A.A. Tseng, Recent developments in nanofabrication using ion projection lithography. Small **1**(6), 594–608 (2005)
16. W.H. Bruenger, C. Dzionk, R. Berger, H. Grimm, A.H. Dietzel, F. Letzkus, R. Springer, Evaluation of ion projection using heavy ions suitable for resistless patterning of thin magnetic films. Microelectr. Eng. **61–62**, 295–300 (2002)

17. J. Butschke, A. Ehrmann, B. Höfflinger, M. Irmscher, R. Käsmaier, F. Letzkus, H. Löschnerb, J. Mathuni, C. Reuter, C. Schomburg, R. Springer, SOI wafer flow process for stencil mask fabrication. Microelectr. Eng. **46**(1–4), 473–476 (1999)
18. F. Letzkus, J. Butschke, B. Höfflinger, M. Irmscher, C. Reuter, R. Springer, A. Ehrmann, J. Mathuni, Dry etch improvements in the SOI wafer flow rocess for IPL stencil mask fabrication. Microelectr. Eng. **53**(1–4), 609–612 (2000)
19. F. Ante, Contact effects in organic transistors. Ph.D. dissertation, Swiss Federal Institute of Technology in Lausanne, Lausanne, Switzerland (2011)
20. H.C. Shah, R.W. Davis Jr, The effect of the roughness of the glass substrate on the roughness of the barrier layer used during fabrication of poly-Si TFTs. Soc. Inf. Display Symp. Tech. Digest **34**(1), 1516–1519 (2003)
21. U. Zschieschang, R. Hofmockel, R. Rödel, U. Kraft, M.J. Kang, K. Takimiya, T. Zaki, F. Letzkus, J. Butschke, H. Richter, J.N. Burghartz, H. Klauk, Megahertz operation of flexible low-voltage organic thin-film transistors. Org. Electron. **14**(6), 1516–1520 (2013)
22. T. Zaki, R. Rödel, F. Letzkus, H. Richter, U. Zschieschang, H. Klauk, J.N. Burghartz, AC characterization of organic thin-film transistors with asymmetric gate-to-source and gate-to-drain overlaps. Org. Electron. **14**(5), 1318–1322 (2013)
23. U. Zschieschang, F. Ante, D. Kälblein, T. Yamamoto, K. Takimiya, H. Kuwabara, M. Ikeda, T. Sekitani, T. Someya, J. Blochwitz-Nimoth, H. Klauk, Dinaphtho[2,3-b:2',3'-f]thieno[3,2-b]thiophene (DNTT) thin-film transistors with improved performance and stability. Org. Electron. **12**(8), 1370–1375 (2011)
24. H. Klauk, U. Zschieschang, J. Pflaum, M. Halik, Ultralow-power organic complementary circuits. Nature **445**, 745–748 (2007)
25. U. Zschieschang, F. Ante, M. Schloerholz, M. Schmidt, K. Kern, H. Klauk, Mixed self-assembled monolayer gate dielectric for continuous threshold voltage control in organic transistors and circuits. Adv. Mater. **22**(40), 4489–4493 (2010)
26. T. Zaki, Design of current-steering D/A converters using organic thin-film transistors enabled by high-resolution silicon stencil masks, M.Sc. thesis, University of Stuttgart, Stuttgart, BW, Germany (2010)

Chapter 5
Static Characterization

An analytical model describing the steady-state characteristics of the OTFTs is presented herein. The model captures the physical behavior of the device, which helps to improve the electrical performance, stability and uniformity of OTFTs, as well as the design and optimization of organic integrated circuits. In general, contact effects in OTFTs are significant and they limit the device performance; therefore, source and drain contact resistances are experimentally investigated to analyze and model their behavior. It is found that contact resistances are dependent on the applied biasing potentials and on the dimensions of the OTFTs. The static model is validated by comparing the numerical solutions with measured characteristics of both p- and n-channel OTFTs; a reliable fit of the data is demonstrated.

5.1 Transmission Line Method

The transmission line method (TLM)—introduced by Murrmann and Widmann in 1969 [1]—is used to extract some intrinsic and extrinsic characteristics of the OTFTs, including the contact resistance at the metal/semiconductor interface and the sheet resistance of the organic semiconductor, from which the threshold voltage and the intrinsic carrier mobility of the devices can be calculated [2, 3]. The theory and implementation of this method on our DNTT OTFTs are described in this section. For this method, a set of OTFTs with different channel lengths is required; correspondingly, eleven OTFTs with $W = 200\,\mu\text{m}$, $L_{\text{ov}} = 100\,\mu\text{m}$ and $L = 100$–$4\,\mu\text{m}$ are fabricated. Figure 5.1 shows the measured static drain currents (I_{D}) of six of those OTFTs. The characteristics show that I_{D} increases when L is reduced. The OTFTs have on/off current ratios ($I_{\text{on}}/I_{\text{off}}$) of about 10^6 and subthreshold slopes ($S_{\text{s-th}}$) of 90 mV/decade. From the saturation transfer curves, effective mobilities (μ) of 2.7, 2.4, 2.0, 1.8, 1.6 and 1.2 cm^2/Vs, and transconductances (g_{m}) of 5, 12, 20, 30, 40 and 60 μA/V (at $V_{\text{GS}} = -3$ V) are extracted for $L = 100$, 40, 20, 12, 8 and 4 μm, respectively. The reduction of μ as L reduces is owed to the increase in the relative contribution of the contact resistances to the total device resistance [4].

T. Zaki, *Short-Channel Organic Thin-Film Transistors*, Springer Theses,
DOI 10.1007/978-3-319-18896-6_5

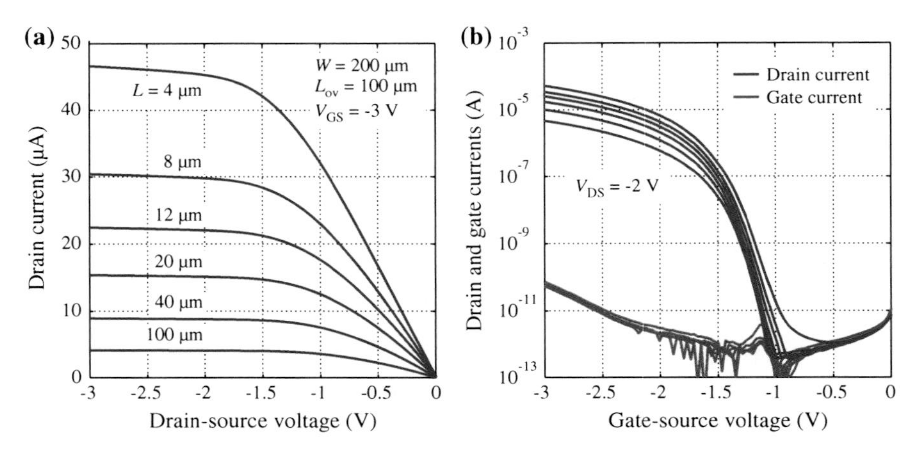

Fig. 5.1 Measured static I–V characteristics of inverted-staggered DNTT OTFTs with $W = 200$ μm, $L_{ov} = 100$ μm, and $L = 100, 40, 20, 12, 8$ and 4 μm. **a** Output characteristics at $V_{GS} = -3$ V. **b** Transfer characteristics at $V_{DS} = -2$ V

Since the output characteristics of the DNTT OTFTs (p-channel) shown in Fig. 5.1 do not exhibit a non-linearity at low drain-source voltages, it can be assumed in the following analysis that the OTFTs have ohmic contact resistances; however, it is demonstrated in Sect. 5.3 that this is not the case for aggressively-scaled DNTT OTFTs (with $L < 1$ μm) and for $F_{16}CuPc$ OTFTs (n-channel) as well. Accordingly, the total resistance of the OTFT (R_{tot}) is given by the sum of the source/drain contact resistances ($2R_C$; assuming also that the OTFT is symmetrical) and the channel resistance ($R_{ch} = 1/g_{ch}$): $R_{tot} = 2R_C + R_{ch}$. By substituting for $R_{ch} = 1/g_{ch}$ using (3.10) at low drain-source voltages ($V_{DS} \ll V_{GS} - V_{TH}$), we obtain

$$R_{tot}W = 2R_C W + R_{sheet}L. \tag{5.1}$$

Figure 5.2a depicts the extracted width-normalized total resistance ($R_{tot}W$) of all the eleven OTFTs as a function of L at various gate-source voltages above threshold and at a low drain-source voltage of -0.1 V. At each V_{GS}, the resistances R_{sheet} and R_C can be calculated from the slope and the intersection with the y-axis (when $L = 0$) of the linear least-square fit, respectively.

As described in Chap. 3, the drain current (I_D) when considering ohmic contact resistances for symmetrical OTFTs can be written as

$$I_D = \mu_o C_I \frac{W}{L} \cdot (V_{DS} - 2I_D R_C) \cdot \left(V_{GS} - V_{TH} - \frac{V_{DS}}{2} \right). \tag{5.2}$$

At a sufficiently small drain-source voltage ($V_{DS} \ll V_{GS} - V_{TH}$), I_D can be simplified to the following expression:

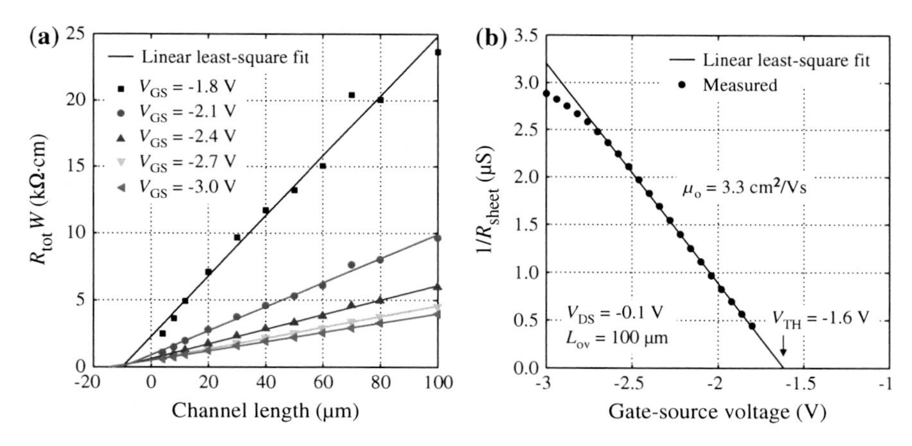

Fig. 5.2 Analysis of the intrinsic and extrinsic properties of OTFTs with $W = 200$ μm, $L = 100$–4 μm and $L_{ov} = 100$ μm using the TLM. **a** Measured width-normalized total resistance ($R_{tot}W$) as a function of L for different V_{GS} at $V_{DS} = -0.1$ V. From the linear least-square fit at $V_{GS} = -3$ V, a width-normalized contact resistance ($R_C W$) of 0.25 kΩcm, a sheet resistance (R_{sheet}) of 347 kΩ/□ and a transfer length (L_T) of 7.3 μm are extracted. All linear least-square fits (except for $V_{GS} = -1.8$ V) meet at a single point, which defines a characteristic length ($-l_0$) of about 8.5 μm and a gate-voltage independent parasitic resistance ($R_{C,0}W$) of 0.2 kΩcm. **b** The reciprocal of the extracted sheet resistance ($1/R_{sheet}$) as a function of V_{GS}. From the linear least-square fit, an intrinsic carrier mobility (μ_o) of 3.3 cm²/Vs and a threshold voltage (V_{TH}) of -1.6 V are determined

$$I_D = \mu_o C_I \frac{W}{L} \cdot (V_{DS} - 2 I_D R_C) \cdot (V_{GS} - V_{TH}). \tag{5.3}$$

Given that $R_{tot} = V_{DS}/I_D$, the width-normalized total OTFT resistance can be derived as follows:

$$R_{tot}W = 2R_C W + \frac{L}{\mu_o C_I (V_{GS} - V_{TH})}. \tag{5.4}$$

From (5.1) and (5.4), the intrinsic carrier mobility (μ_o) and the threshold voltage (V_{TH}) of the OTFTs can therefore be deduced by plotting the reciprocal of the extracted sheet resistance ($1/R_{sheet}$) as a function of V_{GS}, as shown in Fig. 5.2b. Correspondingly, it is found that the p-channel DNTT OTFTs have $\mu_o = 3.3$ cm²/Vs and $V_{TH} = -1.6$ V. Using the same procedure for n-channel $F_{16}CuPc$ OTFTs (having $I_{on}/I_{off} = 10^5$), $\mu_o = 0.06$ cm²/Vs and $V_{TH} = 1.1$ V are determined. The extracted $\mu_o = 3.3$ cm²/Vs is larger than $\mu = 2.7$ cm²/Vs of the DNTT OTFT with $L = 100$ μm. This is because the gate-overlap length (L_{ov}) of these OTFTs is taken to be as large as 100 μm. Indeed, if L is designed to be much larger than L_{ov}, the impact of the parasitic resistances on the carrier mobility will be minimal.

As depicted in Fig. 5.2b, the reciprocal of the sheet resistance slightly deviates from the linear least-square fit at high gate bias ($|V_{GS}| \geq 2.7$ V), which implies that the OTFTs have about 10 % lower carrier mobility at this biasing regime. This can be

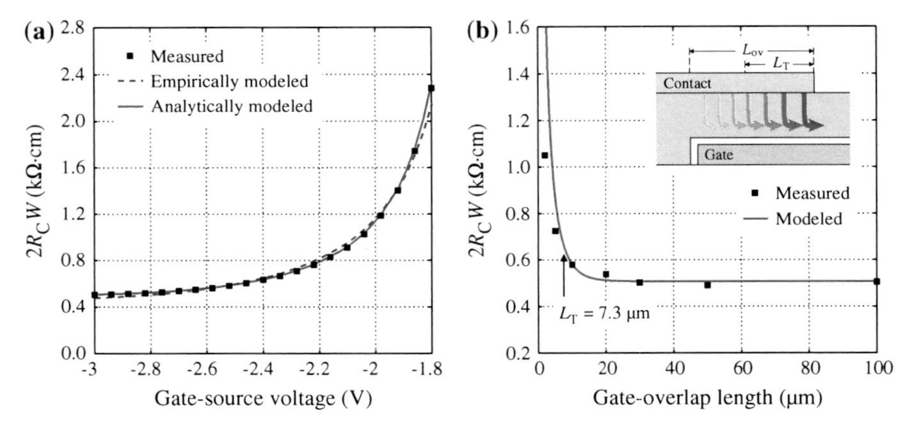

Fig. 5.3 **a** Measured and modeled $2R_C W$ as a function of V_{GS} for the OTFTs with $W = 200$ μm, $L = 100$–4 μm and $L_{ov} = 100$ μm. The *dashed line* represents the empirical model given by (5.5) with $\mu_o = 3.3$–2.9 cm^2/Vs (as a function of V_{GS}), $l_0 = 8.5$ μm and $R_{C,0}W = 2.2$ kΩcm. The *solid line* represents the analytical model given by (5.6) and (3.10) with constants $L_T = 7.3$ μm and $V_{TH} = -1.6$ V. (b) Measured and modeled $2R_C W$ as a function of L_{ov}. The solid line represents the analytical model given by (5.6) with $L_T = 7.3$ μm and $R_{sheet} = 347$ kΩ/□. The figure depicts the excellent agreement between the measured and modeled R_C characteristics with the parameters extracted using the TLM. The *inset* shows a schematic diagram illustrating the current distribution underneath the source or drain contacts of the OTFTs (adopted from [6]), assuming that $L_{ov} > L_T$, where L_T is the characteristic transfer length over which 63 % of the current flows from the top-metal contact to the organic semiconductor or vice versa

attributed to bias-stress effects. Even though the OTFTs exhibit very low hysteresis when being measured in the forward and backward directions (not shown here), a similar effect is observed in the linear transfer characteristics, i.e., the measured drain current is slightly lower in the backward bias stream than in the forward bias stream due to the lower carrier mobility caused by bias stress at high gate-source voltages.

Furthermore, Fig. 5.3a shows the extracted width-normalized contact resistances ($2R_C W$) as a function of the gate voltage. Apparently, the contact resistance is largely modulated by the gate voltage due to the presence of the overlaps between the source/drain contacts and the gate electrode; the contact resistance of the OTFTs decreases with increasing the absolute gate voltage. An empirical model for this effect can be easily deduced from the TLM data [5]. As shown in Fig. 5.2a, the $R_{tot}W$ versus L curves merge at $l_0 \cong 8.5$ μm and has a residual value of a gate-voltage independent parasitic resistance ($R_{C,0}W$) of 0.2 kΩcm. From (5.4), the width-normalized contact resistances ($2R_C W$) must therefore satisfy the following relation [5]:

$$2R_C W = R_{C,0}W + \frac{l_0}{\mu_0 C_I (V_{GS} - V_{TH})}. \tag{5.5}$$

That is, the contact resistance can be considered as an accumulation channel resistance of length l_0 in series with a minimum effective contact resistance. The output of the model with $\mu_0 = 3.3$–2.9 cm^2/Vs (as a function of V_{GS}), $C_I = 700$ nF/cm^2,

$l_0 = 8.5\ \mu m$ and $R_{C,0}W = 2.2\ k\Omega cm$ is shown as a *dashed line* in Fig. 5.3a; the model fitted quite well with the extracted data.

Only the channel length dependence of the OTFT performance with a fixed gate-overlap length (L_{ov}) of 100 μm is considered so far. Based on the discussions above, R_C should depend also on L_{ov} because of the current distribution under the contacts. To investigate this effect, other similar six sets of OTFTs but with $L_{ov} = 50$, 30, 20, 10, 5 and 2 μm are fabricated on the same substrate. Figure 5.3b shows the width-normalized contact resistances ($2R_C W$) extracted using the TLM for each OTFT set at $V_{GS} = -3$ V. An accurate evaluation of the contact resistance for these inverted-staggered (bottom-gate, top-contact) OTFTs requires the use of a distributed resistance network underneath the source/drain contacts as derived in Appendix C.2; accordingly, R_C is analytically described by [2]

$$2R_C W = 2R_{sheet} L_T \coth\left(\frac{L_{ov}}{L_T}\right), \tag{5.6}$$

where L_T is a characteristic transfer length over which 63 % of the current flows from the semiconductor into the metal or from the metal into the semiconductor [6]. As defined in Appendix C.2, the transfer length (L_T) is given by

$$L_T = \sqrt{\frac{\rho_c}{R_{sheet}}}, \tag{5.7}$$

where ρ_c (Ωm^2) is the contact resistivity multiplied by the effective distance between the source/drain contacts and the accumulated channel. Moreover, the expression of $R_C W$ given by (5.6) can be simplified to the following two conditions [2]:

$$R_C W \cong \begin{cases} \rho_c/L_{ov} & \text{for} L_{ov} \ll L_T, \\ \rho_c/L_T & \text{for} L_{ov} \gg L_T. \end{cases} \tag{5.8}$$

If $L_{ov} \ll L_T$, the *effective* contact area is the *actual* contact area of $L_{ov}W$. If $L_{ov} \gg L_T$, however, the effective contact area is $L_T W$, which is smaller than the actual contact area, leading to an increased current density than if the entire contact area was active.

In principle, the transfer length (L_T) can be extracted from the $R_{tot}W$ versus L curves shown in Fig. 5.2a. By substituting (5.7) and (5.8) in (5.1) for long gate-overlap lengths ($L_{ov} \gg L_T$), the width-normalized total OTFT resistance can be expressed as

$$R_{tot}W = R_{sheet}(2L_T + L). \tag{5.9}$$

Therefore, L_T can be calculated at each V_{GS} from the intersection of the linear least-square fit with the x-axis (when $R_{tot}W = 0$). Referring to Fig. 5.2a, the OTFTs with $L_{ov} = 100\ \mu m$ have $R_{sheet} = 347\ k\Omega/\square$ and $L_T = 7.3\ \mu m$ (extracted at $V_{GS} = -3$ V). Using these values, the analytical model of $R_C W$ given by (5.6) yielded an excellent fit to the measured results, as illustrated by the solid line in Fig. 5.3b. Moreover, a precise fit is also achieved in Fig. 5.3a by substituting in (5.6) for R_{sheet} using (3.10).

5.2 Compact Charge-Drift Model

The steady-state characteristics of the OTFTs have been extensively studied and several static models have been proposed in literature [7–12]. Brown et al. [7] observed that the charge carrier mobility in amorphous organic transistors is bias dependent, i.e., the mobility increases when the gate voltage increases. According to that perceived effect, Vissenberg and Matters [8] and Meijer et al. [9] developed analytical models based on charge drift in the presence of variable-range hopping[1] (VRH). However, these models are limited to the linear operating regime. Calvetti et al. [10] and Marinov and Deen [11] derived analytical models based on the same common points, but on the other hand, the models cover both the linear and the saturation operating regimes with a single formulation. Each of these models and many others [13–17] were proposed to explain the conduction mechanism of their specific devices while taking into consideration the mobility enhancement behavior at higher gate bias.

In general, it is quite challenging to define a unified compact DC model for OTFTs due to various reasons, among them are the following [11]: (i) there is a wide range of OTFT technologies, including the use of different materials, processing steps and diverse transistor configurations; and (ii) there is significant variability in experimental data for similar devices in literature. Consequently, many models are developed to reflect the unique features for the particular OTFTs.

As shown in Fig. 5.2b, the extracted intrinsic carrier mobility of our p-channel DNTT OTFTs does not increase with increasing gate bias; in fact, it is slightly reduced due to bias-stress effects. Note that the same property is observed for our n-channel OTFTs as well. Therefore, the aforementioned models are not viable to describe the real physical behavior of our OTFTs, without which they may produce misleading device characteristics. Since our OTFTs behave to a large degree as the conventional FETs, the simple charge-drift model presented by Borkan and Weimer in [18] can be adopted for our simulations. As derived in Chap. 3, the drain current (I_D) is accordingly expressed above threshold ($V_{GS} > V_{TH}$) as

$$I_D = \begin{cases} \mu C_I \dfrac{W}{L}\left[(V_{GS} - V_{TH})V_{DS} - \dfrac{V_{DS}^2}{2}\right] & \text{for } V_{DS} < V_{GS} - V_{TH}, \\ \mu C_I \dfrac{W}{2L}(V_{GS} - V_{TH})^2 & \text{for } V_{DS} \geq V_{GS} - V_{TH}. \end{cases} \tag{5.10}$$

[1] The variable-range hopping (VRH) is a model describing the conduction mechanism in a strongly disordered system, assuming the charge-carrier transport as hopping between localized electronic states. These states can be considered as charge traps induced by chemical impurities, structural defects or grain boundaries. The hopping rate between two localized states, which corresponds to the carrier mobility, is thermally activated or tunneling assisted and the relaxation takes place under phonon emission [3].

This generic charge-drift model formulates the base for developing a reliable OTFT compact DC model. In the following, we explore the details of modifying this generic model to make it valid in all operation regimes of the device with a single expression and also to include non-ideal effects, namely subthreshold current, channel length modulation and parasitic resistances.

The representation of the drain current (I_D) given by (5.10) in the linear operation regime ($V_{DSi} < V_{GSi} - V_{TH}$) of an intrinsic OTFT (excluding the parasitic contact resistances) can be rewritten as

$$I_D = \mu_o C_I \frac{W}{2L} \left[(V_G - V_{TH} - V_{Si})^2 - (V_G - V_{TH} - V_{Di})^2 \right]. \tag{5.11}$$

The subscript i designates the internal source and drain potentials of the intrinsic OTFT, which corresponds to the portion of the inverted-staggered OTFT that is above the gate electrode and between the source and drain contacts. In fact, the source (V_{Si}) and drain (V_{Di}) contacts are interchangeable in the model, which reflect the symmetrical structure of the OTFTs.

To make expression (5.11) valid in both linear and saturation operation regimes, the term ($V_G - V_{TH} - V_{Di}$) has to be taken only with positive values in the linear regime (when $V_{Di} < V_G - V_{TH}$) and it has to be very close to zero in the saturation regime (when $V_{Di} \geq V_G - V_{TH}$). Moreover, expression (5.11) is valid only above threshold (when $V_{GSi} > V_{TH}$). However, a non-zero current is typically measured when the gate voltage is above the turn-on voltage (V_{ON}) and below the threshold voltage (V_{TH}), which is referred to as the subthreshold regime (see Fig. 5.1). This regime is particularly important in low-power digital circuits because it describes how fast the switches are turned on and off [19]. Therefore, to include the subthreshold regime in the static model, I_D has to be proportional to $\exp(V_G - V_{TH} - V_{Si})$ when $V_{GSi} < V_{TH}$. These conditions can be easily incorporated into the generic charge-drift model by replacing the terms ($V_G - V_{TH} - V_{Si}$) and ($V_G - V_{TH} - V_{Di}$) in (5.11) with an asymptotically interpolation function [$V_{eff}(V)$, where $V = V_{Si}$ or V_{Di}] that is given by [11]

$$V_{eff}(V) = V_{SS} \ln \left[1 + \exp\left(\frac{V_G - V_{TH} - V}{V_{SS}} \right) \right] \tag{5.12}$$

$$\cong \begin{cases} V_{SS} \exp\left(\dfrac{V_G - V_{TH} - V}{V_{SS}} \right) \\ \quad \text{for subthreshold regime,} \\ \quad \text{when } V_G - V_{TH} - V < -V_{SS}, \\ V_G - V_{TH} - V \\ \quad \text{for above-threshold regime,} \\ \quad \text{when } V_G - V_{TH} - V > V_{SS}, \end{cases} \tag{5.13}$$

where V_{SS} is related to the steepness of the subthreshold characteristics of the OTFT. Note that $V_{eff}(V)$ is very close to zero when $V > V_G - V_{TH}$, which is a necessary condition as mentioned above to make the model valid in the linear and saturation regimes using a single formulation. A similar interpolation has been also used for MOSFETs [20] and polysilicon TFTs [21].

In the saturation regime, the drain current is determined by the channel length (L) and the gate-overdrive voltage ($V_{GS} - V_{TH}$). The saturation regime starts at the pinch-off point when $V_{DS} = V_{GS} - V_{TH}$. Once the pinch-off point occurs, any further increase in the drain potential results in shortening of the accumulated channel, thus increasing the drain current correspondingly. This effect is referred to as channel length modulation. A simple modification to (5.11) can be implemented to incorporate this phenomenon by replacing the channel length L with $L - \Delta L$, given that [11]

$$L\left(1 - \frac{\Delta L}{L}\right) \cong \frac{L}{1 + \lambda |V_{Di} - V_{Si}|}, \tag{5.14}$$

where λ is the channel-length modulation coefficient. By taking the absolute value of V_{DSi}, the inherent symmetry of the model is preserved.

The final representation of the drain current (I_D) of the intrinsic OTFT is determined by substituting for the subthreshold region using (5.12) and the channel length modulation using (5.14) in (5.11) to obtain

$$I_D = \mu_o C_I \frac{W}{2L} \cdot (1 + \lambda |V_{Di} - V_{Si}|) \cdot \left\{ [V_{eff}(V_{Si})]^2 - [V_{eff}(V_{Di})]^2 \right\}. \tag{5.15}$$

All expressions derived so far are for n-channel OTFTs. Similar characteristic equations are valid for p-channel OTFTs, for which the polarities of the voltages and currents have to be inverted.

As discussed in the previous section, the parasitic contact resistances ($2R_C$) of OTFTs contribute substantially to the total device resistance and have a considerable impact on the drain currents. In fact, measurements show that R_C is dependent on the gate bias as demonstrated in Fig. 5.3. Consequently, the model of R_C given by (5.6) is used in the static OTFT model to account for this field dependence.

Ideally, the model is geometry scalable. However, the variability in the characteristics of OTFTs is relatively large and in practice there is always some discrepancy between idealized assumptions and experimentally acquired data [11]. In addition, some parameters, such as the contact resistance, might have different scaling rules. Thus, one should allow for some variability in the values of μ, V_{TH}, λ and V_{SS} for OTFTs with different dimensions, even if the they are fabricated on the same substrate.

The model is implemented in a SPICE simulator. Figure 5.4 shows the measured versus simulated data for the I–V output and transfer characteristics of a p-channel DNTT OTFT with $W = 100$ µm, $L = 4$ µm and $L_{ov} = 20$ µm. The accurate agreement of the simulated to the measured data justifies the reliability of the proposed model, which is needed to carry out more complex simulations for integrated

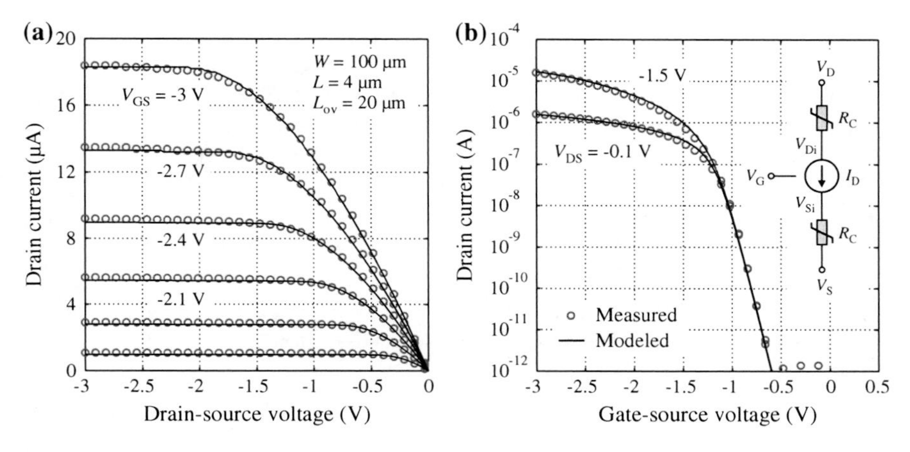

Fig. 5.4 Measured versus simulated I–V characteristics of a p-channel DNTT OTFT with $W = 100$ µm, $L = 4$ µm and $L_{ov} = 20$ µm. **a** Output characteristics. **b** Linear and saturation transfer characteristics. The *inset* shows the OTFT DC model schematic. The model parameters are $K_p = \mu C_I = 478$ nA/V^2, $V_{TH} = -1.08$ V, $\lambda = 0.5$ %/V and $V_{SS} = 90$ mV

circuits based on OTFTs. Note that the OTFT, the characteristics of which are shown in Fig. 5.4, is located on the same sample as the digital-to-analog converters presented in Chap. 8. The model is tested on several p- and n-channel OTFTs with different dimensions and a quite good agreement between the measured and simulated data is generally achieved. However, it is observed that the output characteristics of aggressively-scaled p-channel (only those that are freshly fabricated) and n-channel OTFTs exhibit a pronounced non-linearity at low drain-source voltages, i.e., the contact resistances are dependent not only on the gate bias but also on the drain bias. This eventually necessitated the implementation of a simple modification to the OTFT model to consider this non-ideal contact effect, as discussed hereinafter.

5.3 Non-linear Contact Resistance

Non-ideal contacts in TFTs strongly affect the device characteristics and performance, especially for short channel devices, where these effects become dominant [22]. In the previous section, it is demonstrated that the characteristics of p-channel DNTT OTFTs with channel lengths down to 4 µm can be accurately modeled with ohmic contact resistances that are dependent only on the gate bias. However, measurements of aggressively-scaled OTFTs with channel lengths below 1 µm show pronounced non-linearity of the output characteristics at low drain-source voltages ($|V_{DS}| < |V_{GS} - V_{TH}|$), as depicted in Fig. 5.5. This implies that these transistors exhibit non-linear parasitic contact resistances, which are not ohmic because they are not only dependent on the gate bias but also on the drain bias. A similar effect is measured also for the n-channel $F_{16}CuPc$ OTFTs even with a channel length of

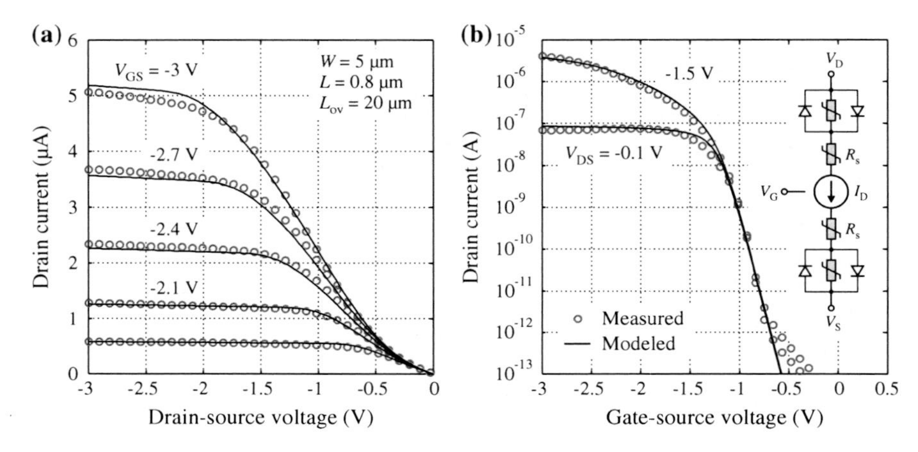

Fig. 5.5 Measured versus simulated I–V characteristics of a p-channel DNTT OTFT with $W = 5$ μm, $L = 0.8$ μm and $L_{\mathrm{ov}} = 20$ μm. **a** Output characteristics. **b** Linear and saturation transfer characteristics. The *inset* shows the OTFT DC model schematic. The model parameters are $K_{\mathrm{p}} = \mu C_{\mathrm{I}} = 830$ nA/V^2, $V_{\mathrm{TH}} = -1.16$ V, $\lambda = 4$ %/V, $V_{\mathrm{SS}} = 95$ mV, $I_{\mathrm{SS}} = 1.6$ μA and $\eta = 3.5$. Note that these characteristics are measured immediately after the fabrication and the high non-linearity observed in (**a**) at low $|V_{\mathrm{DS}}|$ had disappeared after 5 months [3]

4 μm, as shown in Fig. 5.6. It is important to note that the non-linear drain current increase observed in the aggressively-scaled DNTT OTFTs had disappeared 5 months after fabrication [3]; this suggests that their contact resistances had improved over time. In principle, the non-ohmic contact resistances severely limit the effective carrier mobility of the OTFTs and might also lead to frequency dispersion of their current–voltage and capacitance–voltage characteristics [22]. Therefore, we present in the following a modification to the compact model allowing to simulate the highly non-linear source and drain contact resistances for the said devices.

Referring to Figs. 5.5 and 5.6, the non-linear regions of the output characteristics at $|V_{\mathrm{DS}}| < |V_{\mathrm{GS}} - V_{\mathrm{TH}}|$ indicate that the contact resistances decrease with $|V_{\mathrm{DS}}|$. In order to simulate this contact effect, we propose an equivalent circuit that consists of a pair of anti-parallel diodes, a parallel resistance (R_{p}) and a series resistance (R_{s}) for each contact. The corresponding equivalent circuit representing a complete OTFT compact model that reflects the non-ohmic nature of the source and drain contacts is depicted in the *inset* of Figs. 5.5 and 5.6. A similar model is used by Necliudov et al. [23, 24] for inverted-coplanar (bottom-gate, bottom-contact) OTFTs, of which the contact resistances are also dependent on the gate and drain biases.

A pair of anti-parallel diodes is used at each contact to preserve the symmetry of the current–voltage characteristics of the OTFTs. The diode current (I_{diode}) at a given voltage (V_{diode}) is given by

$$I_{\mathrm{diode}} = I_{\mathrm{SS}} \left[\exp\left(\frac{q V_{\mathrm{diode}}}{\eta k_{\mathrm{B}} T} \right) - 1 \right], \tag{5.16}$$

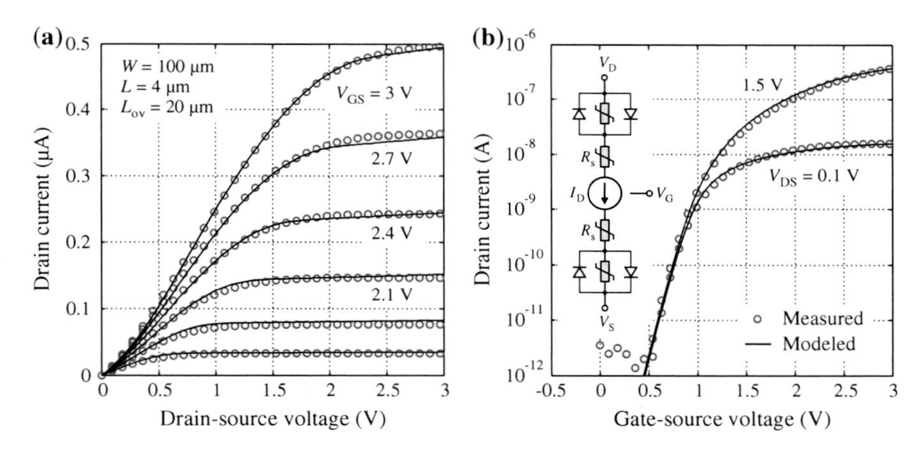

Fig. 5.6 Measured versus simulated I–V characteristics of an n-channel $F_{16}CuPc$ OTFT with $W = 100$ μm, $L = 4$ μm and $L_{ov} = 20$ μm. **a** Output characteristics. **b** Linear and saturation transfer characteristics. The *inset* shows the OTFT DC model schematic. The model parameters are $K_n = \mu C_I = 11$ nA/V^2, $V_{TH} = 0.92$ V, $\lambda = 4$ %/V, $V_{SS} = 125$ mV, $I_{SS} = 3$ nA and $\eta = 1.6$

where I_{SS} is the reverse bias saturation current, also referred to as the scale current, k is the Boltzmann constant, T is the temperature and η is the ideality factor. The scale current (I_{SS}) and the ideality factor (η), which are responsible for the steepness of the current–voltage characteristic, are used as fitting parameters in the model. At very low drain-source voltages $|V_{DS}| \ll |V_{GS} - V_{TH}|$, each diode contributes little to the current conduction. Thus, the total device resistance is viewed in this case as a pair of contact resistors R_s and R_p in series with R_{ch}, i.e., $R_{tot} = 2R_s + 2R_p + R_{ch}$. Using the TLM, similar to that shown in Fig. 5.2, one can extract the contact resistance ($R_C = R_s + R_p$) as well as the intrinsic channel resistance (R_{ch}) as a function of the gate-source voltage (V_{GS}). However, when $|V_{DS}|$ exceeds the diode turn-on voltage, the impact of the parallel resistance (R_p) diminishes; therefore, the total device resistance is given in this case by $R_{tot} = 2R_s + R_{ch}$, meaning that the contact resistance is reduced to $R_C = R_s$ [24].

The model is implemented in a SPICE simulator; comparisons of the simulation with the experimental data shown in Figs. 5.5 and 5.6 illustrate the good fit at all operating regimes. Note that all measurements presented in this chapter are performed in ambient air and at room temperature. Moreover, an IC-CAP PEL code is written to control the test setup and to automate the data collection process; the code is listed in Appendix D.2.

5.4 Summary

Using the TLM, static intrinsic and extrinsic properties of the OTFTs are characterized. The p- and n-channel OTFTs based on DNTT and $F_{16}CuPc$ organic semiconductors have intrinsic carrier mobilities (μ_o) of 3.3 and 0.06 cm^2/Vs, respectively. The source/drain parasitic resistances are experimentally studied for different L and L_{ov}; a distributed resistive network is employed to accurately model their behavior. Furthermore, the static $I-V$ characteristics of the OTFTs are well described by a simple compact model, in which the channel length modulation, subthreshold current and parasitic resistances are considered. It is found that n-channel and aggressively-scaled p-channel (pristine) OTFTs exhibit pronounced non-linearities in their output characteristics at low $|V_{DS}|$; this non-ohmic parasitic effect is modeled using an equivalent circuit incorporating a pair of diodes at each contact.

References

1. H. Murrmann, D. Widmann, Current crowding on metal contacts to planar devices. IEEE Trans. Electron Devices **16**(12), 1022–1024 (1969)
2. D.K. Schroder, *Semiconductor Material and Device Characterization*, 3rd edn. (Wiley-IEEE Press, Hoboken, 2006)
3. F. Ante, Contact effects in organic transistors. Ph.D. dissertation, Swiss Federal Institute of Technology in Lausanne, Lausanne, Switzerland (2011)
4. F. Ante, D. Kälblein, T. Zaki, U. Zschieschang, K. Takimiya, M. Ikeda, T. Sekitani, T. Someya, J.N. Burghartz, K. Kern, H. Klauk, Contact resistance and megahertz operation of aggressively scaled organic transistors. Small **8**(1), 73–79 (2012)
5. S. Luan, G.W. Neudeck, An experimental study of the source/drain parasitic resistance effects in amorphous silicon thin film transistors. J. Appl. Phys. **72**(2), 766–772 (2004)
6. T.J. Richards, H. Sirringhaus, Analysis of the contact resistance in staggered, top-gate organic field-effect transistors. J. Appl. Phys. **102**(9), 094510-1–094510-6 (2007)
7. A.R. Brown, C.P. Jarrett, D.M. de Leeuw, M. Matters, Field-effect transistors made from solution-processed organic semiconductors. Synth. Metals **88**(1), 37–55 (1997)
8. M.C.J.M. Vissenberg, M. Matters, Theory of the field-effect mobility in amorphous organic transistors. Phys. Rev. B **57**(20), 12964–12967 (1998)
9. E.J. Meijer, C. Tanase, P.W.M. Blom, E. van Veenendaal, B.-H. Huisman, D.M. de Leeuw, T.M. Klapwijk, Switch-on voltage in disordered organic field-effect transistors. Appl. Phys. Lett. **80**(20), 3838–3840 (2002)
10. E. Calvetti, L. Colalongo, Z.M. Kovács-Vajna, Organic thin film transistors: a DC/dynamic analytical model. Solid-State Electron. **49**(4), 567–577 (2005)
11. O. Marinov, M.J. Deen, U. Zschieschang, H. Klauk, Organic thin-film transistors: part I-Compact DC modeling. IEEE Trans. Electron Devices **56**(12), 2952–2961 (2009)
12. C.-W. Sohn, T.-U. Rim, G.-B. Choi, Y.-H. Jeong, Analysis of contact effects in inverted-staggered organic thin-film transistors based on anisotropic conduction. IEEE Trans. Electron Devices **57**(5), 986–994 (2019)
13. M. Estrada, A. Cerdeira, J. Puigdollers, L. Reséndiz, J. Pallares, L. F. Marsal, C. Voz, B. Iñíguez, Accurate modeling and parameter extraction method for organic TFTs. Solid-State Electron. **49**(6), 1009–1016 (2005)
14. M. Fadlallah, G. Billiot, W. Eccleston, D. Barclay, DC/AC unified OTFT compact modeling and circuit design for RFID applications. Solid-State Electron. **51**(7), 1047–1051 (2007)

15. B. Iñíguez, R. Picos, D. Veksler, A. Koudymov, M. S. Shur, T. Ytterdal, W. Jackson, Universal compact model for long- and short-channel thin-film transistors. Solid-State Electron. **52**(3), 400–405 (2008)
16. F. Torricelli, Z.M. Kovács-Vajna, A charge-based OTT model for circuit simulation. IEEE Trans. Electron Devices **56**(1), 20–30 (2009)
17. C.H. Kim, A. Castro-Carranza, M. Estrada, A. Cerdeira, Y. Bonnassieux, G. Horrowitz, B. Iñíguez, A compact model for organic field-effect transistors with improved output asymptotic behaviors. IEEE Trans. Electron Devices **60**(3), 1136–1141 (2013)
18. H. Borkan, P.K. Weimer, An analysis of the characteristics of insulated-gate thin-film transistors. RCA Rev. **24**, 153–165 (1963)
19. S.M. Sze, *Semiconductor Devices Physics and Technology*, 2nd edn. (Wiley, Hoboken, 2002)
20. Y. Tsividis, *Operation and Modeling of the MOS Transistors*, 2nd edn. (McGraw-Hill, New York, 1999)
21. M. Jacunski, M. Shur, A. Owsu, T. Ytterdal, M. Hack, B. Iñíguez, A short-channel DC SPICE model for polysilicon thin-film transistors including temperature effects. IEEE Trans. Electron Devices **46**(6), 1146–1158 (1999)
22. M.S. Shur, D. Veksler, V. Chivukula, A. Koudymov, T. Ytterdal, B. Iñíguez, W. Jackson, Modeling of thin film transistors with non-ideal contacts. ECS Trans. **8**(1), 165–170 (2007)
23. P.V. Necliudov, M.S. Shur, D.J. Gundlach, T.N. Jackson, Modeling of organic thin film transistors of different designs. J. Appl. Phys. **88**(11), 6594–6597 (2000)
24. P.V. Necliudov, M.S. Shur, D.J. Gundlach, T.N. Jackson, Contact resistance extraction in pentacene thin film transistors. Solid-State Electron. **47**(2), 259–262 (2003)

Chapter 6
Admittance Characterization

An accurate modeling of the dynamic response of OTFT-based circuits requires an analysis of the charge storage behavior in the transistors. Accordingly, this chapter presents an experimental study of the OTFT admittance at different biasing potentials and validates the results with accurate modeling as well as 2-D device simulations. Effects induced by the parasitic elements that extend beyond the periphery of the intrinsic transistor, including fringe current and contact impedance, are carefully considered. Furthermore, a small-signal model is built to characterize both the resistive and the reactive parts of the measured device admittance. Finally, the implications of the measurements are also discussed relating to the OTFTs dynamic performance, particularly the cutoff frequency and the charge response time.

6.1 Parameter Extraction

For admittance measurements, fully-patterned OTFTs with channel widths (W) of 400 μm and channel lengths (L) of 200, 160, 140, 120, 100, 80, 50 and 30 μm are fabricated. To minimize and precisely control the parasitic capacitances, the top-contact layer is carefully aligned with the patterned bottom-gate layer. Considering the good alignment capability of the high-resolution silicon stencil masks, the gate-overlap length is symmetrically laid out at 10 μm for the source and drain electrodes. Moreover, the organic semiconductor layer extends beyond the periphery of the intrinsic OTFTs by 30 μm on each side (also called fringe regions).

In the beginning, static current–voltage ($I-V$) measurements are performed using a DC source/monitor unit (HP 4141B) on the chip, in the dark, in ambient air and at room temperature. Using the transmission line method (TLM) analysis, which is described in detail in the previous chapter, at a drain-source voltage (V_{DS}) of -0.1 V, a set of parameters are extracted; an intrinsic charge carrier mobility (μ_o) of 2.1 cm^2/Vs and a threshold voltage (V_{TH}) of -1.04 V are calculated, and at a gate-source voltage (V_{GS}) of -3 V, a sheet resistance (R_{sheet}) of 420 kΩ/□ and a contact resistance ($R_C W$) of 0.1 kΩ cm (this value is equal to half of the value reported

T. Zaki, *Short-Channel Organic Thin-Film Transistors*, Springer Theses,
DOI 10.1007/978-3-319-18896-6_6

in [1] due to the different notation used) are found. The I–V output and transfer characteristics of the OTFT with $L = 200$ μm is shown previously in Fig. 3.4. Furthermore, the simulation and modeling parameters are given in Tables 3.1 and 3.2, noting that some of the values are extracted from admittance measurements as demonstrated further below. It is observed from the measurements that the OTFTs have very little hysteresis (not shown in Fig. 3.4) indicating a low interface trap density and good electrical stability of the DNTT layer on the SAM/AlO_x gate dielectric. Since the transistor has a relatively long channel ($L = 200$ μm $\gg L_{ov} = 10$ μm), contact effects are insignificant; accordingly, the effective charge carrier mobility in both the saturation and linear operation regimes are very close to the intrinsic mobility ($\mu_o = 2.1$ cm^2/Vs) and independent of the gate voltage.

The frequency response analysis by means of admittance (Y) measurements is subsequently conducted to provide much insight with regard to the limitation factors of the device performance and their origin. For example, the admittance of OTFTs has been investigate in literature to obtain the doping concentration and charge carrier mobility [2, 3]. Moreover, it has been a useful tool to understand the charge transport mechanism in the polycrystalline organic semiconducting films that include many grain boundaries and trap sites [4], also to study the charge injection mechanism through the contact interfaces between the metallic source/drain contacts and the organic semiconductor [5, 6]. The admittance (Y) measurements are carried out using an LCR meter (HP 4284A) at frequencies from 100 Hz up to 1 MHz and at different bias potentials ranging from zero to −3 V under the same atmospheric conditions as in the static measurements. The IC-CAP PEL codes written to control the LCR meter and to automate data collection are given in Appendix D.2. The capacitance (C) and the conductance (G) are calculated using equations $C = \mathrm{Im}(Y/\omega)$ and $G = \mathrm{Re}(Y)$, respectively, where ω is the angular frequency.

To initially extract the gate insulator capacitance per unit area (C_I), dedicated metal-insulator-metal (MIM) devices with different dimensions (200 × 200, 100 × 100, 50 × 50 and 10 × 10 μm^2) are fabricated on the same substrate. In principle, the MIM is a two-terminal pendant of the OTFT sharing the same layer structure except for the semiconductor. Figure 6.1a shows the measured capacitance–voltage (C–V) curves of the MIM structures at frequencies ranging from 500 Hz upto 1 MHz. The average value measured for the MIMs with different areas is found to be $C_I = 560$ nF/cm^2 (Tables 3.1 and 3.2). With the hybrid insulator thickness of about $d_i = 5.3$ nm, an effective value of $\varepsilon_i = 3.37$ resulted for the relative dielectric constant.

Figure 6.1b shows also the C–V curves of the OTFT with $L = 200$ μm calculated from the measured admittance [1]. For this setup, the source and drain contacts are electrically shorted and connected to the low terminal (virtual ground) of the LCR meter, while the gate electrode is connected to the high terminal. A DC potential sweep from zero to −3 V along with a superimposed AC voltage (V_m) of ±100 mV is applied to the gate. The C–V curves show a typical step-like transition from depletion (at low absolute bias potential) to accumulation (at higher negative bias), determined by the doping concentration of the semiconductor (Table 3.1) [2, 4–8]. As the frequency (f) increases, the accumulated charges in the channel

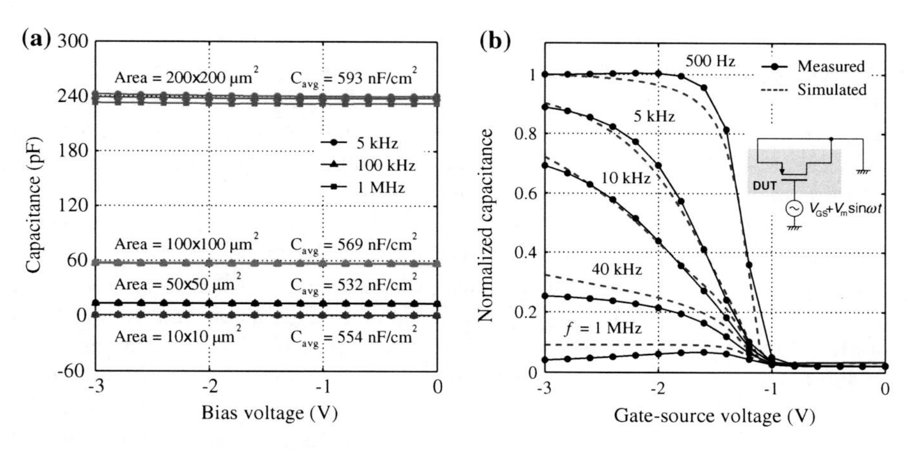

Fig. 6.1 **a** Measured capacitance–voltage (C–V) characteristics of MIM devices with different dimensions (200×200, 100×100, 50×50 and 10×10 μm^2) as a function of the bias voltage and at different frequencies. The results show that the capacitance is dependent neither on the bias voltage nor on the frequency. An average dielectric capacitance per unit area (C_I) of 560 nF/cm^2 for all the structures is extracted. **b** Measured versus simulated C–V characteristics of an OTFT ($W = 400$ μm, $L = 200$ μm and $L_{ov} = 10$ μm) as a function of gate bias and at different frequencies. The inset shows the used measurement setup, where the device under test (DUT) is the p-channel OTFT. The *dashed line* represents the simulations, assuming a mobility of $\mu = 2.0$ cm^2/Vs and a contribution of the fringe part of about 13%. The simulations also fit well to the conductance-voltage (G–V) characteristics (not shown here) [1]

cannot follow the signal. Considering the measurement at $f = 500$ Hz, a minimum capacitance of $C_{min} = 13$ pF (in depletion) and a maximum capacitance of $C_{max} = 566$ pF (in accumulation) are measured. At $V_{GS} > V_{TH} \simeq -1$ V, no channel exists and the measured C_{min} corresponds to the overlap capacitances ($2 \cdot C_{ov}$) between the gate and the source/drain contacts, and can be modeled as a series sum of two dielectrics (with $\varepsilon_i = 3.37$ for the hybrid gate dielectric and $\varepsilon_{DNTT} = 2.84$ for the fully depleted layer of DNTT). On the other hand, at sufficiently large negative bias ($V_{GS} \lesssim -2$ V), the measured C_{max} is equal to the geometrical capacitance between the gate and the accumulated channel. It is simply given by the hybrid dielectric capacitance per unit area (C_I), while considering that the gate-induced channel extends beyond the periphery of the intrinsic OTFT through the overlaps and fringe regions.

The usual lack of a self-alignment OTFT process makes the gate-to-contact overlap capacitances inevitable [7, 9–11]. As a result, the gate electrode is patterned to minimize the impact of the parasitics. The pronounced ratio between the measured capacitances in the depletion and accumulation modes (44 times) is due to the patterning of the gate layer and the use of a relatively small overlap length ($L_{ov} = 10$ μm $\ll L = 200$ μm). This helps in modeling the OTFT intrinsic capacitance accurately. Note that this change in the capacitance is more evident than previously reported works ($\lesssim$6 times) [2, 4–8] owing to the better accuracy and alignment capability of the high-resolution silicon stencil masks [12–15].

Furthermore, Fig. 6.1b depicts the results of the 2-D device simulation, which are performed by our partners at the University of Ilmenau on Sentaurus Device simulator from Synopsys [1]. At the frequency of $f = 500$ Hz, the transition slope between the accumulation and depletion modes is determined only by an unintentional doping in the organic semiconductor (N_A). From the minimum capacitance, the effective layer thickness (d) of the organic semiconductor is estimated. At this frequency, the measured curve is well described with $N_A = 10^{16}$ cm^{-3} and $d = 11$ nm (Table 3.1) with the exception of the transition near $V_{GS} = -1$ V; however, the error in this region is less than 10 % and probably caused by the large voltage step taken in the measurements. To simulate correctly the flat band voltage, a concentration of fixed interface charges of $N_{if} = 1.1 \times 10^{12}$ cm^{-2} is assumed. A slightly larger N_{if} is used in the dynamic simulation compared to the value used in the static simulation ($N_{if} = 5 \times 10^{11}$ cm^{-2} given in Table 3.1) because of the negative bias stress caused by $V_{GS} = -3$ V during the measurements and resulted in a negative shift of the threshold voltage (V_{TH}). The frequency dependence of the C–V curves is well simulated by considering a mobility of $\mu = 2.0$ cm^2/Vs for the injected holes from the source/drain electrodes. As for the highest frequencies (40 KHz and 1 MHz), a smaller contribution of the fringe part should be assumed to reduce the error. Nevertheless, the value of the mobility corresponds very well with the aforementioned TLM measurements and static simulation.

The sheet charge density of the channel (Q_{ch}) in the OTFT can be calculated from the C–V measurements; accordingly, the charge carrier mobility (μ) can be easily extracted from the corresponding linear I–V characteristics as follows [3]. The carrier mobility (μ) is defined as the drift velocity (v_d) of the charge carriers divided by the electric field in the channel (E_{ch}). Given that $v = I_D/(W Q_{ch})$, the charge carrier mobility can be written as

$$\mu = \frac{I_D}{W Q_{ch} E_{ch}}. \tag{6.1}$$

At sufficiently low drain-source bias ($V_{DS} \ll V_{GS} - V_{TH}$; I_D in the linear operation regime), the gate-induced accumulated charges at the semiconductor/dielectric interface are assumed to be uniformly distributed along the channel length (L); consequently, the electric field across the entire channel is approximately constant ($E_{ch} = V_{DS}/L$). Moreover, the induced charge density in the channel at different gate-source voltages can be determined from the C–V measurements at a low drain-source voltage ($V_{DS} = -0.1$ V) using the following expression [3]:

$$Q_{ch} = \int_{+\infty}^{V_{GS}} \left(\frac{C_{ch}}{WL} \right) dV, \tag{6.2}$$

where C_{ch} is the channel capacitance and is obtained by subtracting the overlap capacitances ($2 \cdot C_{ov}$) between the gate and the source/drain contacts from the total measured OTFT capacitance.

Large contact resistances in the OTFT may introduce error into the calculation of the mobility [3]. Therefore, to minimize the effects of the contact resistances, the measured C–V and I–V (Figs. 6.1b and 3.4) characteristics of the longest channel ($L = 200$ μm) are considered here. From the TLM, it is found that the contact resistances are significantly smaller ($< 3\%$) than the channel resistance for the OTFT with $W = 400$ μm and $L = 200$ μm. Subtracting the measured minimum capacitance ($C_{min} = 13$ pF) at $V_{GS} = -0.8$–0 V, which corresponds to the overlap capacitances ($2 \cdot C_{ov}$), from the measured C–V characteristics at the frequency of $f = 500$ Hz (Fig. 6.1b) yields the desired channel capacitance (C_{ch}). A sheet charge density (Q_{ch}) of 1.23 μC/cm^2 is therefore obtained at $V_{GS} = -3$ V by substituting C_{ch} in (6.2). At $V_{DS} = -0.1$ V, a linear drain current ($|I_D|$) of 0.46 μA is measured and a constant electric field ($|E_{ch}|$) of 500 V/m is calculated. Finally, a charge carrier mobility (μ) of 1.9 cm^2/Vs is extracted from (6.1). This value corresponds very well with the aforementioned TLM measurements and static simulations.

6.2 Frequency Response Analysis

To further investigate the charge response behavior, the capacitance (C) and the loss (G/ω) of the OTFTs are measured as a function of the frequency. This helps to clarify the limitation factors for the AC characteristics of the OTFTs. Similar to the previous setup shown in the inset of Fig. 6.1b, the results include the intrinsic as well as the parasitic OTFT components, i.e., the fringe regions that extend beyond the periphery of the intrinsic OTFT and the overlap regions between the gate and the source/drain electrodes.

Figure 6.2 show the admittance measurements of the OTFT with $W = 400$ μm, $L = 200$ μm and $L_{ov} = 10$ μm [1]. Above threshold ($V_{TH} = -1.08$ V), the holes appear to respond well to the applied small signal and the measured capacitance is proportional to the channel length. This is valid up to a certain frequency depending on the applied gate potential; however, their responsiveness starts to degrade above this frequency due to limited lateral flow of holes [7]. At the said frequency, also referred to as relaxation or cutoff frequency (f_T), the measured capacitance (C) decreases and the loss (G/ω) reaches a maximum. This kind of dispersion occurs as a result of the frequency- and voltage-dependent accumulation layer, while another dispersion is expected to occur above 1 MHz due to the overlap regions. For the latter case, the measured capacitances fall to zero. For a smaller V_{GS}, the cutoff frequency (f_T) decreases indicating a lower mobility of the holes. In other words, the mobility of the holes increases by filling up the trap inside the DNTT [7].

As mentioned above, the admittance of OTFTs is investigated to obtain information about the carrier concentration and the interface trap properties [3, 16, 17]. It is also a useful tool to build a small-signal model because both the resistive and the reactive characteristics of a device can be obtained from the measurements [18]. Therefore, the limitations of the AC characteristics by the significant parasitic impedance can be quantitatively explained by means of a simple equivalent

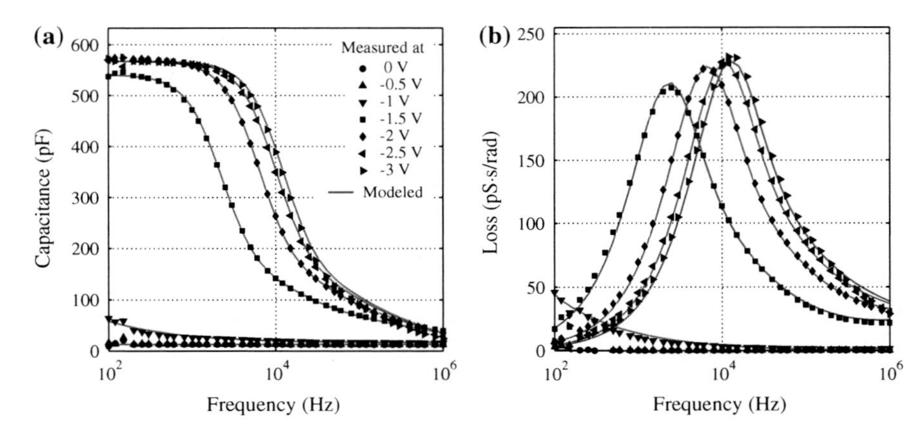

Fig. 6.2 Measured versus modeled admittance of an OTFT ($W = 400$ μm, $L = 200$ μm and $L_{ov} = 10$ μm) as a function of frequency and at different gate bias. **a** Capacitance–frequency (C–f) characteristics, where $C = \text{Im}(Y/\omega)$. **b** Loss–frequency (G/ω–f), where $G = \text{Re}(Y)$. The OTFT has a maximum cutoff frequency (f_T) of about 13 kHz (estimated from the peak of G/ω) at $V_{GS} = -3$ V. The results show that f_T increases with increasing the V_{GS} [1]

circuit [4]. Contact effects in an OTFT are generally dependent on the gate bias and related to the organic semiconductor property, particularly in inverted-staggered (bottom-gate, top-contact) OTFT structure [19]. It is possible to characterize the gate-overlap admittance by fabricating a dedicated MIS structure as in [4, 18]. However, as mentioned in the previous section, the overlap capacitances can be extracted directly from the C–V measurements, obviating the need for special MIS structures. In this case, the admittance of the intrinsic and fringe regions can be obtained from the total admittance by subtracting the estimated gate-overlap admittance.

A small-signal equivalent circuit is developed to explain the conduction mechanism and quantitatively evaluate the impact of the parasitic impedance [1]. As depicted in Fig. 6.3, the model consists of a distributed circuit with parasitic impedances at the contact [4, 5, 7, 18]. The channel as well as the fringe regions act as a resistance-capacitance (RC) transmission line owing to the distributed coupling between the gate electrode and the semiconductor. From the geometry, the contribution of the fringe parts is found to be about 16.5%. In principle, parasitic effects are more dominant in shorter channel OTFTs and they considerably limit the cutoff frequency.

It is reported in [6] that parasitic impedance accounting for a depletion region near the metal/semiconductor interface is necessary to fit the measured channel capacitance [7]. Furthermore, another group has shown in [4] that this parasitic impedance at the metal/semiconductor interface can be suppressed by contact doping; this is supposed to originate from the reduction of the current injection barrier and interface dipole. In this work, however, the contact regions are not doped and no parasitic impedances at the metal/semiconductor interface are considered. It is necessary here

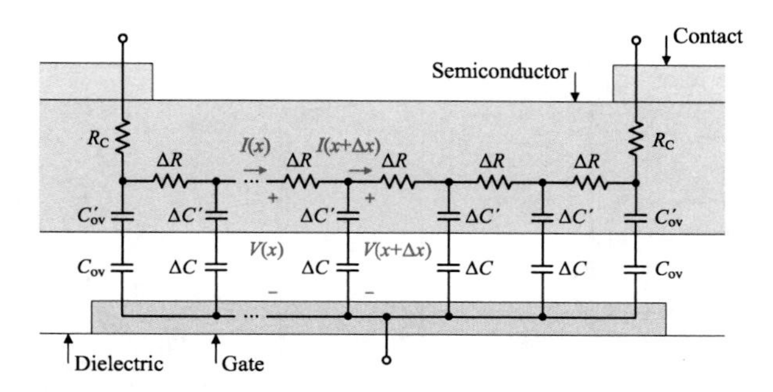

Fig. 6.3 Schematic cross section of an inverted-staggered (bottom-gate, top-contact) OTFT. The small-signal equivalent circuit of the intrinsic as well as the extrinsic (overlap) parts of the OTFT using a distributed RC equivalent or transmission line model is depicted inside the schematic. A particular section of the transmission line of length Δx has the resistance $\Delta R = r_{ch}\Delta x$, and the capacitances $\Delta C = c_i \Delta x$ and $\Delta C' = c_s \Delta x$. The overlap region comprises source/drain contact resistances $R_{S,D} = R_C$, and overlap capacitances $\Delta C_{ov} = c_i L_{ov}$ and $\Delta C'_{ov} = c_s L_{ov}$. Note that the capacitances $\Delta C'$ and $\Delta C'_{ov}$ govern the change in the depletion layer thickness depending on the applied gate voltage. The resistance in the gate electrode is assumed to be negligible

to include only contact resistances and overlap capacitances into the model to properly simulate the measured data.

Referring to Fig. 6.3, the model consists of five parameters, namely source/drain contact resistances ($R_{S,D}$), channel resistance per unit length (r_{ch}), dielectric capacitance per unit length (c_i) and semiconductor capacitance per unit length (c_s) [1]. The resistance in the gate electrode is neglected because the resistance in the channel is large enough in this experiment. The resistances $R_{S,D}$ (Ω) and r_{ch} (Ω/cm) are directly obtained from the TLM, where $R_{S,D} = R_C$ and $r_{ch} = R_{sheet}/W$. Note that the expression of R_C for our inverted-staggered OTFTs is derived in Appendix C.2. Since there is no potential difference between the drain and source electrodes in this setup, r_{ch} is assumed to be uniform along the channel. The accumulated charges of the intrinsic gate region of the OTFT are, however, non-uniformly distributed along the channel when a drain-source bias is applied. Furthermore, r_{ch} is assumed to be uniform along the channel also because the amount of accumulated holes does not change considerably during the admittance measurements by the AC small signal under high negative DC gate-source bias [7]. In depletion, the capacitance c_s (F/cm) governs the change in the depletion layer thickness depending on the applied gate voltage. In accumulation, the influence of the semiconductor capacitance is negligible ($c_s \gg c_i$). Finally, c_i (F/cm) is acquired from the capacitance measurements of the MIM structure (Fig. 6.1a), where $c_i = C_I \cdot W$.

At $V_{GS} = -3$ V, the following values are used: $R_C \cdot W = 0.12$ kΩ cm, $R_{sheet} = 500$ k$\Omega/\Box$ and $C_I = 560$ nF/cm^2. There is only a slight difference (less than 20%) to the values extracted from the TLM. Table 6.1 summarizes all the modeling parameters used at the different gate-source voltages. As depicted in Fig. 6.2, the model shows an excellent agreement with both the capacitance and loss of the measured admittance over the complete frequency range (100 Hz–1 MHz).

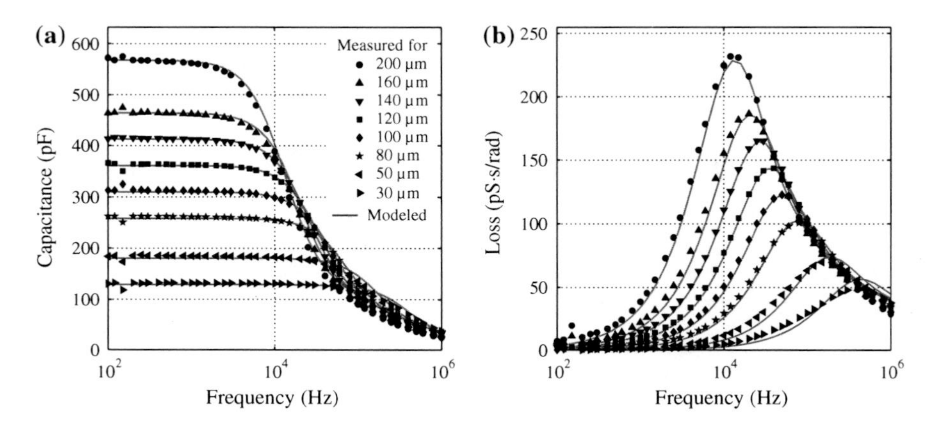

Fig. 6.4 Measured versus modeled admittance of several OTFTs ($W = 400$ μm, $L_{ov} = 10$ μm, and $L = 200, 160, 140, 120, 100, 80, 50$ and 30 μm) as a function of frequency and at $V_{GS} = -3$ V. **a** Capacitance–frequency (C–f) characteristics, where $C = \mathrm{Im}(Y/\omega)$. **b** Loss–frequency ($G/\omega$–$f$), where $G = \mathrm{Re}(Y)$. Note that the model (shown in Fig. 6.3) here used a constant set of parameters (Table 6.1) except for the channel length (L). A maximum cutoff frequency (f_T) of about 0.4 MHz (estimated from the peak of G/ω) at $V_{GS} = -3$ V is extracted for the OTFT with $L = 30$ μm. The results show that f_T increases with decreasing L [1]

Table 6.1 List of model parameters used in the OTFT ($W = 400$ μm and $L_{ov} = 10$ μm) small-signal equivalent circuit

Model parameter	Gate-source voltage [V_{GS} (V)]						
	0	−0.5	−1.0	−1.5	−2.0	−2.5	−3.0
$R_{sheet} = r_{ch} \cdot W$ (kΩ/□)	⇑	⇑	⇑	3100	1000	620	500
$C_S = c_s/W$ (μF/cm^2)[a]	0.2	0.23	0.35	10.5	⇑	⇑	⇑

The parameters listed above, namely the sheet resistance (R_{sheet}) and semiconductor capacitance (C_S), are those that are dependent on the gate-source voltage (V_{GS}). In addition, $R_C \cdot W = 0.12$ kΩ cm and $C_I = 560$ nF/cm^2 are included in the model

[a] In depletion, C_S governs the change in the depletion layer thickness

To evaluate the scalability of the model with reducing the device dimensions, Fig. 6.4 shows the measured and simulated admittance at $V_{GS} = -3$ V for all the fabricated OTFTs with $W = 400$ μm, $L_{ov} = 10$ μm, and $L = 200, 160, 140, 120, 100, 80, 50$ and 30 μm. The same set of model parameters is used. A reliable and precise fit of the small-signal equivalent circuit to the experimental data is obtained.

Following the method presented in [4], the extracted capacitances can be divided into two regions, namely a constant region at low frequencies and a decreasing region at high frequencies. In the high frequency region, the measured capacitances follow the power law ($C \propto f^{-p}$). Figure 6.5a shows an example for the OTFT with $L = 200$ μm. The values of the exponent p are between 0.59 and 0.68. The cutoff frequency (f_T) can be estimated as the intersection point of the extrapolated fitting lines (using power law fit) and the capacitance in the low frequency region (lateral line fit) [4]. This method is useful if no conductance measurements are available.

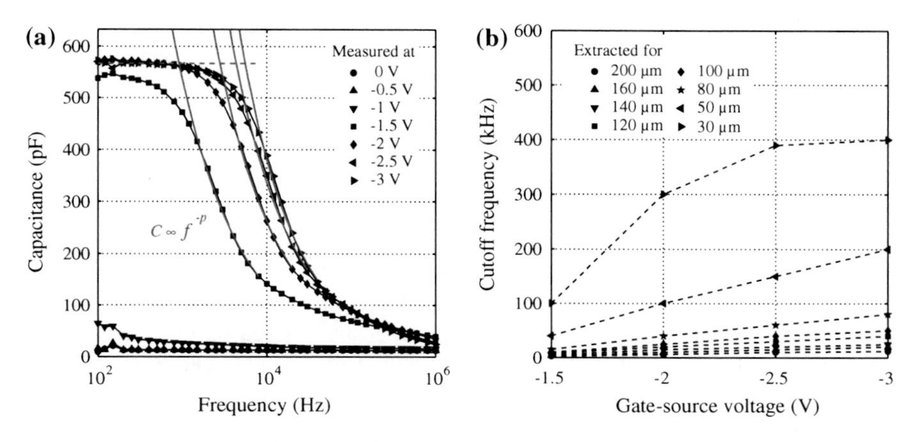

Fig. 6.5 **a** Measured capacitance–frequency (C–f) characteristics of an OTFT with ($W = 400\,\mu$m, $L = 200\,\mu$m and $L_{ov} = 10\,\mu$m) as a function of frequency and at different gate bias. At $V_{GS} > V_{TH}$, the capacitance follow the power law ($C \propto f^{-p}$) in the high frequency region and is nearly constant in the low frequency region, noting that the cutoff frequency can be roughly estimated from the intersection point between the two fitting lines [4]. The values of the exponent p are 0.59, 0.6, 0.62 and 0.68 at $V_{GS} = -1.5, -2.0, -2.5$ and -3.0 V, respectively. **b** Extracted cutoff frequency (f_T) of all the OTFTs ($W = 400\,\mu$m, $L_{ov} = 10\,\mu$m, and $L = 200, 160, 140, 120, 100, 80, 50$ and 30 μm) from the peak of G/ω. The results verify that f_T approximately complies with (3.12), i.e., f_T increases with decreasing L and with increasing V_{GS}

However, a more accurate approach is to estimate f_T from the peak of G/ω as depicted in Fig. 6.5b. The results verify that the cutoff frequency increases with decreasing channel length and with increasing gate bias. This approximately conforms with the expression (3.12).

6.3 Effective Delay of Gate-Induced Charges

The charges in the OTFT channel cannot be induced instantaneously with the applied gate-source voltage; therefore, the charge response time can be a bottleneck in the high frequency operation of the device [18]. A quantitative evaluation of the effective RC delay time (τ), which is needed to make the charges induced in the channel, requires first to derive expressions for the gate-to-channel capacitance (C_{ch}) and loss (G_{ch}/ω) of the intrinsic OTFT. Referring to Fig. 6.3, a transmission line model (RC network) is used for the calculation of the capacitance. Assuming the transistor is operated in the accumulation regime, the influence of the capacitances $\Delta C'$ and $\Delta C'_{ov}$ that govern the change in the depletion layer thickness depending on the gate voltage are negligible, i.e., $\Delta C \gg \Delta C'$ and $\Delta C_{ov} \gg \Delta C'_{ov}$. Considering only the intrinsic part of the OTFT (valid when $L \gg L_{ov}$), a particular section of the transmission line of length Δx has the resistance $\Delta R = R_{sheet}\Delta x/W$ and the capacitance $\Delta C = C_I W \Delta x$. By following a procedure similar to that used in [20] and [21], we apply Kirchhoff's voltage and current laws and take the limit $\Delta x \rightarrow 0$ to obtain

$$\frac{\partial I}{\partial x} = -j\omega C_{\mathrm{I}} W V(x), \tag{6.3}$$

$$\frac{\partial V}{\partial x} = -\frac{R_{\mathrm{sheet}}}{W} I(x). \tag{6.4}$$

We then combine (6.3) and (6.4) and solve the resulting second-order linear differential equation to get expressions for the current $I(x)$ and the voltage $V(x)$. After a considerable amount of algebra, it can be deduced that the gate-to-channel capacitance (C_{ch}) and loss (G_{ch}/ω) are expressed by the following [1, 18, 20]:

$$C_{\mathrm{ch}} = \frac{1}{\omega} \mathrm{Im}\left[\frac{2I(x=0)}{V(x=0)}\right] = \frac{C_{\mathrm{I}} W L}{\alpha}\left(\frac{\sinh\alpha + \sin\alpha}{\cosh\alpha + \cos\alpha}\right), \tag{6.5}$$

$$\frac{G_{\mathrm{ch}}}{\omega} = \mathrm{Re}\left[\frac{2I(x=0)}{V(x=0)}\right] = \frac{C_{\mathrm{I}} W L}{\alpha}\left(\frac{\sinh\alpha - \sin\alpha}{\cosh\alpha + \cos\alpha}\right), \tag{6.6}$$

where $\alpha = \sqrt{\omega R_{\mathrm{sheet}} C_{\mathrm{I}} L^2/2}$. Detailed steps for the derivation of C_{ch} and G_{ch}/ω are given in Appendix C.1.

When $\alpha = 1$, the channel capacitance (C_{ch}) given in (6.5) is equal $0.97 \times C_{\mathrm{I}} W L$. Therefore, the delay time ($\tau = 1/\omega$) that is needed for 97% of the total charges in the channel to be induced and effectively responding, is accordingly determined as

$$\tau = \frac{R_{\mathrm{sheet}} C_{\mathrm{I}} L^2}{2}. \tag{6.7}$$

This relation is slightly different from the one reported in [18] because of the different device structure used. Figure 6.6a shows the calculated τ as a function of L using the same modeling parameters: $R_{\mathrm{sheet}} = 500\ \mathrm{k\Omega/\Box}$ and $C_{\mathrm{I}} = 560\ \mathrm{nF/cm^2}$ [1]. Such a plot can be used to deduce information about the maximum frequency achieved for a given minimum channel length and/or maximum channel length needed for a given frequency. For example, one can estimate from Fig. 6.6a that for a successful operation at $f = 10$ kHz, the OTFT should have $L \leq 106$ μm. To verify this estimation, normalized C–V characteristics of the OTFTs with $L = 200$, 160, 140, 120, 100, 80, 50 and 30 μm measured at 10 kHz are depicted in Fig. 6.6b. The normalization is implemented by dividing the measured C–V data by $1.165\ C_{\mathrm{I}} W(L + 2L_{\mathrm{ov}})$ to account for the parasitic components, i.e., the fringe (16.5%) and overlap regions ($L_{\mathrm{ov}} = 10$ μm). There is an error of less than 3% because of this assumption; when $L = 100$ μm, the normalized channel capacitance almost reaches unity at $V_{\mathrm{GS}} = -3$ V, but longer devices cannot fully induce the charges at this frequency. The result correspond closely to what is estimated from (6.7).

Equation (6.7) suggests that there are three ways to reduce the effective charge response time (τ) of an OTFT [18]. First, the sheet resistance (R_{sheet}) can be reduced by using an organic semiconductor with a larger mobility or by applying a higher

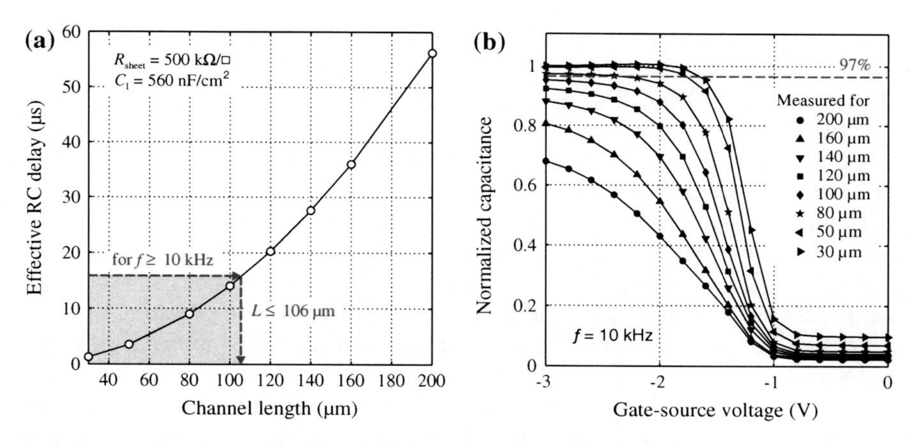

Fig. 6.6 **a** Calculated effective delay (τ) for 97% of the total charges in the OTFT channel to be induced and effectively responding. The expression of τ is represented by (6.7), considering the same values for the sheet resistance (R_{sheet}) and the gate dielectric capacitance per unit area (C_I), which are used for the small-signal equivalent circuit model. Information about the maximum f achieved for a given minimum L and/or maximum L needed for a given f can be deduced from this relation. For instance, it can be estimated that for a successful operation at $f = 10$ kHz, the OTFT should have $L \leq 106$ μm. **b** Normalized C–V measurements at $f = 10$ kHz. The contribution of the fringe part is assumed to be constant (16.5%) for all channel lengths and the normalization is done by dividing the measurement results by $1.165\, C_\text{I} W(L + 2L_{\text{ov}})$. There is an error of less than 3% because of this assumption [1]

voltage. However, the evolution of the mobility (Fig. 2.2b) in recent years does not offer a promising order-of-magnitude improvement in the near future [22]. Furthermore, applying a higher voltage would increase the power consumption of the device and this is not favorable for battery-powered or frequency-coupled portable devices. In addition, this may avert the possibility to integrate hybrid solutions combining large-area organic electronics with high-performance thin silicon chips. Second, a smaller dielectric capacitance (C_I) can be employed, but this may cause low on-current (I_{on}) or transconductance (g_m). Finally, the most reasonable approach is to reduce τ by scaling down the channel length (L) of the OTFT.

6.4 Meyer's Capacitance Model

For the simulation of the dynamic performance of OTFTs, the variations in the stored charges of the devices have to be considered. In an OTFT, there are stored charges in the gate electrode and the conducting accumulated channel. In a simplified manner, the variation in these stored charges can be expressed through different capacitance elements as discussed herein. One should distinguish between the parasitic and intrinsic elements of the OTFT. As mentioned before, the parasitics include the fringe regions that extend beyond the periphery of the intrinsic OTFT, and the overlap regions between the gate electrode and the source/drain contacts. In general, applying a drain-source bias to the OTFT results in a non-uniform distribution of charges along the channel. An accurate description of this effect requires a distributed RC network as an equivalent circuit model, which can in practice be simplified into lumped capacitive elements between the source, drain and gate terminals. In this case, the resulting errors in circuit simulations are typically small [23]. In contrast to DC models, the development of capacitance compact models for OTFTs were not verified by experimental data since measurements of OTFTs in the quasistatic regime were not available [7, 24–28]. In this section, to the best of our knowledge, we present the first experimental analysis of the OTFT intrinsic capacitances at different biasing potentials and validate the results with accurate modeling as well as 2-D device simulations [1].

Using a similar approach that is developed for silicon-based MOSFETs, analysis of how the accumulated charges in an OTFT are distributed between the source, drain and gate electrode at different bias potentials is discussed in the following [23]. Ward and Dutton have reported in 1978 a MOSFET capacitance model, which comprises a set of charge-conserving and nonreciprocal capacitances between the different intrinsic terminals of the device [29]. Note that nonreciprocity here means that $C_{\mathrm{ij}} \neq C_{\mathrm{ji}}$, where i and j denote source, drain, gate or substrate in the case of MOSFETs. In a simplified analysis by Meyer in 1971, a set of reciprocal capacitances are obtained as derivatives of the total gate charges with respect to the various terminal voltages [30]. In principle, Meyer's model does not guarantee charge conservation because it is only a subset of Ward and Dutton model. Hence, there have been attempts in [24] and [25] to derive OTFT capacitance models based on Ward and Dutton analysis. However, we prefer to use here the simple Meyer's model as the resulting errors in circuit simulations are usually small. Unlike MOSFETs, the OTFT layers are grown on an insulating substrate, most notably on a glass substrate. Therefore, there is no bulk connection/terminal and there are no charges stored in the substrate.

In the following, only the intrinsic part of the OTFT is considered, i.e., the overlap and fringe regions are excluded. However, the effects induced by the parasitic elements on the capacitance measurements are investigated hereinafter. According to the Meyer's capacitance model, the distributed intrinsic capacitances of an OTFT can be split into the following two lumped capacitances, namely gate-source (C_{gs}) and gate-drain (C_{gd}) capacitances, which are expressed as

$$C_{\mathrm{gs}} = \left.\frac{\partial Q_{\mathrm{G}}}{\partial V_{\mathrm{GS}}}\right|_{V_{\mathrm{GD}}}, \tag{6.8}$$

$$C_{\mathrm{gd}} = \left.\frac{\partial Q_{\mathrm{G}}}{\partial V_{\mathrm{GD}}}\right|_{V_{\mathrm{GS}}}, \tag{6.9}$$

where Q_{G} is the total intrinsic gate charges [23]. The capacitances C_{gs} (F) and C_{gd} (F) are assumed to be dominated by the gate-induced accumulated charges. Thus, the contribution of the accumulated charges to the gate charges is determined by integrating the sheet charge density of the channel (Q_{ch}) over the active gate area:

$$Q_G = C_I W \int_0^L (V_G - V_{TH} - V_x)\, dx, \tag{6.10}$$

where V_x is the voltage at an arbitrary distance x from the source. Using the well-established concept of charge drift, the drain current (I_D) in an OTFT can be deduced from (6.1) as

$$I_D = \mu C_I W (V_G - V_{TH} - V_x) \frac{\partial V_x}{\partial x}. \tag{6.11}$$

The current is constant along the channel and $dV_x = (\partial V_x/\partial x)\, dx$. By substituting for dx using (6.11) in (6.10) and changing the integral limits, we obtain

$$\begin{aligned} Q_G &= \frac{\mu C_I^2 W^2}{I_D} \int_{V_S}^{V_D} (V_G - V_{TH} - V_x)^2\, dV_x \\ &= \frac{\mu C_I^2 W^2}{3 I_D} \left[(V_{GS} - V_{TH})^3 - (V_{GD} - V_{TH})^3 \right]. \end{aligned} \tag{6.12}$$

Using the expression (3.4) for the drain current (I_D) in the linear operation regime and replacing V_{DS} by $V_{GS} - V_{GD}$, we get

$$\begin{aligned} I_D &= \frac{\mu C_I W}{L} \left[(V_{GS} - V_{TH}) V_{DS} - \frac{V_{DS}^2}{2} \right] \\ &= \frac{\mu C_I W}{2L} V_{DS} \left[(V_{GS} - V_{TH}) + (V_{GD} - V_{TH}) \right] \\ &= \frac{\mu C_I W}{L} \left[(V_{GS} - V_{TH})^2 - (V_{GD} - V_{TH})^2 \right]. \end{aligned} \tag{6.13}$$

Substituting (6.13) in (6.12), the total intrinsic gate charges (Q_G) can therefore be expressed as

$$Q_G = \frac{2}{3} C_I W L \frac{(V_{GS} - V_{TH})^3 - (V_{GD} - V_{TH})^3}{(V_{GS} - V_{TH})^2 - (V_{GD} - V_{TH})^2}. \tag{6.14}$$

Finally, The gate-source (C_{gs}) and gate-drain (C_{gd}) capacitances can be obtained by substituting (6.14) in (6.8) and (6.9), respectively:

$$C_{gs} = \frac{2}{3} C_I W L \left[1 - \left(\frac{V_{GT} - V_{DS}}{2 V_{GT} - V_{DS}} \right)^2 \right], \tag{6.15}$$

$$C_{gd} = \frac{2}{3} C_I W L \left[1 - \left(\frac{V_{GT}}{2 V_{GT} - V_{DS}} \right)^2 \right], \tag{6.16}$$

where $V_{GT} = V_{GS} - V_{TH}$, also called the gate-overdrive voltage. At $V_{DS} = 0$ V, both the gate-source and gate-drain capacitances are equal to $1/2 \cdot C_I W L$. In saturation regime ($V_{DS} \geq V_{GS} - V_{TH}$), the drain-source voltage is given by $V_{DS} = V_{GT}$, and thus, C_{gs} and C_{gd} are equal to $2/3 \cdot C_I WL$ and zero, respectively. This indicates that a small change in the applied drain-source voltage (V_{DS}) when the channel is pinched off during the saturation regime does not have an impact on the gate or channel charges; however, the channel is completely assigned to the source terminal, resulting in a maximum value for the gate-source capacitance ($C_{gs,max} = 2/3 \cdot C_I WL$) [23].

Discontinuities in the derivatives of the Meyer capacitances given by (6.15) and (6.16) occur at the onset of saturation. Such discontinuities should be avoided in the device models since they give rise to increased simulation time and conversion problems in circuit simulators [23]. A smooth transition between the nonsaturated (linear or off regimes) and the saturated regimes is assured by replacing V_{DS} in (6.15) and (6.16) by V_{DSe}, where V_{DSe} is an effective drain-source voltage that is equal to V_{DS} for $V_{DS} < V_{GT}$ and is equal to V_{GT} for $V_{DS} > V_{GT}$. This is achieved by the following asymptotic interpolation function for the effective drain-source voltage (V_{DSe}) [23]:

$$V_{DSe} = \frac{1}{2}\left[V_{DS} + V_{GT} - \sqrt{V_\delta^2 + (V_{DS} - V_{GT})^2}\right], \tag{6.17}$$

where V_δ is a constant voltage that determines the width of the transition region; It is a fitting parameter that is extracted from the measurements.

We have considered so far admittance measurements with shorted source and drain contacts, however, a different setup is used here to account for the variations in the stored charges of the OTFT at different drain-source (V_{DS}) and gate-source (V_{GS}) voltages. Accordingly, the gate electrode is connected to the low terminal (virtual ground) of the LCR meter, while the source and drain contacts are connected alternatively to the high terminal of the LCR meter (HP 4284A precision LCR meter) and a DC voltage source (Keithley 230 programmable voltage source) [1, 7]. The DC voltages V_{DS} and V_{GS} are both swept from zero to -3 V and an AC signal of ± 100 mV with $f = 500$ Hz is applied to either the source or the drain.

Figure 6.7 shows the measured gate-source (C_{gs}) and gate-drain (C_{gd}) capacitances along with modeled and simulated results for the OTFT with $W = 400$ µm, $L = 200$ µm and $L_{ov} = 10$ µm [1]. The measured data include both the intrinsic and parasitic components. Similar to the previous measurements, the parasitic component is composed of about 16.5% for the fringe region and a constant $C_{ov} = 6.5$ pF for each of the gate-source and gate-drain overlap areas. By excluding the parasitic components from the measured data, it is found that $C_{gs} = C_{gd} = 1/2 \cdot C_I WL$ at $V_{DS} = 0$ V, while C_{gs} and C_{gd} approach $2/3 \cdot C_I WL$ and zero, respectively, at $V_{DS} \geq V_{GS} - V_{TH}$ (saturation regime). This charge storage effect complies with the Meyer's capacitance model described above.

Using $C_I = 580$ nF/cm^2, $V_{TH} = -1.18$ V and $V_\delta = 0.1$ V in addition to multiplying both (6.15) and (6.16) by 1.165 for the fringe region and adding 6.5 pF for the overlap capacitance (C_{ov}), a precise fit between the modeled and measured data is accomplished as demonstrated in Fig. 6.7. Furthermore, the measured and modeled characteristics are compared to 2-D device simulation results. Using the same

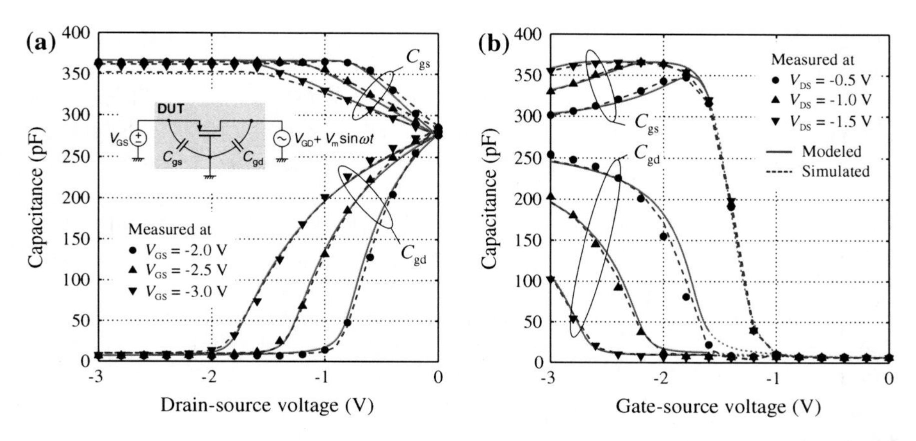

Fig. 6.7 Measured gate-source (C_{gs}) and gate-drain (C_{gd}) capacitances along with modeled and simulated results of an OTFT ($W = 400$ µm, $L = 200$ µm and $L_{ov} = 10$ µm) at $f = 500$ Hz. **a** C_{gs} and C_{gd} as a function of V_{DS}. **b** C_{gs} and C_{gd} as a function of V_{GS}. Note that the modeled curves are represented by (6.15) and (6.16) along with the interpolation function (6.17), which are valid above threshold (*solid lines*). The voltage-dependent fringe factor of the simulated curves is calculated by dividing the measured value of C_{gs} by the simulated intrinsic value. The inset shows the used measurement setup, where the gate electrode is connected to the low terminal (virtual ground) of the LCR meter, and the source and drain contacts are connected alternatively to the high terminal of the LCR meter and the DC voltage source [1]

simulation parameters given in Table 3.1, an excellent agreement for all voltage regions is obtained for both C_{gs} and C_{gd}. It is necessary here to include a voltage-dependent fringe factor because otherwise the results would not be well described. The reason is simply the smaller influence of fringe effects for small intrinsic capacitances, which occurs in the case of the subthreshold or the linear operation regions. The fringe factor is calculated by dividing the measured C_{gs}–V_{GS} by the simulated intrinsic values at the different drain-source voltages. The results do not show a strong dependency of the fringe factor on the drain-source voltage but it changes considerable depending on the gate-source voltage ranging from about 0.6 at $V_{GS} = -1.3$ V to 1.19 at $V_{GS} = -3$ V.

6.5 Summary

Frequency response analysis on inverted-staggered (bottom-gate, top-contact) OTFTs based on DNTT organic semiconductor by means of admittance measurements is performed. The results confirm that the cutoff frequency (f_T) of the OTFTs increases with decreasing channel length (L) and with increasing gate-source bias (V_{GS}). A small-signal equivalent circuit based on a distributed RC network or a transmission line model is used to produce a precise and reliable fit to the data for different operating frequencies (100 Hz–1 MHz), biasing potentials (−3 V to zero) and channel lengths (30 to 200 μm). A quantitative evaluation of the effective RC delay time (τ), which is needed to make the charges induced in the OTFT channel, is demonstrated. Furthermore, the charge storage behavior in the OTFT is very well described by compact modeling and verified against 2-D device simulations. The dependence of the intrinsic gate-source (C_{gs}) and gate-drain (C_{gd}) capacitances on the applied gate-source (V_{GS}) and gate-drain (V_{GD}) voltages shows an excellent agreement with Meyer's capacitance model, which can in principle be easily incorporated into a SPICE simulator. This is the first study to undertake experimental analysis of OTFT intrinsic capacitances at different biasing potentials and validate the results with accurate modeling and simulations. The material parameters used for all the dynamic characterization correspond closely to those extracted from the static measurements, with the exception of the density of fixed interface states (N_{if}). This is because of the negative bias stress imposed during the dynamic measurements, which results in a negative shift of the threshold voltage (V_{TH}).

References

1. T. Zaki, S. Scheinert, I. Hörselmann, R. Rödel, F. Letzkus, H. Richter, U. Zschieschang, H. Klauk, J.N. Burghartz, Accurate capacitance modeling and characterization of organic thin-film transistors. IEEE Trans. Electron Devices **61**(1), 98–104 (2014)
2. S. Grecu, M. Bronner, A. Opitz, W. Brütting, Characterization of polymeric metal-insulator-semiconductor diodes. Synth. Met. **146**(3), 359–363 (2004)
3. K. Ryu, I. Kymissis, V. Bulović, C.G. Sodini, Direct extraction of mobility in pentacene OFETs using *C-V* and *I-V* measurements. IEEE Electron Device Lett. **26**(10), 716–718 (2005)
4. T. Miyadera, T. Minari, K. Tsukagoshi, H. Ito, Y. Aoyagi, Frequency response analysis of pentacene thin-film transistors with low impedance contact by interface molecular doping. Appl. Phys. Lett. **91**(1), 013512-1–013512-3 (2007)
5. D.M. Taylor, N. Alves, Separating interface state response from parasitic effects in conductance measurements on organic metal-insulator-semiconductor capacitors. J. Appl. Phys. **103**(5), 054509-1–054509-6 (2008)
6. B.H. Hamadani, C.A. Richter, J.S. Suehle, D.J. Gundlach, Insights into the characterization of polymer-based organic thin-film transistors using capacitance-voltage analysis. Appl. Phys. Lett. **92**(20), 203303-1–203303-3 (2008)
7. K. Kim, Y. Kim, Intrinsic capacitance characterization of top-contact organic thin-film transistors. IEEE Trans. Electron Devices **57**(9), 2344–2346 (2010)
8. E. Itoh, K. Miyairi, Interfacial charge phenomena at the semiconductor/gate insulator interface in organic field effect transistors. Thin Solid Films **499**(1–2), 95–103 (2006)
9. T. Zaki, R. Rödel, F. Letzkus, H. Richter, U. Zschieschang, H. Klauk, J.N. Burghartz, S-parameter characterization of submicrometer low-voltage organic thin-film transistors. IEEE Electron Device Lett. **34**(4), 520–522 (2013)
10. Y.-Y. Noh, N. Zhao, M. Caironi, H. Sirringhaus, Downscaling of self-aligned, all-printed polymer thin-film transistors. Nat. Nanotechnol. **2**, 784–789 (2007)
11. M. Caironi, E. Gili, T. Sakanoue, X. Cheng, H. Sirringhaus, High yield, single droplet electrode arrays for nanoscale printed electronics. ACS Nano **4**(3), 1451–1456 (2010)
12. F. Letzkus, T. Zaki, F. Ante, J. Butschke, H. Richter, H. Klauk, J.N. Burghartz, Si stencil masks for organic thin film transistor fabrication, in *Proceedings of the SPIE Photomask Technology*, pp. 81662B-1–81662B-12 (2011)

13. F. Letzkus, T. Zaki, F. Ante, J. Butschke, H. Richter, H. Klauk, J.N. Burghartz, Si Stencil-Masken für die Herstellung organischer Dünnschichttransistoren, in *MikroSystemTechnik Kongress*, pp. 181–184 (2011) (in German)
14. F. Ante, F. Letzkus, J. Butschke, U. Zschieschang, K. Kern, J.N. Burghartz, H. Klauk, Submicron low-voltage organic transistors and circuits enabled by high-resolution silicon stencil masks, in IEEE International Solid-State Circuits Conference Technical Digest, pp. 21.6.1-21.6.4 (2010)
15. F. Ante, F. Letzkus, J. Butschke, U. Zschieschang, J.N. Burghartz, K. Kern, H. Klauk, Top-contact organic transistors and complementary circuits fabricated using high-resolution silicon stencil masks, in Device Research Conference, pp. 175–176 (2010)
16. C.A. Mills, D.M. Taylor, A. Riul, A.P. Lee, Effects of space charge at the conjugated polymer/electrode interface. J. Appl. Phys. **91**(8), 5182–5189 (2002)
17. I. Torres, D.M. Taylor, Interface states in polymer metal-insulator-semiconductor devices. J. Appl. Phys. **98**(7), 073710-1–073710-9 (2005)
18. K.-D. Jung, C.A. Lee, D.-W. Park, B.-G. Park, H. Shin, J.D. Lee, Admittance measurements on OFET channel and its modeling with *R*-*C* network. IEEE Electron Device Lett. **28**(3), 204–206 (2007)
19. C.-W. Sohn, T.-U. Rim, G.-B. Choi, Y.-H. Jeong, Analysis of contact effects in inverted-staggered organic thin-film transistors based on anisotropic conduction. IEEE Trans. Electron Devices **57**(5), 986–994 (2019)
20. D.W. Greve, V.R. Hay, Interpretation of capacitance-voltage characteristics of polycrystalline silicon thinfilm transistors. J. Appl. Phys. **61**(3), 1176–1180 (1987)
21. D.W. Greve, Programming mechanism of polysilicon resistor fuses. IEEE Trans. Electron Devices **29**(4), 719–724 (1982)
22. F. Ante, "Contact effects in organic transistors," Ph.D. dissertation, Swiss Federal Institute of Technology in Lausanne, Lausanne, Switzerland (2011)
23. T. Ytterdal, Y. Cheng, T. Fjeldly, *Device Modeling for Analog and RF CMOS Circuit Design* (Wiley, Chichester, 2003)
24. A. Castro-Carranza, M. Estrada, J.C. Nolasco, A. Cerdeira, L.F. Marsal, B. Iñíguez, J. Pallarès, "Organic thin-film transistor bias-dependent capacitance compact model in accumulation regime,". IET Circ. Dev. Syst. **6**(2), 130–135 (2012)
25. O. Marinov, M.J. Deen, Quasistatic compact modelling of organic thin-film transistors. Org. Electron. **14**(1), 295–311 (2013)
26. E. Calvetti, L. Colalongo, Z.M. Kovács-Vajna, Organic thin film transistors: a DC/dynamic analytical model. Solid-State Electron. **49**(4), 567–577 (2005)
27. M. Fadlallah, G. Billiot, W. Eccleston, D. Barclay, DC/AC unified OTFT compact modeling and circuit design for RFID applications. Solid-State Electron. **51**(7), 1047–1051 (2007)
28. F. Torricelli, Z.M. Kovács-Vajna, L. Colalongo, A charge-based OTFT model for circuit simulation. IEEE Trans. Electron Devices **56**(1), 20–30 (2009)
29. D.E. Ward, R.W. Dutton, A charge-oriented model for MOS transistor capacitance. IEEE J. Solid-State Circ. **13**(5), 703–708 (1978)
30. J.E. Meyer, MOS models and circuit simulation. RCA Rev. **32**(1), 42–63 (1971)

Chapter 7
Scattering Parameter Characterization

Recent developments of the performance of OTFTs in the past few years have led to a renewed interest in characterizing their AC electrical properties using a self-contained method. Accordingly, the first comprehensive experimental study of the frequency response of OTFTs using scattering S-parameter measurements is introduced in this chapter. A small-signal model, which includes intrinsic as well as extrinsic components, derived from the physical behavior of the device is presented. An excellent fit between measured and simulated S-parameters is demonstrated. The frequency performance of the model is examined in terms of its current-gain cutoff frequency. The channel length dependence of the cutoff frequency is described in a compact model and a close agreement to the measured data of OTFTs with variable device dimensions is shown. Moreover, the correspondence between static and dynamic characterization is discussed. For this study, low-voltage OTFTs based on the air-stable DNTT organic semiconductor having various channel and gate-overlap lengths are utilized. Furthermore, the impact of misalignment between the source/drain contacts and the patterned gate on the dynamic TFT performance is explored and a simple method to estimate the degree of misalignment from the measured S-parameters is proposed. The intentional asymmetry between gate-source and gate-drain overlaps is precisely controlled by the use of high-resolution silicon stencil masks.

7.1 Measurement Setup

Analysis of the frequency response of OTFTs is of great interest to gain deeper insight into the physics of the device and for comparing different structures as well as assessing materials-related limitations [1]. In addition, it helps to predict the maximum operating frequency of which circuits based on OTFTs, such as an amplifier, can work satisfactorily without reduction of the gain [2]. So far, the frequency limit has been derived from the propagation delay of a ring oscillator, which only provides

T. Zaki, *Short-Channel Organic Thin-Film Transistors*, Springer Theses,
DOI 10.1007/978-3-319-18896-6_7

an average delay figure. As presented in Sect. 3.4, the minimum measured stage delays are 0.2 μs for unipolar and 17 μs for complementary ring oscillators, where both comprised air-stable organic semiconductors, fabricated using high-resolution silicon stencil masks and operated at supply voltages below 5 V [3, 4]. It is also mentioned that simultaneous achievement of low operation voltage and small propagation delay is very difficult for OTFTs-based ring oscillators because of the limited carrier mobility, particularly for the air-stable organic semiconductors, and the large parasitic capacitances and contact resistances [5]. Nevertheless, in view of the low supply voltage and air stability, these results mark a record accomplishment in the OTFT technology. Given the variability of state-of-the-art OTFTs and the design variations of the ring oscillators, this characterization method is not fully satisfactory.

To fairly compare different OTFTs with respect to their frequency response, the bandwidth has to be measured in a consistent way [6]. It is shown in the previous chapter that AC admittance measurements on stand-alone OTFTs, under certain DC bias, can be used to give information about the cutoff frequency and/or charge response time within the channel. This characterization is essential for estimating all the circuit elements present in the device and in particular for constructing an equivalent circuit model [7]; however, it is only appropriate for relatively large devices to ensure accurate measurements. For this reason, others have demonstrated the possibility to extract the cutoff frequency on individual OTFTs using direct measurement of the gate and drain modulation currents [6–9]. In this setup, the gate and drain are biased by a DC voltage source, while a superimposed AC signal is applied to the gate through a bias tee using a function generator. The input and output AC current components are measured by special current probes (Integrated Sensor Technologies 711S standard) without electrical connection and monitored using a digital oscilloscope [8, 9]. An alternative approach is to measure the input and output AC currents by a lock-in amplifier [6, 7]. In both cases, the frequency scan is done by reprogramming (computer-controlled) the function generator. Such user-configured measurement setup, involving multiple test instruments, is cumbersome and makes the extraction of device parasitics very difficult. Therefore, a self-contained method for characterizing the cutoff frequency of OTFTs by means of S-parameter measurements is highly desirable, as demonstrated on other field-effect transistor types such as ZnO TFTs [10], graphene FETs [11], heterojunction bipolar transistors (HBTs) [12] and radio-frequency MOSFETs [13]. In this work, to the best of our knowledge, we present the first report on S-parameter measurements of OTFTs [1].

Unlike previous techniques, the use of a self-contained method by measuring S-parameters is superior, since it only requires a vector network analyzer (VNA) to characterize the entire AC electrical properties of the OTFTs [14]. This simple and accurate approach is well suited for wafer probing systems because it is very fast and is performed at relatively low frequencies (100 kHz–5 MHz) [15]. Therefore, a large amount of data directly connected with the design or the process of the OTFTs can be easily obtained. Note that the S-parameter measurements also include the intrinsic as well as the extrinsic components of the OTFTs. As a result, appropriate analysis of the parasitic effects can be determined. This allows to build an OTFT

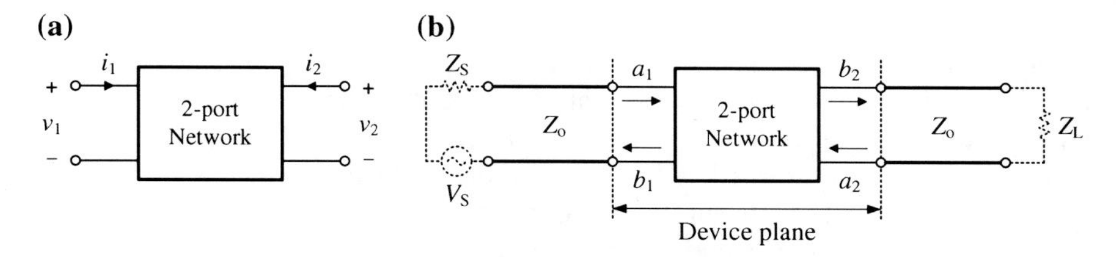

Fig. 7.1 Block diagram of two-port network/device with defined **a** voltages and currents, and **b** traveling waves. In (**b**), the network/device is connected to a VNA using two line segments (transmission lines) with a characteristic impedance Z_o. Only the forward direction is illustrated when the source and load are connected to port 1 and port 2, respectively. The source (Z_S) and load (Z_L) impedances are configured by the VNA. When perfect terminations ($Z_S = Z_L = Z_o$) are applied in this configuration, the input reflection (S_{11}) and forward transmission (S_{21}) coefficients can be measured. In addition, the connections can be reversed by the VNA to measured the full two-port S-parameters [17]

small-signal equivalent circuit in a more direct and straightforward manner without possible uncertainties due to user-configured measurement setup.

Full two-port S-parameter measurements that relate the AC currents and voltages between the drain and the gate contacts of the OTFTs are carried out using an HP 3577A VNA at frequencies up to 5 MHz. A MATLAB code is written to handle and automate this process; the code is presented in Appendix D.3. All the measurements that are demonstrated herein are performed on-wafer, in ambient air and at room temperature. A steady-state bias is applied to the gate and drain electrodes superimposed by additional radio-frequency (RF) excitations, while the source electrode is connected to a common (virtual) ground.

In principle, two-port parameter sets for transistors are useful design aids that are provided by manufacturers [16]. As mentioned above, they can be used to extract the well-defined unity-gain bandwidth of the transistor. For a general two-port network/device (e.g. an OTFT) with voltages (v_1 and v_2) and currents (i_1 and i_2) applied at the input and output ports as indicated in Fig. 7.1a, it can be described using any of the following relations:

$$\begin{aligned} v_1 &= Z_{11} \cdot i_1 + Z_{12} \cdot i_2 \\ v_2 &= Z_{21} \cdot i_1 + Z_{22} \cdot i_2 \end{aligned} \qquad \begin{bmatrix} v_1 \\ v_2 \end{bmatrix} = \begin{bmatrix} Z_{11} & Z_{12} \\ Z_{21} & Z_{22} \end{bmatrix} \begin{bmatrix} i_1 \\ i_2 \end{bmatrix}, \tag{7.1}$$

$$\begin{aligned} i_1 &= Y_{11} \cdot v_1 + Y_{12} \cdot v_2 \\ i_2 &= Y_{21} \cdot v_1 + Y_{22} \cdot v_2 \end{aligned} \qquad \begin{bmatrix} i_1 \\ i_2 \end{bmatrix} = \begin{bmatrix} Y_{11} & Y_{12} \\ Y_{21} & Y_{22} \end{bmatrix} \begin{bmatrix} v_1 \\ v_2 \end{bmatrix}, \tag{7.2}$$

$$\begin{aligned} v_1 &= h_{11} \cdot i_1 + h_{12} \cdot v_2 \\ i_2 &= h_{21} \cdot i_1 + h_{22} \cdot v_2 \end{aligned} \qquad \begin{bmatrix} v_1 \\ i_2 \end{bmatrix} = \begin{bmatrix} h_{11} & h_{12} \\ h_{21} & h_{22} \end{bmatrix} \begin{bmatrix} i_1 \\ v_2 \end{bmatrix}, \tag{7.3}$$

$$\begin{aligned} v_1 &= A \cdot v_2 - B \cdot i_2 \\ i_1 &= C \cdot v_2 - D \cdot i_2 \end{aligned} \qquad \begin{bmatrix} v_1 \\ i_1 \end{bmatrix} = \begin{bmatrix} A & B \\ C & D \end{bmatrix} \begin{bmatrix} v_2 \\ -i_2 \end{bmatrix}, \tag{7.4}$$

where Z_{ij}, Y_{ij}, h_{ij} and ABCD are the impedance, admittance, hybrid and cascade (chain) parameters, respectively. Referring to Fig. 7.1a, it can be deduced from (7.1) that $Z_{11} = v_1/i_1$ when $i_2 = 0$, meaning that port 2 is open circuit. On the other hand, it can be deduced from (7.2) that $Y_{11} = i_1/v_1$ when $v_2 = 0$, meaning that port 2 is short circuit. Same applies to all the other parameters in the four two-port parameter sets. Moreover, the h-parameters are often used for the description of the active devices such as transistors [16]. In fact, the h_{21}-parameter is the short-circuit current gain ($h_{21} = i_2/i_1$ when $v_2 = 0$) of the device and its absolute value is used to characterize the cutoff frequency (f_T) when $|h_{21}| = 1$. More details about the current gain and the cutoff frequency of a transistor are presented in the following two sections. Finally, the ABCD-parameters are very useful for cascaded circuit topologies as they allow matrix multiplications of single circuit/network components, which are connected in series.

In order to measure the two-port parameter sets, namely Z-, Y-, h- and ABCD-parameters, open and short terminations are required. These terminations, however, create some problems when measured at high frequencies (especially at radio frequencies, which are relevant to this work) [16]. At high frequencies, these terminations are difficult to obtain due to stray inductances and capacitances. Furthermore, it is problematic to directly measure voltages and currents in a test setup because they depend on the length of the cables used to connect the DUT to the measurement equipment. Accordingly, the measured values depend on the position along the cables. Therefore, in order to avoid these drawbacks, S-parameter set can be used to characterize a two-port network, which is related to the scattering and reflection of the incident waves on the network rather than the total voltages and currents [18]. Instead of the short and open terminations, the ports are terminated in this case by a cable of a certain characteristic impedance[1] (Z_o), usually 50 Ω, and a matched load with an impedance of $Z_L = Z_o$ to compute the individual S-parameters. Moreover, assuming a lossless transmission line is used, the traveling waves do not vary in magnitude along the line.

Referring to Fig. 7.1, the traveling waves a_1 and b_1 at port 1, and a_2 and b_2 at port 2 are defined in terms of v_1, v_2, i_1, i_2 and the real-valued positive characteristic impedance Z_o as follows [17]:

$$a_1 = \frac{v_1 + Z_o i_1}{2\sqrt{Z_o}}, \qquad a_2 = \frac{v_2 + Z_o i_2}{2\sqrt{Z_o}},$$
$$b_1 = \frac{v_1 - Z_o i_1}{2\sqrt{Z_o}}, \qquad b_2 = \frac{v_2 - Z_o i_2}{2\sqrt{Z_o}}.$$

The square of the magnitude of these variables (a_1, a_2, b_1 and b_2) has the unit watt. Hence, these incoming and outgoing waves can be considered to be in the form of power waves travelling in both directions along the lines. Assuming that the used transmission lines (cables) are uniform in cross section; therefore, the transmission line (Fig. 7.1b) can be modeled by a general equivalent circuit with a series impedance

[1] A characteristic impedance (Z_o) of a cable, i.e., a uniform transmission line, is the ratio of the amplitudes of the voltage and the current of a single wave propagating forward along the line. It is determined by the geometry and the materials of the cable, but it is independent of the cable length. Typical values are $Z_o = 50, 75, 90$ or 300 Ω.

and a shunt admittance per unit length. Accordingly, the characteristic impedance (Z_o) is expressed generally as [18]

$$Z_o = \sqrt{\frac{R + j\omega L}{G + j\omega C}}. \tag{7.5}$$

For a lossless line, the series resistance (R) and the shunt conductance (G) are both zero, and thus, the expression of the characteristic impedance given by (7.5) can be simplified to $Z_o = \sqrt{L/C}$.

The S-parameters completely define the network/device characteristics; in addition, they provide a clear physical interpretation of the transmission and reflection performance of the network/device [18]. For the two-port network shown in Fig. 7.1b, the S-parameters are defined using the reflected or emanating waves (b_1 and b_2) as the dependent variables, and the incident waves (a_1 and a_2) as the independent variables [18]. The general expressions for these waves as a function of the S-parameters are given by

$$\begin{aligned} b_1 &= S_{11} \cdot a_1 + S_{12} \cdot a_2 \\ b_2 &= S_{21} \cdot a_1 + S_{22} \cdot a_2 \end{aligned} \qquad \begin{bmatrix} b_1 \\ b_2 \end{bmatrix} = \begin{bmatrix} S_{11} & S_{12} \\ S_{21} & S_{22} \end{bmatrix} \begin{bmatrix} a_1 \\ a_2 \end{bmatrix}. \tag{7.6}$$

The parameters S_{11} and S_{22} are the complex-valued input and output reflection coefficients, whereas S_{21} and S_{12} are the complex-valued forward and reverse transmission coefficients, respectively. Using (7.6), the individual S-parameters can be determined by taking the ratio of the reflected or transmitted wave to the incident wave with a perfect termination placed at the corresponding port. For example, when a matched load ($Z_L = Z_o$) is applied to port 2 as depicted in Fig. 7.1b, the input reflection and forward transmission coefficients can be calculated using $S_{11} = b_1/a_1$ and $S_{21} = b_2/a_1$, respectively. The matched termination means that the load impedance (Z_L) is equal to the characteristic impedance (Z_o) of the line (not the network); consequently, any wave traveling through the line towards the load would be totally absorbed. This guarantees that $a_2 = 0$ since there is no reflection at port 2. The same applies for the output reflection and reverse transmission coefficients; $S_{22} = b_2/a_2$ and $S_{12} = b_1/a_2$ when a matched load is applied to port 1 ($a_1 = 0$).

Accurate characterization of the DUT, i.e., an OTFT on a glass wafer, requires that the parasitics associated with the measurement setup and on-wafer pads to be removed from the measured results. The employed VNA (HP 3577A) has 50 Ω RPC-3.5 interfaces, which are connected to high precision flexible cables using adapters. The cables are mounted directly to the probe tip holders that are connected to the OTFT. At this point, a distinction between the desired OTFT characteristics and the measured extrinsic parameters has to be made. There are typically differences between the measured S-parameters and the simulation results, which are mostly introduced by systematical errors due to the measurement setup. In principle, systematical errors are repeatable errors that result from non-idealities in the measurement system such as mismatches between the test system and the DUT, directivity

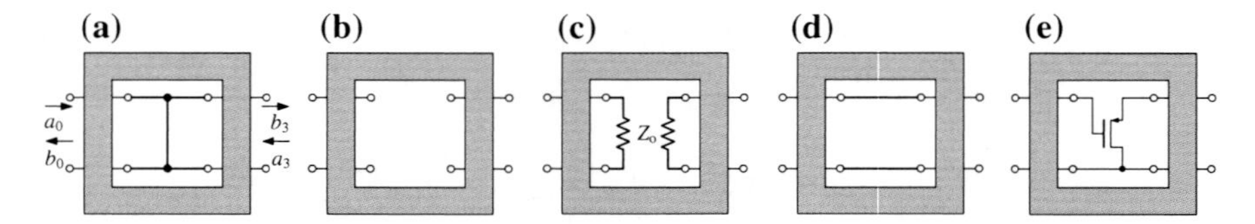

Fig. 7.2 Schematic diagram of several devices under test (DUT) with a fictitious two-port error adapter (*red*-colored blocks). The error adapter separates the DUT from an ideal measurement system, thus creating a distinction between the intrinsic (actual; related to the traveling waves a_1, a_2, b_1 and b_2 shown in Fig. 7.1b) and extrinsic (measured; related to the traveling waves a_0, a_3, b_0 and b_3 shown in this figure) S-parameters. In the used model, the error adapter contains twelve error terms as described in Appendix B. The devices/terminations depicted in the figure are the standard **a** short, **b** open, **c** through, and **d** matched (Z_o) loads used for the calibration/de-embedding procedure in addition to **e** an OTFT as an active load. The schematic is adopted from [24]

effects in the couplers, and cable losses [19]. Therefore, the test setup needs to be calibrated at the point where the DUT is connected, also referred to as the *device plane* as shown in Fig. 7.1b.

The process of calibrating the system and removing the undesired extrinsic effects due to the physical network placed between the VNA and the OTFT is called *de-embedding*. The de-embedding procedure uses a model of the test fixture and mathematically removes the undesired characteristics from the overall measurement [19]. For a typical two-port network, the test system can be modeled as having twelve errors that can be corrected; this 12-term error correction model is implemented in this work and is described in Appendix B [19–21]. Because of the variety of the test fixtures, there are also other error correction models, such as 8-term and 16-term models, which can be used [22, 23]. However, the 12-term error correction model produced promising results that are sufficient for our purpose as presented hereinafter. Accordingly, the standard short, open, through and matched loads are utilized to remove the parasitics and to obtain the desired S-parameters at the device terminals (see Fig. 7.2).

Furthermore, the ground-signal-ground (GSG) pads configuration is used by our test structures to conform with the microwave wafer probes employed during the measurements. The layout of an OTFT and the calibration structures with the GSG pads are presented in Appendix A. The open, short and through calibration structures are fabricated on the same glass substrate, but the matched load (with an impedance of 50 Ω) is not realizable with our OTFT technology. As a result, a standard calibration substrate with well-characterized terminations (short, open, through and matched loads that are necessary for the 12-term error correction model) at the desired frequency range (≤5 MHz) is employed. Note that the contact pads can also incorporate parasitic capacitance and inductance to the system, and in fact, the accuracy of the de-embedding procedure strongly depends on the quality as well as the layout of the terminations. Therefore, identical layout of the active device (including the contact pads) with that of the calibration structures is used to achieve high fidelity in the de-embedding process [11].

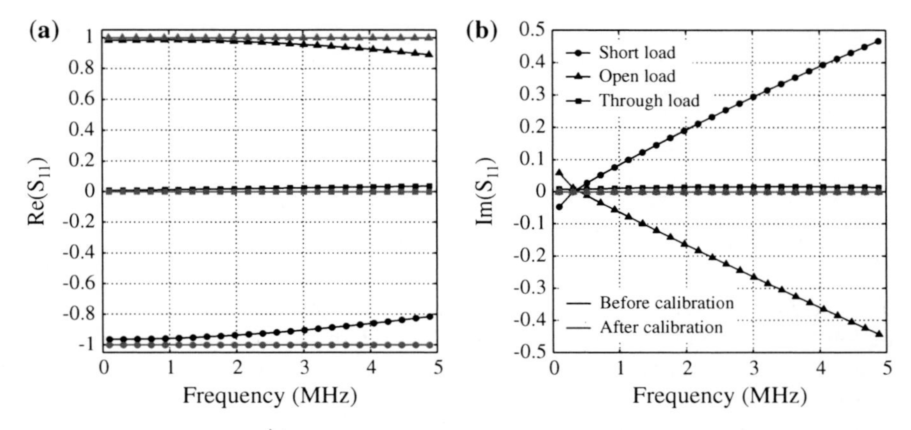

Fig. 7.3 Comparison between the measured input reflection coefficient (S_{11}) of the standard short, open and through loads before and after calibration. **a** The real part of S_{11} [Re(S_{11})]. **b** The imaginary part of S_{11} [Im(S_{11})]. The calibration procedure is implemented by de-embedding the parasitics associated with the measurement setup and on-wafer pads using the 12-term error correction model as explained in Appendix B. The same calibration accuracy is obtained for S_{12}, S_{21} as well as S_{22} (not shown here)

There are basically two main approaches to implement the de-embedding process [19]. The first is to measured the OTFT and the calibration structures beforehand and then mathematically remove the undesired parasitics from the measurements using any error correction model. The second uses the VNA itself to directly perform the de-embedding calculations, given that the error correction model is incorporated in the VNA. The latter approach allows the user to examine the de-embedded S-parameters in real-time. However, the used HP 3577A VNA offers only one-port calibration (3-term error correction model), which is not appropriate for our OTFT measurements. Therefore, the first approach using the 12-term error correction model, whose equations are described in Appendix B, is implemented on MATLAB to calculate the desired de-embedded S-parameters.

The calibration steps are as follows. First, the standard short, open, through and matched loads are measured, from which the twelve error terms are extracted using MATLAB. Second, the full two-port S-parameters of an OTFT are measured. Finally, the de-embedded S-parameters of the OTFT are calculated using MATLAB. As long as the measurement system and the temperature are stable over time, the same twelve error terms can be used for de-embedding all subsequent measurements [19].

Figure 7.3 shows the measured input reflection coefficient (S_{11}) of the short, open and through loads before and after calibration. The differences between the measured and expected values of the real and imaginary components of S_{11} result from the systematical errors, of which the impacts are successfully removed by the de-embedding process over the complete frequency range (0.1–5 MHz) as depicted in the figure. For example, the measured S_{11} of a short load before calibration are between $-0.96 - 0.05j$ and $-0.82 + 0.47j$ at $f = 0.1$ and 5 MHz, respectively; however, the expected value of $S_{11} = -1$ is achieved after calibration for the entire frequency range. Same applies for the other S-parameters, namely S_{12}, S_{21} and S_{22}, and all load structures. This verifies the AC measurement and the calibration procedure used here for studying the frequency response of OTFTs.

7.2 Small-Signal Model

Over the past few years, tremendous research efforts have been devoted to increase the field-effect mobility in organic materials by choosing the proper composite or improving the deposition conditions. As a result of these efforts, many OTFTs with high mobilities exceeding $1\,\mathrm{cm^2/Vs}$ have been reported [2, 8]. In particular, high mobilities of more than $5\,\mathrm{cm^2/Vs}$ in OTFTs based on pentacene and fullerene C_{60} organic semiconductors have been demonstrated [8]; this surpasses the mobilities of the conventional a-Si:H TFTs by about ten times. Besides high field-effect mobility, other ways of improving the dynamic performance of OTFTs, such as reducing the lateral dimensions of the device, has also attracted considerable attention [2, 8]. Downscaling the channel length results in an increase of the transconductance; furthermore, reducing the gate-overlaps results in a decrease of the parasitic capacitances [25]. In principle, high frequency operation is expected in the planar OTFTs with high mobility as well as small lateral dimensions. However, this can be difficult because in many cases the mobilities of OTFTs decreases with the channel length as a result of the parasitic contact resistances [8]. This problem can be resolved by choosing an appropriate device configuration (e.g. inverted-staggered topology) or by suppressing the parasitic resistance (e.g. by contact doping or choosing a proper material for the contact metal with a work function that matches the LUMO/HOMO energy level of the organic semiconductor) [26].

As described in Chap. 3, OTFTs are commonly fabricated in the inverted configuration, i.e., inverted-coplanar or inverted-staggered topologies (Fig. 3.1). It depends on weather the source/drain contacts are deposited on top of the organic semiconductor film or on the gate dielectric prior to the deposition of the organic semiconductor film, they are referred to as top-contact or bottom-contact OTFTs, respectively [2]. The bottom-contact OTFTs has the advantage that the standard photolithography and spin coating techniques, which are used in the established IC and display industries, can be used to obtain short channel devices. In this case, the organic semiconductor is deposited merely at the end of the process to avoid degradation of the conjugated organic compound due to solvents or elevated temperatures. Nevertheless, the bottom-contact OTFTs suffer from higher contact resistance compared to that of the top-contact OTFTs [27, 28]; the adhesion layers for the bottom contacts and the carrier injection barrier into the organic material typically cause the large contact resistance [8]. This results in a significant reduction of the field-effect mobility, which contradicts with the requirement of high mobility for the high-frequency operation. On the other hand, the top-contact OTFTs have been commonly fabricated using plastic shadow masks, which result in a relatively large minimum channel lengths of about 10–20 μm [25]. However, we are able to fabricate top-contact OTFTs with submicrometer channel lengths (down to 0.6 μm as demonstrated further below) and very small contact resistance without exposing the semiconductor layer to potentially harmful solvents during device processing using high-resolution silicon stencil masks [1].

Furthermore, it is demonstrated in [2] that top-contact OTFTs have smaller gate capacitance compared to that of the bottom-contact OTFTs, resulting in improved frequency performance. If one ignores the artifacts introduced by the contact resistance, the gate-source capacitance is found to be similar for both configurations. However, the gate-drain capacitance is found to be lower in top-contact OTFTs, particularly in the saturation operation regime, because the absence of the accumulation layer beneath the contact area makes the organic semiconductor film to behave as a dielectric layer. As a result, a smaller gate-drain overlap capacitance and a larger bandwidth are expected for top-contact OTFTs.

In view of the promising advantages of the top-contact configuration and the advances offered by the new OTFT technology process developed at IMS CHIPS and MPI-SSR, the top-contact OTFTs fabricated by high-resolution silicon stencil masks becomes the most attractive choice for this study. Besides the differences between the bottom- and top-contact OTFT configurations mentioned above, there are also other important aspects that relate to the organic semiconductor film thickness, of which implication with respect to the device performance is not discussed so far. According to the device simulations presented in [2], significantly higher device speed can be obtained just by choosing an optimal thickness of the organic semiconductor film. In principle, there is a tradeoff when choosing the organic semiconductor film thickness, i.e., a larger thickness would result in an increase of the contact resistance but a decrease of the overlap capacitance. A thickness of 20 nm for the organic semiconductor film is chosen for our OTFTs.

All the OTFTs fabricated for this study are prepared on the same substrate in order to minimize the device-to-device variations. Moreover, the gate electrodes are patterned to minimize the parasitic capacitances associated with overlap between the gate and the source/drain contacts. The channel width (W) of all the OTFTs is 100 μm, the channel length (L) is 10, 2, 1 or 0.6 μm, and the gate-overlap length (L_{ov}) is 20 or 5 μm. To ensure proper fabrication yield, each transistor is dissected to four parallel OTFTs with $W = 25$ μm sharing the same semiconductor layer; this is to comply with the layout design rules constrained by the required mechanical stability of the silicon stencil masks (Table 4.7). The layout of the OTFTs with the dissected top contacts and connected to the aforementioned GSG pads is shown in Appendix A. Despite the fact that there are OTFTs, such as those based on pentacene and fullerene C_{60} organic semiconductors, offering high mobilities of more than 5 cm^2/Vs, we prefer to use the DNTT owing to its better air stability. Furthermore, the DNTT-based OTFTs feature promising shelf-life, bias-stress stability, matching and hysteresis behavior as demonstrated thoroughly in [29] and [25]. An intrinsic mobility (μ_o) of $\sim$2.5 cm^2/Vs and a contact resistance ($2R_C W$) of 0.1 kΩ cm are extracted for the fabricated DNTT-based OTFTs using the TLM. For comparison, this measured value of the contact resistance is substantially smaller than that of

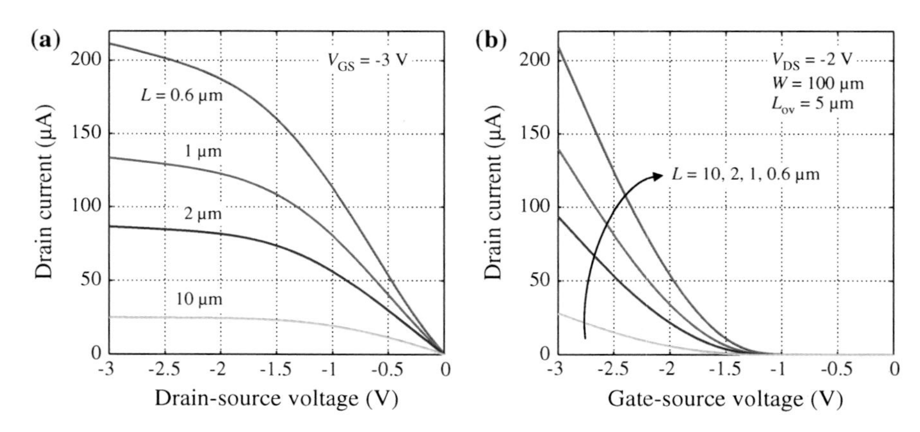

Fig. 7.4 Measured static I–V characteristics of OTFTs with $W = 100$ µm, $L = 10$, 2, 1 and 0.6 µm, and $L_{ov} = 5$ µm. **a** Output characteristics. **b** Saturation transfer characteristics. The figure depicts that by reducing the channel length of the TFTs, the drain current is increased. From the transfer characteristics, effective saturation mobilities (μ) of 1.9, 1.2, 0.87 and 0.75 cm^2/Vs are extracted for channel lengths of 10, 2, 1 and 0.6 µm, respectively. In addition, a threshold voltage (V_{TH}) of -1.1 ± 0.1 V is extracted for all the TFTs [1]

the values reported for the OTFTs based on pentacene ($2R_C W = 0.94$ kΩ cm) and fullerene C_{60} ($2R_C W = 3$ kΩ cm) organic semiconductors [8].

Measured static drain currents of the OTFTs with the various channel lengths and the gate overlap of 5 µm are shown in Fig. 7.4. The on/off current ratios of the OTFTs are found to be 10^7–10^5 for channel lengths of 10–0.6 µm, respectively. The measurements reveal that the extracted effective saturation mobility (μ) is decreased as the channel length is reduced. As discussed above, this is because of the increase in the relative contribution of the contact resistances to the total device resistance [3]. The relationship between these parameters is explained further below. Emphasis is also made on the OTFT dynamic performance given the advanced material (air-stable DNTT-based OTFTs with small contact resistances) and the aggressively scaled lateral dimensions ($L = 0.6$ µm and $L_{ov} = 5$ µm, which are enabled by the OTFT fabrication process based on high-resolution silicon stencil masks).

In literature, large-signal OTFT models combining the I–V characteristics and device parasitics have been thoroughly studied [30, 31]. This is necessary when analyzing for example digital circuits in which the bias points of transistors are significantly disturbed by the input signals [32]. Conversely, in most analog circuits the perturbation in bias conditions is small. Accordingly, a small-signal model is used to simplify the design and analysis. Furthermore, it is essential for estimating all the circuit elements and extracting various figures of merit of the device. In principle, the small-signal model is just an approximation of the large-signal model around an operating point [33]. This has been demonstrated for OLEDs and OPVCs to give deeper insight into the transport properties and operational mechanism in these devices [7, 34, 35]. Therefore, the knowledge of the small-signal model of an

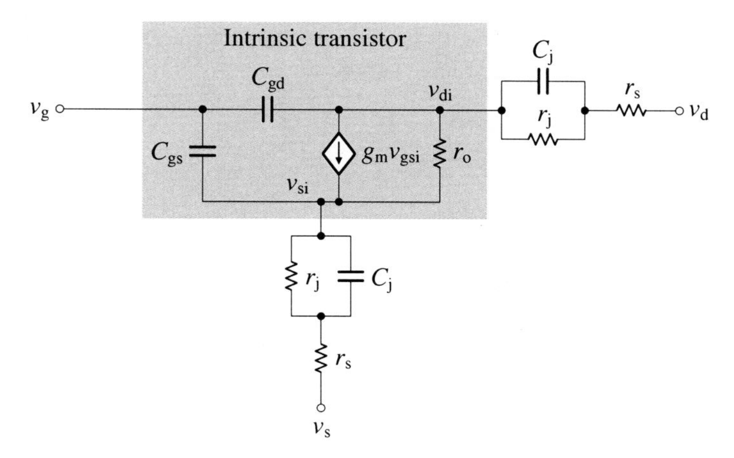

Fig. 7.5 Schematic diagram of the OTFT small-signal model including both intrinsic (highlighted) and extrinsic (not highlighted) components. The subscript i designates the internal potential of the intrinsic OTFT, which corresponds to the portion of the OTFT above the gate and between the source/drain contacts. Though the capacitances C_{gs} and C_{gd} are highlighted, each includes a constant overlap capacitance (C_{ov}), which is a contribution of the extrinsic part of the OTFT [14]

OTFT is very beneficial in many aspects: (i) simplification of the analyses and the design of analog/hybrid circuits, (ii) determination of the device performance, (iii) verification of the large signal model, (iv) characterization of the technology process, and (v) extraction of the material properties. In the following, a physics-based small-signal OTFT model and a verification of the model to S-parameter measurements are presented. The model is simple to allow convenient upgradability of further effects and it includes no empirical fitting parameters.

Since most transistors in analog circuitry are biased in the saturation operation regime; correspondingly, Fig. 7.5 shows a proposed OTFT small-signal equivalent circuit including both intrinsic and extrinsic components. Nevertheless, in the linear regime the transistors operate as a switch, which can be roughly modeled by a linear resistor with an equivalent voltage-dependent resistance of $L/(\mu C_I W(V_{GS} - V_{TH}))$ together with the device extrinsic components. The values of the small-signal parameters are constant with the change in the frequency but they depend on the large-signal parameters and the DC quiescent point as described hereinafter.

As depicted in Fig. 7.5, the OTFT small-signal model contains circuit elements similar to the classical small-signal models of field-effect transistors (FETs). The intrinsic part of the model, which corresponds to the portion of the OTFT above the gate and between the source/drain contacts, is composed of a voltage dependent current source that is equal to $g_m v_{gsi}$ (where v_{gsi} is for the internal gate-source potential; the subscript i denotes the internal nodes of the intrinsic OTFT), output resistance ($r_o = 1/g_o$), and gate-source (C_{gs}) and gate-drain (C_{gs}) capacitances. The transconductance (g_m) is the change in the drain current divided by the change in the gate-source voltage as given by (3.7). Owing to the channel length modulation effect,

r_o is placed in the model and is given by the change in the drain current divided by the change in the drain-source voltage as follows:

$$r_o = \left.\frac{\partial I_D}{\partial V_{DS}}\right|_{V_{GS}=\text{const.}} = \frac{1}{\lambda I_D}, \quad (7.7)$$

where λ represents the relative variation in the channel length for a given change in the drain-source voltage. Although the use of fully patterned OTFTs makes the overlap capacitances minimized, the usual lack of a self-alignment OTFT process makes the overlap capacitances inevitable. According to the experimental AC analysis presented in the previous chapter using admittance measurements, the capacitances C_{gs} and C_{gs} are found at the bias of interest (in the saturation regime) to be

$$C_{gs} = C_{ov} + (2/3)C_{ch} = C_I W[L_{ov} + (2/3)L], \quad (7.8)$$
$$C_{gd} = C_{ov} = C_I W L_{ov}. \quad (7.9)$$

Note that C_{ov} is a contribution of the extrinsic part, i.e., parasitic overlap capacitance, of the device.

Since the intrinsic OTFT exhibits a hybrid-π topology (Fig. 7.5), it is convenient to use the admittance (Y) parameters to characterize the electrical properties [15]. Accordingly, the Y-parameters used to describe the small-signal model of the intrinsic OTFT can be derived as follows:

$$Y_{11i} = j\omega(C_{gs} + C_{gd}), \quad (7.10)$$
$$Y_{12i} = -j\omega C_{gd}, \quad (7.11)$$
$$Y_{21i} = g_m - j\omega C_{gd}, \quad (7.12)$$
$$Y_{22i} = g_o + j\omega C_{gd}. \quad (7.13)$$

From which the transfer function of the current gain, defined as the ratio of the small-signal output current to the input current of the transistor with short-circuited output ($h_{21i} = i_{di}/i_{gi} = Y_{21i}/Y_{11i}$), is found to be

$$h_{21i} = \frac{g_m - j\omega C_{gd}}{j\omega(C_{gs} + C_{gd})}. \quad (7.14)$$

Note that the current gain (h_{21}) of a transistor is often denoted in literature by the symbol β [1, 14]; however, this distinction is not made in this work for the sake of clarification.

It is shown so far that the small-signal model of the OTFTs is similar in composition to those of conventional FETs, i.e., consisting of g_m, r_o, C_{gs} and C_{gd}. Furthermore, the dynamic performance of the OTFTs is strongly limited by the contact interface between the metallic source/drain electrodes and the semiconductor layer [36]. Detailed analyses of the capacitance-voltage (C–V) characteristics and frequency responses of metal-insulator-semiconductor (MIS) structures have been presented in literature to describe the current injection mechanism through the contact interfaces of OTFTs [36–38]. As shown in the schematic diagram of the proposed small-signal model, the extrinsic part that is used to describe the non-ohmic source/drain contacts consists of a junction capacitance (C_j), a junction resistance (r_j) and a series resistance (r_s). The origin of the junction capacitance in the contact model is most likely due to the formation of a small depletion region near the metal-semiconductor interface, while the junction resistance governs the injection current to the channel [38]. Both are highly dependent on the processing conditions; for example, they can be considerably reduced when performing local doping into the contact interface [36]. It is demonstrated in the following that the contribution of the junction capacitance in our OTFTs can be neglected. Unlike the MIS structures, the resistances r_j and r_s has to account not only for the vertical but also for the much larger horizontal component of the total OTFT contact resistance (Appendix C.2) [3].

Each terminal (gate, source and drain) of the OTFTs is in fact exhibiting a very small ohmic resistance resulting from the resistivity of the aluminium and gold conductive materials; nevertheless, they are neglected in the above model for simplicity. Moreover, the gate leakage current is also not considered, but it can be easily incorporated into the model by distinctive resistors tied between the gate and source/drain contacts. The proposed model is adequate for most frequency analyses and can be even further simplified by utilizing only the intrinsic part of the model while considering the contact resistances in the calculation of the effective charge carrier mobility using (3.11); the validity of this simplification is demonstrated hereinafter.

Using the test setup described in the previous section, full two-port S-parameter measurements that constitute a complete set of coefficients to describe the input/output behavior of the OTFTs are performed. Figure 7.6a depicts measured and simulated S-parameters of the OTFT with $W = 100$ μm, $L = 1$ μm and $L_{ov} = 20$ μm [14]. The measurements are carried out under various bias conditions, but only the saturation regime (at $V_{GS} = -3$ V and $V_{DS} = -2$ V) is illustrated here, as it is the bias of interest for most analog circuits. A relatively large gate-overlap length of 20 μm is utilized in this experiment in order to minimize the impact of unintentional misalignment on the measurement results, noting that this effect is carefully studied and the results are presented in Sect. 7.4. The corresponding model parameter set used for the small-signal simulation is summarized in Table 7.1. The intrinsic mobility and the contact resistance are initially extracted using the TLM ($\mu_o = 2.5$ cm^2/Vs and $2R_C W = 0.1$ kΩ cm); subsequently the rest of the parameters are calculated from the analytical formulas and/or optimized to fit the simulations to the measurements. Owing to the parts of the organic semiconductor that extend beyond the periphery of the intrinsic transistor and cause additional fringe currents, the model has to account for an additional induced parsitic capacitance by adding $C_I A_{fringe}$ to (7.8), where $A_{fringe} \simeq 480$ μm^2 (about 12% of the total overlap area). As depicted in the figure, the simulation results agree well with the measured S-parameters over the entire frequency range from 100 kHz to 1 MHz. The relative mean square errors (rMSEs) are found to be <0.03 %, 10.4 %, 6.9 % and <0.03 % for S_{11}, S_{12}, S_{21} and S_{22}, respectively. The equation describing the calculation of the errors is given by [39]

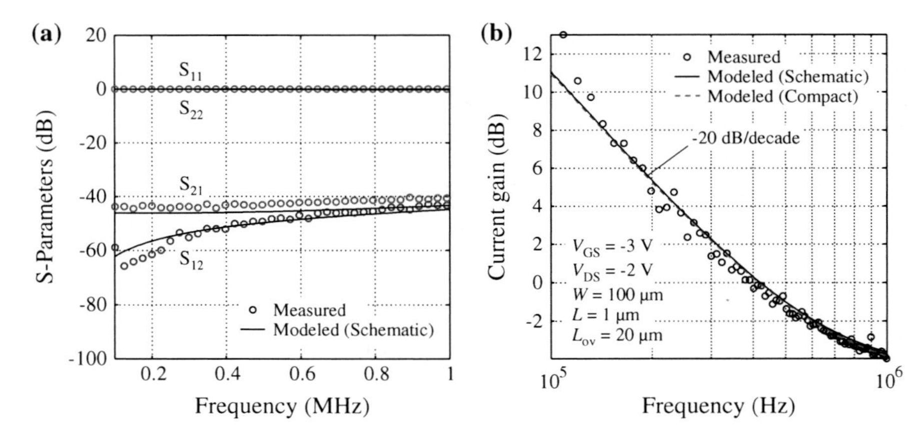

Fig. 7.6 **a** Full two-port measured and simulated S-parameters of the OTFT with $W = 100$ µm, $L = 1$ µm and $L_{ov} = 20$ µm operated in the saturation regime. **b** Extracted short-circuit current gain of the OTFT. The *solid(black) lines* represent the schematic simulations, while the *dashed(red)line*represents the simplified compact model when ohmic contact resistances are considered. The results demonstrate the reliability of the proposed OTFT small-signal model [1, 14]

Table 7.1 Summary of the small-signal model parameters used in the simulation of the p-channel OTFT with $W = 100$ µm, $L = 1$ µm and $L_{ov} = 20$ µm [14]

Parameter	Notation	Value	Unit
Intrinsic mobility	μ_o	~2.5	cm²/Vs
Threshold voltage	V_{TH}	−1.2	V
Insulator capacitance per unit area[a]	C_I	900	nF/cm²
Junction capacitance per unit area	C_j	2000	nF/cm²
Normalized junction resistance	r_j	60	Ω cm
Normalized series resistance	r_s	9	Ω cm
Normalized output resistance	r_o	10	kΩ cm

[a]The slight difference of C_I from the previously reported values is most probably due to wafer-to-wafer variations

$$E_{ij} = \frac{1}{n}\sum_{k=1}^{n}\left(\frac{|S_{ij,sim}| - |S_{ij,meas}|}{|S_{ij,meas}|}\right)^2, \tag{7.15}$$

where n is the number of the data points and the subscripts i and j are equal to 1 and/or 2.

Figure 7.6b shows the extracted and simulated current gains (h_{21}) [14]. The expression used to calculate h_{21} from the measured full two-port de-embedded S-parameters is given by [40]

$$h_{21} = \frac{-2S_{21}}{(1 - S_{11})(1 + S_{22}) + S_{12}S_{21}}. \tag{7.16}$$

The current gain decays with increasing frequency as expected for a FET-like device, following the $1/f$ slope (−20 dB/decade) resulting from the total gate capacitance ($C_G = C_{gs}+C_{gd}$) given in the denominator of (7.14). This validates the measurement and the de-embedding procedures employed here to study the frequency response of the OTFTs. The current-gain cutoff frequency (f_T) of this OTFT, defined as the frequency at which the current gain is unity ($|h_{21}| = 1$), is found to be about 0.4 MHz. The characterization of the current-gain cutoff frequency is discussed in detail in the following section. Using the same parameter set given in Table 7.1, the solid line in the Fig. 7.6b represents the simulated current gain of the small-signal equivalent circuit model. The good correspondence between the simulations and the measurements verifies that the proposed model is accurate and reliable. The small difference between the measured and simulated data near 100 kHz, which is the lowest frequency limit of the employed VNA, is owed to the accuracy of both the measurement and the calibration procedure.

Given the relatively large value of the junction capacitance ($C_j = 2000$ nF/cm^2), the model can be further simplified as follows. Assuming an ohmic contact resistance (i.e., no junction capacitance) with $2R_C W = 0.1$ kΩ cm (extracted using the TLM), the corresponding effective g_m and μ represented in the saturation regime for symmetrical OTFTs by (3.7) and (3.11), respectively, can be calculated. Accordingly, the short-circuit current gain (h_{21}) is computed using the analytical expression given by (7.14) for the classical hybrid-π small-signal model, while taking into consideration the impact of the contact resistance on the mobility. An excellent agreement is achieved as demonstrated by the dashed line in Fig. 7.6b. However, it is important to note that these two models do not represent any different physical operation or performance of the OTFT; they are just different representations of the same device.

7.3 Current-Gain Cutoff Frequency

In principle, the dynamic performance of the OTFTs can be improved by reducing the lateral dimensions. As a measure of the device speed, the parameter current-gain cutoff frequency is selected for this study. To investigate the dependence of the lateral dimensions on the frequency response, the OTFTs with various channel and gate-overlap lengths down to 0.6 and 5 µm, respectively, are characterized herein. Correspondingly, Fig. 7.7 shows the short-circuit small-signal current gain (h_{21}) derived from the measured S-parameters of the OTFTs with $W = 100$ µm, $L = 10$, 2, 1 and 0.6 µm, and $L_{ov} = 20$ and 5 µm at a gate-source voltage (V_{GS}) of −3 V and a drain-source voltage (V_{DS}) of −2 V [1]. For all the transistors, the de-embedded current gain is inversely proportional to the frequency (f), following the $1/f$ slope expected for a conventional FET [11]. Referring to (7.14), this $1/f$ frequency dependence of h_{21}, which is equivalent to a decay slope of −20 dB/decade, results from the total gate capacitance ($C_G = C_{gs} + C_{gd}$). This observation is significant because it not only validates the used characterization approach but also verifies the regular FET-like behavior of the OTFTs. At high frequencies, however,

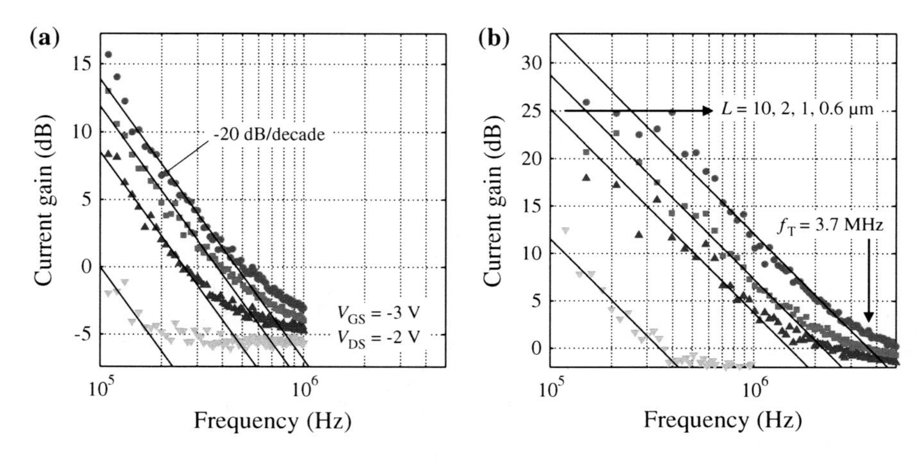

Fig. 7.7 Short-circuit current gain (h_{21}) derived from the de-embedded full two-port S-parameter measurements of OTFTs with $W = 100$ μm, $L = 10$, 2, 1 and 0.6 μm, **a** $L_{ov} = 20$ μm, and **b** $L_{ov} = 5$ μm. The current gain decreases with increasing frequency, following the $1/f$ slope as expected for a FET-like device. This decay slope is represented by the -20 dB/decade solid lines. At high frequencies, the decay levels off due to the zero in the transfer function h_{21}. The cutoff frequency (f_T) of each OTFT is extracted from the point at which the extrapolated roll-off slope (-20 dB/decade) crosses the x-axis (when $|h_{21}| = 1$) [1]

the decay levels off as depicted in Fig. 7.7; this is due to the zero in the transfer function h_{21}, which denotes a current flow through the relatively large gate-drain capacitance (C_{gd}).

In the following, the capacitance C_{gd} is neglected during the derivation of the current-gain cutoff frequency, but its impact is discussed in detail in the next section. For a typical FET, the input current (gate current) induced by an applied AC gate voltage increases linearly with the frequency due to the finite gate capacitance ($i_G = j2\pi f C_G v_{GS}$), assuming that the independent DC contribution of the gate leakage current is insignificant. Furthermore, the output current (drain current) is mainly given by the transistor's DC characteristics and a small superimposed AC component ($i_D = g_m v_{GS}$). For the unity-gain bandwidth, the upper frequency limit of the device is reached if the input and output AC levels exhibit the same amplitude. Beyond the said frequency, also referred to as the cutoff or transition frequency (f_T), the transistor is not able to drive another load transistor of the same kind. Hence, no circuitry that is composed of such transistors can operate at a frequency beyond f_T [6]. Accordingly, the cutoff frequency (f_T) is considered as one of the most important figures of merit for characterizing the dynamic performance of the transistors. In this connection, f_T given by (3.12) can be rewritten in the saturation regime as

$$f_T = \frac{g_m}{2\pi(C_{gs} + C_{gd})} = \frac{\mu(V_{GS} - V_{TH})}{2\pi L[(2/3)L + 2L_{ov}]}. \tag{7.17}$$

Therefore, the current-gain cutoff frequency results from an interplay between the transconductance and the device capacitances [2]. To attain high cutoff frequency, a small variation in gate voltage should create a huge drain current modulation, i.e., a large transconductance (g_m); furthermore, the total gate capacitance (C_G) should be minimized to reduce the gate current [6]. In other words, a high bandwidth is achieved by a transistor with high mobility, short channel length, low threshold voltage and low parasitic capacitances [8]. It is important to note that for transistors with short channels and large gate overlaps, the total gate capacitance in (7.17) has to account for the additional fringe region that extend beyond the periphery of the intrinsic transistor.

In practice, f_T is extracted from the measured de-embedded current-gain (h_{21}) at the point at which the extrapolated roll-off slope (−20 dB/decade) due to the pole in the transfer function h_{21} crosses the x-axis (when $|h_{21}| = 1$, which is equivalent to zero decibel) as illustrated by the solid lines in Fig. 7.7. Correspondingly, the extracted f_T of all the tested OTFTs are plotted in Fig. 7.8a to investigate the impact of scaling down the channel and gate-overlap lengths on the dynamic performance. In the figure, there is an additional point for an OTFT with $W = 100\ \mu m$, $L = 0.8\ \mu m$ and $L_{ov} = 5\ \mu m$, of which the measured current gain is not shown in Fig. 7.7. As expected, the cutoff frequency is found to increase consistently with reduced channel and/or gate-overlap lengths. A maximum $f_T = 3.7$ MHz is obtained for the OTFT with $W = 100\ \mu m$, $L = 0.6\ \mu m$ and $L_{ov} = 5\ \mu m$. To the best of our knowledge, this is the highest value measured for low-voltage, air-stable OTFTs to date.

In order to check the validity of the method, a comparison between the extracted and simulated cutoff frequencies is carried out. The solid and dashed lines in Fig. 7.8a represent f_T calculated using (7.17) and two different representations for the effective saturation mobility (μ). It is shown that for a constant μ (dashed lines) of 2.5 cm^2/Vs, which is equal to the intrinsic carrier mobility (μ_o), the results are overestimated for shorter channels. This is attributed to the contact resistance (R_C), the influence of which becomes more dominant for smaller channel lengths. Therefore, an ohmic contact resistance had to be considered in the calculation of the effective saturation mobility as given by (3.11). A reliable and good matching of the model to the experimental data of the OTFTs with different device geometries is obtained with $V_{TH} \cong -1.2$ V and $2R_C W \cong 0.1\ k\Omega\,cm$. This result suggests that a higher transistor bandwidth can be also achieved by reducing the contact resistance at the metal/semiconductor interface.

The correspondence between the DC and AC characteristics is examined. Given that $\mu = 0.8\ cm^2/Vs$ and $V_{TH} = -1.1$ V from the static transfer characteristics of the OTFT with $W = 100\ \mu m$, $L = 1\ \mu m$ and $L_{ov} = 5\ \mu m$, $2R_C W$ can be similarly reproduced from (7.17) to be $\sim 0.1\ k\Omega\,cm$. The extracted value is identical to the aforementioned result determined from the S-parameter measurements.

Furthermore, the DC transconductance (g_m) extracted from the static transfer characteristics (Fig. 7.4) along with the cutoff frequency (f_T) extracted from the S-parameter measurements (multiplied by a constant $2\pi C_G$ where $C_G = 8.5$ pF) at various gate-source voltages of the same OTFT is plotted in Fig. 7.8b. This is to verify the proportionality between g_m and f_T as given by (7.17). The figure illustrates that

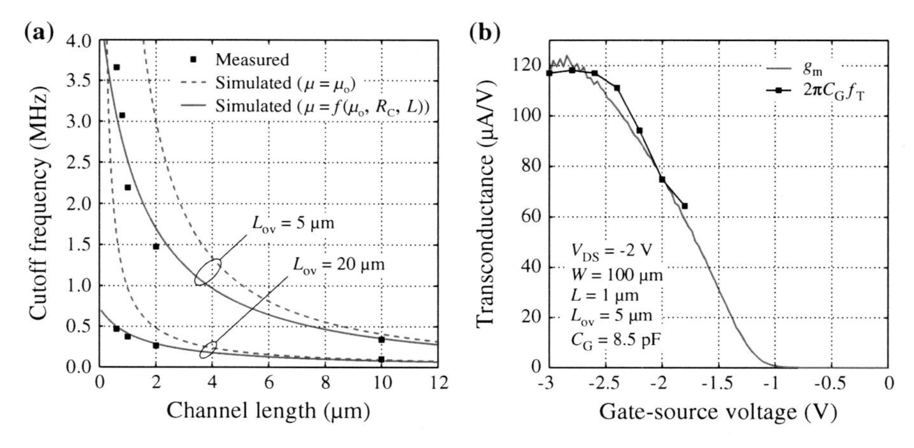

Fig. 7.8 **a** Current-gain cutoff frequency (f_T) extracted for the OTFTs with the different L and L_{ov}. Note that there is an additional point for an OTFT with $W = 100$ μm, $L = 0.8$ μm and $L_{ov} = 5$ μm, of which the short-circuit current gain (h_{21}) characteristics is not shown in Fig. 7.7. The dashed lines represent simulated f_T with a constant effective mobility (μ) that is equal to the intrinsic mobility (μ_o). The solid lines represent the simulated f_T with an effective mobility (μ) that is channel-length dependent as given by (3.11), providing excellent agreement with the measured data. **b** Extracted f_T and g_m as a function of V_{GS}. A total gate capacitance (C_G) of 8.5 pF is extracted from the linear relation between f_T and g_m as given by (7.17). The figure confirms the close correspondence between the DC and AC measurements [41]

the frequency response of the OTFT is highly dependent on the DC bias conditions. In principle, the performance of the transistor is mainly determined by g_m, which is defined as the ratio of the drain current to an applied gate-source voltage. The strong correlation between the two quantities as a function of the gate-source voltage is clearly observed with a maximum f_T of 2.2 MHz corresponding to a peak g_m of about 120 μA/V at $V_{GS} = -2.8$ V. In addition, C_G can be independently estimated from the device geometry. Given that $C_I = 900$ nF/cm^2, $W = 100$ μm, $L = 1$ μm and $L_{ov} = 5$ μm, a total gate capacitance (C_G) of 9.6 pF is calculated, which is in close agreement with the extracted value from f_T and g_m measurements. The small difference between the calculated and extracted values of C_G can be attributed to an accumulation of errors that are encountered in converting the original measured S-parameters to h_{21}, from which f_T is extracted.

The OTFTs in practical applications require simultaneously low-voltage operation, high bandwidth, small parasitics and good air stability. Recent results available in literature are cutoff frequencies of 27.7 MHz for fullerene C_{60} (n-channel; not air-stable) TFTs and 11.4 MHz for pentacene (p-channel) TFTs having channel lengths of 2 μm and operating, however, at a large supply voltage (V_{DD}) of 20 V [8]. Others have reported cutoff frequencies of 2 MHz for poly(3-hexylthiophene) (P3HT) (p-channel) TFTs with a channel length of 480 nm operating at 10 V and 1.6 MHz for poly(2,5-bis(3-alkylthiophen-2-yl)thieno[3,2-b]thiophene) (pBTTT) (p-channel) TFTs with a channel length of 200 nm operating at 8 V [6, 42]. As demonstrated

Table 7.2 Measured current-gain cutoff frequency (f_T) benchmarked against data reported in literature for individual OTFTs

Material	Type	L (nm)	V_{DD} (V)	f_T (MHz)	Reference
Fullerene C_{60}[a]	n-channel	2000	20	27.7	[8]
Pentacene	p-channel	2000	20	11.4	[8]
Pentacene[b]	p-channel	1000	50	0.9	[44]
P3HT	p-channel	480	10	2	[6]
pBTTT	p-channel	200	8	1.6	[42]
Sexithiophene	p-channel	30	2	0.02	[45]
DNTT	p-channel	600	3	3.7	This work

[a] Fullerene C_{60} is not an air-stable organic semiconductor
[b] This OTFT is implemented using the vertical-channel configuration

above, a peak cutoff frequency of 3.7 MHz for DNTT TFTs with a channel length of 0.6 μm, a gate overlap of 5 μm and operating at a supply voltage of 3 V is measured [1]. In fact, an important advantage of the organic semiconductor DNTT over pentacene and poly(3-hexylthiophene) is its better air stability [29, 43]. Table 7.2 summarizes the cutoff frequencies (measured on individual transistors) reported for the state-of-the-art OTFTs. In view of the low supply voltage and good air stability, the results presented in this work mark a record performance achievement in the OTFT technology.

7.4 Asymmetric Organic Transistors

Regardless of the device structure (staggered or coplanar) and the fabrication technique (printing or vacuum evaporation), the usual lack of a self-alignment process typically necessitates the design of non-negligible overlaps between the source/drain contacts and the gate electrode [46]. Since these overlaps have a significant impact on the static and/or dynamic TFT characteristics, it is necessary to include these contact effects in all the models presented in this work. Furthermore, the limited alignment capabilities of most OTFT technologies, especially printing techniques, make it very difficult to avoid a mismatch between the gate-source and gate-drain overlaps, which has a direct influence on the device performance and is particularly detrimental for OTFTs with small feature sizes [14]. Therefore, a detailed frequency analysis of aggressively scaled asymmetric OTFTs is presented herein.

To study the impact of asymmetry between the gate-source (L_{gs}) and gate-drain (L_{gd}) overlap lengths, where $L_{gs} + L_{gd} = 2L_{ov}$, the OTFT with $W = 100$ μm, $L = 1$ μm and $L_{ov} = 5$ μm is duplicated and intentionally misaligned on the mask level in order to realize several OTFTs with well-defined $L_{gs} = 1, 2, \ldots, 7$ μm, while keeping constant $L_{gs} + L_{gd} = 10$ μm. The seven OTFTs are placed in close proximity to minimize the device-to-device variations and ensure proper device uniformity.

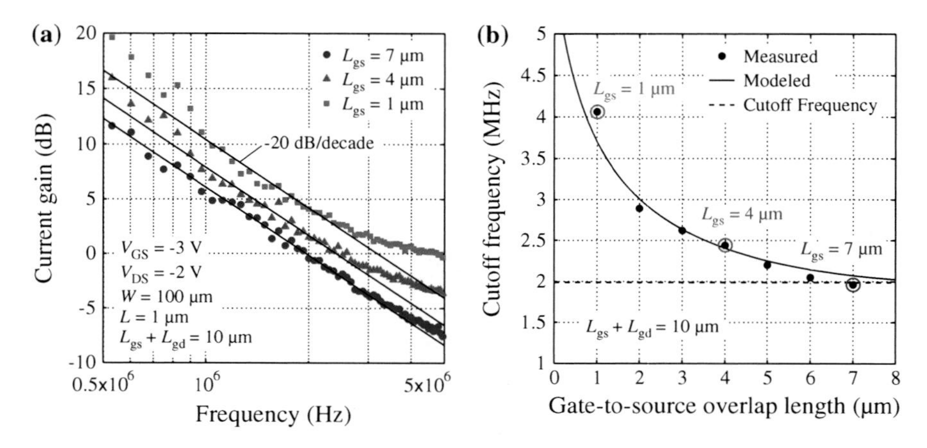

Fig. 7.9 Short-circuit current gain (h_{21}) derived from the de-embedded full two-port S-parameter measurements of asymmetric OTFTs with $W = 100$ μm, $L = 1$ μm and $L_{gs} + L_{gd} = 10$ μm. The -20 dB/decade slopes starts to level off at lower frequencies for OTFTs with smaller L_{gs}. (b) Extracted unity-gain cutoff frequency $f(|h_{21}| = 1)$ for all the OTFTs with $L_{gs} = 1, 2, \ldots, 7$ μm, while keeping constant $L_{gs} + L_{gd} = 10$ μm. As a reference, the dashed line represents the simulated prevalent FET-based cutoff frequency (f_T) calculated from (7.17). The solid line, however, represents the modeled $f(|h_{21}| = 1)$ given by (7.18), showing an accurate and reliable fit to the measured data [14]

The utilization of the OTFT process based on high-resolution silicon stencil masks enabled the fabrication of such small dimensions. Vernier structures, which are originally used to resolve alignment errors more accurately than the minimum feature size of a given technology, are employed to measure the particular degree of misalignment.

Figure 7.9a shows the current gain of the asymmetric OTFTs with $W = 100$ μm, $L = 1$ μm, $L_{gs} = 1$, 4 and 7 μm, and $L_{gs} + L_{gd} = 10$ μm [14]. It is found that the $1/f$ slopes (-20 dB/decade) level off at lower frequencies for smaller L_{gs}. This occurs due to a current flow through the relatively large gate-drain capacitance (C_{gd}), which can be understood mathematically from the zero in the current-gain transfer function (h_{21}) given by (7.14). In other words, as C_{gd} decreases for smaller L_{gd}, the decay in the measured current gain levels off at higher frequencies, which implies that the device behaves more like a conventional silicon MOSFET, where $C_{gd} \simeq 0$ and $C_{gd} \ll C_{gs}$ in the saturation regime.

As discussed in the previous section, the current-gain cutoff frequency (f_T) is generally defined as the frequency at which the magnitude of the current gain (h_{21}) of the device drops to unity when a short-circuited load is applied at the output. Therefore, the expression of f_T given by (7.17) is valid when $C_{gd} \ll C_{gs}$. On the other hand, $f(|h_{21}| = 1)$ can be derived from (7.14) and written as

$$f(|h_{21}| = 1) = \frac{g_m}{2\pi C_{gs}} \cdot \frac{1}{\sqrt{1 + 2C_{gd}/C_{gs}}}. \tag{7.18}$$

Fig. 7.9b depicts the extracted frequency at which the current gain is unity ($|h_{21}| = 1$) for each of the asymmetric OTFTs [14]. The solid line represent $f(|h_{21}| = 1)$ calculated using (7.18) with $g_m = 120$ μA/V. Even though the impact of misalignment is not considered in g_m, a reasonable agreement is obtained. This indicates that in this frequency range, the impact of misalignment on the overlap capacitances is more detrimental. Furthermore, the dashed line designates the cutoff frequency (f_T) that is commonly calculated using (7.17). As shown in the figure, the expression of f_T produces a reasonable approximation but is not well-suited for characterizing asymmetric OTFTs.

The result of this study is worthwhile for understanding the device physics and for process corner simulations of OTFT-based circuits. Moreover, it can be deduced from (7.14) that the current gain levels off at $|h_{21}| = L_{gd}/(L_{gs} + L_{gd})$. Given that the sum $L_{gs} + L_{gd} = 10$ μm, one can easily estimate the values of L_{gs} and L_{gd} separately from the measurements, i.e., the degree of misalignment can be determined from the measured S-parameters. For example, the upper (green) curve in Fig. 7.9a levels off at $|h_{21}| \simeq -1$ dB, accordingly, the gate-overlap lengths $L_{gs} = 1$ μm and $L_{gd} = 9$ μm are calculated. These values match properly with the designed dimensions.

7.5 Summary

S-parameter characterization of OTFTs is, for the first time, demonstrated in this work. A physics-based OTFT small-signal model is presented, which is well-suited for the design and analysis of analog/hybrid circuitry. The model includes the contact effects and reflects closely the characteristics of the OTFTs. The correspondence of the S-parameter characterization to prevalent methods, such as the admittance measurements presented in the previous chapter for relatively large OTFT devices, is illustrated. A reliable fit between the theoretical and experimental S-parameters is achieved. The short-circuit current gain, which is calculated from the S-parameter measurements, shows the ideal $1/f$ frequency dependence, confirming the reliability of the characterization technique and the FET-like behavior of the OTFTs. Furthermore, the extracted current-gain cutoff frequency is found to increase with decreasing channel and/or gate-overlap lengths. The deviation of the measured cutoff frequency for short channels owing to the device contacts is physically modeled and a good agreement to the measured data of OTFTs with various channel and gate-overlap lengths is accomplished. Moreover, by varying the gate-source voltage, the linear dependence between the measured cutoff frequency and the DC transconductance is verified. In view of the low 3-V supply voltage, a record cutoff frequency of 3.7 MHz is measured on an air-stable DNTT OTFT with a channel length of 0.6 μm and a gate-overlap length of 5 μm. To the best of our knowledge, this is the highest value reported for low-voltage, air-stable OTFTs so far.

In addition, the impact of misalignment in OTFTs, occurring between the source/drain contacts and the gate electrode, on the TFT dynamic performance is investigated. Due to the limited alignment capabilities of many OTFT technologies, this effect has to be considered during circuit design in process corner simulations. Although OTFTs are similar in many aspects to conventional FETs, the study of asymmetric transistors shows that incorporating assumptions like $C_{gd} \ll C_{gs}$ into the calculation of the cutoff frequency might produce underestimated results. The submicrometer patterning and the precise control of the misalignment are attributed to the OTFT process that is based on the high-resolution silicon stencil masks.

References

1. T. Zaki, R. Rödel, F. Letzkus, H. Richter, U. Zschieschang, H. Klauk, J.N. Burghartz, S-parameter characterization of submicrometer low-voltage organic thin-film transistors. IEEE Electron Device Lett. **34**(4), 520–522 (2013)
2. M.N. Islam, B. Mazhari, Comparative analysis of unity gain frequency of top and bottom-contact organic thin film transistors. Solid-State Electron. **53**(10), 1067–1075 (2009)
3. F. Ante, D. Kälblein, T. Zaki, U. Zschieschang, K. Takimiya, M. Ikeda, T. Sekitani, T. Someya, J.N. Burghartz, K. Kern, H. Klauk, Contact resistance and megahertz operation of aggressively scaled organic transistors. Small **8**(1), 73–79 (2012)
4. R. Rödel, F. Letzkus, T. Zaki, J.N. Burghartz, U. Kraft, U. Zschieschang, K. Kern, H. Klauk, Contact properties of high-mobility, air-stable, low-voltage organic n-channel thin-film transistors based on a naphthalene tetracarboxylic diimide. Appl. Phys. Lett. **102**(233303), 233303-1–233303-5 (2013)
5. K. Fukuda, T. Sekitani, T. Yokota, K. Kuribara, T.-C. Huang, T. Sakurai, U. Zschieschang, H. Klauk, M. Ikeda, H. Kuwabara, T. Yamamoto, K. Takimiya, K.-T. Cheng, T. Someya, Organic pseudo-CMOS circuits for low-voltage large-gain high-speed operation. IEEE Electron Device Lett. **32**(10), 1448–1450 (2011)
6. V. Wagner, P. Wöbkenberg, A. Hoppe, J. Seekamp, Megahertz operation of organic field-effect transistors based on poly(3-hexylthiophene). Appl. Phys. Lett. **89**(24), 243515-1–243515-3 (2006)
7. M. Jaiswal, R. Menon, Equivalent circuit for an organic field-effect transistor from impedance measurements under dc bias. Appl. Phys. Lett. **88**(12), 123504-1–123504-3 (2006)
8. M. Kitamura, Y. Arakawa, High current-gain cutoff frequencies above 10 MHz in n-channel C_{60} and p-channel pentacene thin-film transistors. Jpn. J. Appl. Phys. **50**, 01BC01-1–01BC01-4 (2011)
9. M. Kitamura, Y. Arakawa, Current-gain cutoff frequencies above 10 MHz for organic thin-film transistors with high mobility and low parasitic capacitance. Appl. Phys. Lett. **95**(2), 023503-1–023503-3 (2009)
10. B. Bayraktaroglu, K. Leedy, R. Neidhard, Microwave ZnO thin-film transistors. IEEE Electron Device Lett. **29**(9), 1024–1026 (2008)
11. Y.-M. Lin, K.A. Jenkins, A. Valdes-Garcia, J.P. Small, D.B. Farmer, P. Avouris, Operation of graphene transistors at gigaherz frequencies. Nano Lett. **9**(1), 422–426 (2009)
12. O. Sevimli, A.E. Parker, A.P. Fattorini, J.T. Harvey, Very low frequency S-parameter measurements for transistor noise modeling, in *International Conference on Electromagnetics in Advanced Applications*, Sep. 2010, pp. 386–389
13. S. Lee, Accurate extraction and analysis of intrinsic cutoff frequency of sub-0.1 μm MOSFETs. Electron. Lett. **42**(16), 945–947 (2006)
14. T. Zaki, R. Rödel, F. Letzkus, H. Richter, U. Zschieschang, H. Klauk, J.N. Burghartz, AC characterization of organic thin-film transistors with asymmetric gate-to-source and gate-to-drain overlaps. Org. Electron. **14**(5), 1318–1322 (2013)
15. G. Dambrine, A. Cappy, F. Helidore, E. Playez, A new method for determining the FET small-signal equivalent circuit. IEEE Trans. Microw. Theory Tech. **36**(7), 1151–1159 (1988)

16. S. Wagner, *Small-signal Device and Circuit Simulation*, Ph.D. dissertation, Vienna University of Technology, Vienna, Austria, Apr. 2005
17. S.J. Orfanidis, *Electromagnetic Waves and Antennas* (E-Book, Piscataway, USA), http://www.ece.rutgers.edu/orfanidi/ewa/ (2013)
18. S-parameter design, Application note 154, Hewlett Packard (1990)
19. De-embedding and embedding S-parameter networks using a vector network analyzer, Application note 1364–1, Agilent Technologies (2001)
20. Applying error correction to network analyzer measuerements, Application note 1287–3, Agilent Technologies (1999)
21. D.K. Rytting, Network analyzer error models and calibration methods, Presentation, Agilent Technologies (1998)
22. H. van Hamme, M. VandenBossche, Flexible vector network analyzer calibration with accuracy bounds using an 8-term or a 16-term error correction model. IEEE Trans. Microw. Theory Tech. **42**(6), 976–987 (1994)
23. J.V. Butler, D.K. Rytting, M.F. Iskander, R.D. Pollard, M. Vanden Bossche, 16-term error model and calibration procedure for on-wafer network analysis measurements. IEEE Trans. Microw. Theory Tech. **39**(12), 2211–2217 (1991)
24. L.F. Tiemeijer, R.M.T. Pijper, J.A. van Steenwijk, E. van der Heijden, A new 12-term open-short-load de-embedding method for accurate on-wafer characterization of RF MOSFET structures. IEEE Trans. Microw. Theory Tech. **58**(2), 419–433 (2010)
25. F. Ante, F. Letzkus, J. Butschke, U. Zschieschang, K. Kern, J.N. Burghartz, H. Klauk, Submicron low-voltage organic transistors and circuits enabled by high-resolution silicon stencil masks, in *IEEE International Solid-State Circuits Conference Technical Digest*, Dec. 2010, pp. 21.6.1–21.6.4
26. F. Ante, Contact effects in organic transistors, Ph.D. dissertation, Swiss Federal Institute of Technology in Lausanne, Lausanne, Switzerland, Dec. 2011
27. D.J. Gundlach, L. Zhou, J.A. Nichols, T.N. Jackson, P.V. Necliudov, M.S. Shur, An experimental study of contact effects in organic thin film transistors. J. Appl. Phys. **100**(2), 024509-1–024509-13 (2006)
28. C.H. Shim, F. Maruoka, R. Hattori, Structural analysis on organic thin-film transistor with device simulation. IEEE Trans. Electron Devices **57**(1), 195–200 (2010)
29. U. Zschieschang, F. Ante, D. Kälblein, T. Yamamoto, K. Takimiya, H. Kuwabara, M. Ikeda, T. Sekitani, T. Someya, J. Blochwitz-Nimoth, H. Klauk, Dinaphtho[2,3-b:2',3'-f]thieno[3,2-b]thiophene (DNTT) thin-film transistors with improved performance and stability. Org. Electron. **12**(8), 1370–1375 (2011)
30. M. Fadlallah, G. Billiot, W. Eccleston, D. Barclay, DC/AC unified OTFT compact modeling and circuit design for RFID applications. Solid-State Electron. **51**(7), 1047–1051 (2007)
31. E. Calvetti, L. Colalongo, Z.M. Kovács-Vajna, Organic thin film transistors: a DC/dynamic analytical model. Solid-State Electron. **49**(4), 567–577 (2005)
32. T. Zaki, F. Ante, U. Zschieschang, J. Butschke, F. Letzkus, H. Richter, H. Klauk, J.N. Burghartz, Circuit impact of device and interconnect parasitics in a complementary low-voltage organic thin-film technology, in *Semiconductor Conference Dresden*, Sep. 2011, pp. 1–4
33. B. Razavi, *Design of Analog CMOS Integrated Circuits* (McGraw-Hill, New York, 2001)
34. S.-C. Chang, Y. Yang, F. Wudl, G. He, Y. Li, AC impedance characteristics and modeling of polymer solution light-emitting devices. J. Phys. Chem. B **105**(46), 11419–11423 (2001)
35. L. Han, N. Koide, Y. Chiba, T. Mitate, Modeling of an equivalent circuit for dye-sensitized solar cells. Appl. Phys. Lett. **84**(13), 2433–2435 (2004)
36. T. Miyadera, T. Minari, K. Tsukagoshi, H. Ito, Y. Aoyagi, Frequency response analysis of pentacene thin-film transistors with low impedance contact by interface molecular doping. Appl. Phys. Lett. **91**(1), 013512-1–013512-3 (2007)
37. S. Grecu, M. Bronner, A. Opitz, W. Brütting, Characterization of polymeric metal-insulator-semiconductor diodes. Synth. Met. **146**(3), 359–363 (2004)
38. B.H. Hamadani, C.A. Richter, J.S. Suehle, D.J. Gundlach, Insights into the characterization of polymer-based organic thin-film transistors using capacitance-voltage analysis. Appl. Phys. Lett. **92**(20), 203303-1–203303-3 (2008)

39. D. Lovelace, J. Costa, N. Camilleri, Extracting small-signal model parameters of silicon MOS-FET transistors, in *International Microwave Symposium Digest*, May 1994, pp. 865–868
40. D.A. Frickey, Conversions between S, Z, Y, h, ABCD, and T parameters which are valid for complex source and load impedances. IEEE Trans. Microw. Theory Tech. **42**(2), 205–211 (1994)
41. T. Zaki, S. Scheinert, I. Hörselmann, R. Rödel, F. Letzkus, H. Richter, U. Zschieschang, H. Klauk, J.N. Burghartz, Accurate capacitance modeling and characterization of organic thin-film transistors. IEEE Trans. Electron Devices **61**(1), 98–104 (2014)
42. M. Caironi, Y.-Y. Noh, H. Sirringhaus, Frequency operation of low-voltage, solution-processed organic field-effect transistors. Semicond. Sci. Technol. **26**(3), 034006-1–034006-8 (2011)
43. U. Zschieschang, F. Ante, T. Yamamoto, K. Takimiya, H. Kuwabara, M. Ikeda, T. Sekitani, T. Someya, K. Kern, H. Klauk, Flexible low-voltage organic transistors and circuits based on a high-mobility organic semiconductor with good air stability. Adv. Mater. **22**(9), 982–985 (2010)
44. T. Takano, H. Yamauchi, M. Iizuka, M. Nakamura, K. Kudo, High-speed operation of vertical type organic transistors utilizing step-edge structures. Appl. Phys. Express **2**(7), 071501-1–071501-3 (2009)
45. J. Collet, O. Tharaud, A. Chapoton, D. Vuillaume, Low-voltage, 30 nm channel length, organic transistors with a self-assembled monolayer as gate insulating films. Appl. Phys. Lett. **76**(14), 1941–1943 (2000)
46. Y.-Y. Noh, N. Zhao, M. Caironi, H. Sirringhaus, Downscaling of self-aligned, all-printed polymer thin-film transistors. Nat. Nanotechnol. **2**, 784–789 (2007)

Chapter 8
Digital-to-Analog Converters

The unique processing characteristics and demonstrated performance of OTFTs suggest that they can be competitive candidates for emerging commercial applications such as flexible displays, low-end RFID tags and smart sensors systems. In such applications, data conversion to interface the digital processors with the analog world is an essential necessity. Hereby, the first demonstration of organic-based current-steering digital-to-analog converters (DACs) is reported [1–5]. Given that many state-of-the-art OTFTs suffer from intrinsically low carrier mobility and large behavioral parameter variations, this chapter gives first an overview of the circuit design techniques used in literature to address these challenges. This is followed by a discussion on the operation principal and the different architectures of DACs, with an emphasis on the current-steering topology. Then the design and measurements of a 6-bit binary and a 3-bit unary current-steering DACs are presented. The 6-bit binary current-steering DAC uses p-channel OTFTs; it marks records in speed (100 kS/s) and compactness ($2.6 \times 4.6\,\text{mm}^2$) owing to the OTFT fabrication process that is based on high-resolution silicon stencil masks. The 3-bit unary current-steering DAC, however, uses complementary OTFTs and is employed as a circuit test vehicle in a simulation case study to investigate the impact of parasitics as well as the influence of using the relatively slow n-channel OTFTs on the circuit performance. The study is based on extrapolations from a well calibrated circuit simulation, resulting from optimized DC/AC SPICE models for both p- and n-channel OTFTs. The models are validated by transient measurements of the driving complementary logic of the DAC, i.e., a binary-to-thermometer decoder, up to a maximum operating frequency of 1 kHz.

8.1 Organic Circuit Design Methodologies

Integrated circuits based on OTFTs have recently shown a rapid progressive development towards higher level of integration and better performance, particularly in applications where large-area, low-cost and novel form factors such as flexibility and

T. Zaki, *Short-Channel Organic Thin-Film Transistors*, Springer Theses,
DOI 10.1007/978-3-319-18896-6_8

transparency are important considerations [6]. The technology is being developed for active-matrix backplanes of flexible electrophoretic (e-paper) [7] and OLED [8, 9] displays. Furthermore, it has attracted considerable attention for realizing memories [10–12], RFID tags [13–18], and smart sensor such as biosensors [19–21] and robotic e-skin [22–25]. Most of the OTFT-based circuits developed today are digital, yet the interest is also growing to harness the technology for analog circuits. In this context, OTFTs are beginning to make significant inroads into building basic analog circuit blocks, where the first OTFT-based amplifiers [26–28], digital-to-analog converters (DACs) [1–3, 29] and analog-to-digital converters (ADCs) [30, 31] are successfully demonstrated in experiment.

Circuit design in OTFT technologies is quite challenging because of several reasons, among them are the following. First, most OTFTs, particularly those manufactured using low-cost printing methods, suffer from large process variations and bias-stress degradation, which result in significant performance deterioration and poor long-term reliability [32]. Second, OTFTs are often having high threshold voltages and operate at fairly large supply potentials, which deter the possibility to fabricate low-power circuits. Third, it is difficult and in most cases not possible to realize area-efficient resistors. Fourth, it is observed that electron transport in most available organic semiconductors is not possible in air [33]. This is because molecules possessing an electron on their LUMO are usually very reactive towards oxygen or water vapor such that oxidation of the molecule occurs in ambient air. Furthermore, electron traps due to hydroxyl groups located at the interface between the dielectric and the organic semiconductor immobilize their transport. As a result, most organic circuits today utilize p-channel OTFTs only.

To overcome these limitations, a number of technological enhancements are made in the OTFT technology used in this work as discussed in Chaps. 3 and 4. The surface of the gate insulator is treated with a molecular self-assembled monolayer (SAM) that provide a capacitance approaching 1 $\mu F/cm^2$, thus allowing the OTFTs to operate with voltages of a few volts [34]. Furthermore, the OTFTs employ organic semiconductors, namely DNTT, $F_{16}CuPc$ and NTCDI-Cl_2-$(CH_2C_3F_7)_2$, which are able to operate in ambient air with a relatively good stability [34–36]. These low-voltage and air-stable OTFTs are fabricated using high-resolution silicon stencil masks that provide submicrometer channel length capability and excellent device uniformity [1, 37–39]. Finally, a robust design solution to realize an OTFT-based transimpedance circuit, which imitates the behavior of a linear passive resistor element, is proposed hereinafter.

Figure 8.1 shows several logic design techniques that have been mostly explored by researchers for OTFT technologies. The key factors to choose the appropriate logic family are the output voltage swing, noise margin, power consumption, number of transistors and routing complexity. The complementary logic design is most favorable in that regard, but with the fact that n-channel OTFTs are still in their infancy, designs that make use of p-channel OTFTs only are still mostly employed [33].

The conventional configuration is the biased-load (Fig. 8.1a); depending on the bias voltage (V_{bias}) applied to its gate terminal, the load transistor can be operated in the saturation regime (when V_{bias} is connected to the drain of the load transistor; also

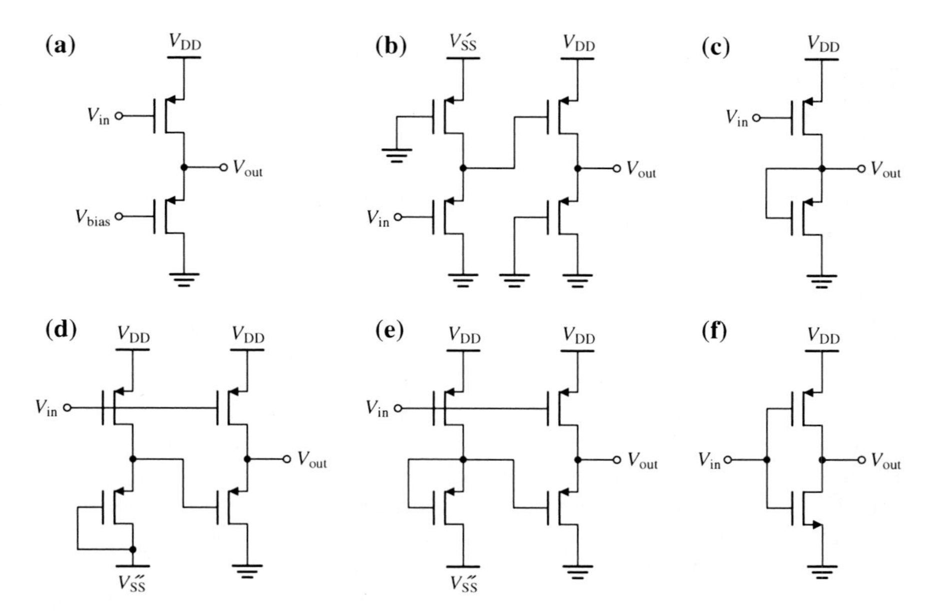

Fig. 8.1 Inverter designs. **a** Biased-load inverter; also referred to as diode-load inverter when V_{bias} is set to zero (when the gate electrode of the load transistor is connected to the ground) [40]. **b** Level-shifted diode-load inverter; the voltage V'_{SS} is typically set higher than V_{DD} to increase the driving potential at the input of the output stage [14, 41]. **c** Zero-V_{GS} inverter; a load TFT with a high V_{TH} (depleted-like device) is usually used in this design [40]. **d** Pseudo-E inverter [32]. **e** Pseudo-D inverter [32]. **f** Complementary inverter [42]

referred to as diode-load inverter [40]) or in the linear regime (when $V_{bias} \leq V_{TH}$) [42]. Both cases have distinct advantages and disadvantages. The saturated biased-load inverter requires a single voltage supply and a relatively simple routing, yet the output low level (V_{OL}) is limited to $|V_{TH}|$. The linear biased-load inverter, however, allows the output node to be completely dissipated to the ground level, resulting in a higher noise margin compared to that of the saturated counterpart. Nevertheless, one should note that there is a tradeoff because as V_{bias} reduces, the load transistor drives larger current. Thus, the output high level (V_{OH}) also reduces, which results in a lower noise margin. The main drawback of the linear biased-load inverter is the use of two separate power supplies, but this can be actually utilized as a built-in post fabrication tool to compensate for process variation and device degradation or can be also employed to switch off the inverter when it is not needed by applying $V_{bias} = V_{DD}$. In general, both saturated and linear biased-load inverters suffer from low gain and relatively high stand-by power dissipation.

In the biased-load inverter, the output swing as well as the trip voltage are dependent on the applied bias voltage V_{bias}. To have a better control over the trip voltage without affecting much the output swing, a level shifter (Fig. 8.1b) can be used at the input [14, 41]. In this special case, a diode-load inverter is used at the output stage to have only one additional power supply V'_{SS} and accordingly reduce the routing complexity. Therefore, by adjusting V'_{SS}, the trip voltage can be shifted towards $V_{DD}/2$. This design, though improves the balance between the high and low noise margins, consumes more power due to the larger flow of current compared to the previous design.

Another approach is by using a zero-V_{GS} load (Fig. 8.1c); this configuration can alleviate the high power dissipation of the level-shifted diode-load inverter at the cost of a slower speed [40]. The zero-V_{GS} load provides a higher output resistance and thus higher gain, sharper voltage transfer characteristic (VTC) and better noise margin. Given that the mobility in OTFTs is dependent on the gate voltage, the load in this case must be significantly wider than the driver to achieve positive noise margins [43]. Not only does this make the area of the inverter larger, it also increases the total gate capacitance, resulting in a slower inverter speed. In order to increase the pull-down force and improve the asymmetric characteristics without area and speed overhead, a zero-V_{GS} OTFT load with a high V_{TH}, i.e., a depletion-like device, can be used. This dual-V_{TH} implementation can reduce the area by 30 times compared to the single-V_{TH} inverter, assuming a trip voltage at $V_{DD}/2$ [43]. However, the problem with the high stand-by power dissipation is still present in this approach because the load transistor always has a conducting channel regardless of the input and output voltage levels.

Organic transistors are mostly not intentionally doped; therefore, the threshold voltages of all OTFTs on a substrate are similar [33]. In principle, electrical doping of organic semiconductors is possible, but this is not favorable as discussed in Chap. 3. Alternatively, several methods have been studied in literature to achieve V_{TH} control in OTFTs: (i) different SAMs in the gate dielectric [44], (ii) floating-gate structure [45], (iii) back-gate bias [46], or (iv) gate materials with different work functions [43, 47]. Despite the successful tuning of the V_{TH}, each of these methods has certain drawbacks for use in practical applications. These methods generally require special process manipulation, add an electrode in each transistor that result in more interconnect lines, create a very small (sometimes negligible) amount of V_{TH} shift relative to the typical supply voltage V_{DD}, or induce substantial adverse effects on the OTFT characteristics including performance and stability. Nevertheless, the control of the threshold voltage after manufacturing using the floating- or back-gate methods is beneficial for many OTFT technologies, since V_{TH} is known to change inadvertently during operation due to bias stress and air exposure [45, 46]. In 2011, the first organic 8-bit microprocessor was introduced, which employed the dual-V_{TH} OTFT architecture with back-gate bias [48, 49]. The microprocessor runs at a clock speed of 40 Hz with a supply voltage of 10 V; however, a very high back-gate voltage of 50 V is needed to tune properly the threshold voltage.

A new logic family called *Pseudo*-CMOS that avoids the process complexities introduced by the dual-V_{TH} technologies while attaining a comparable inverter performance was recently proposed [32, 50, 51]. Pseudo-E (Fig. 8.1d) and Pseudo-D (Fig. 8.1e) are two design variations of this category; they differ in the gate connection of the load transistor of the input stage. Similar to the level-shifted diode-load inverter, four OTFTs are needed to realize an inverter. Moreover, the additional tuning voltage V''_{SS} is also used post fabrication to compensate for the process variations and to keep the device performance in check. The difference to the previous configurations is that the load transistor connected to the output node is switched off when the input is low to prevent a direct current from V_{DD} to ground, resulting in a high output swing. Nevertheless, the load transistor in the input stage always has a conducting channel regardless of the input voltage level.

The operation of all the inverters considered so far is controlled primarily by switching the driver transistor. In addition, a nonzero steady-state current is always drawn from the power supply to the ground when the driver transistor is switched on, which results in an inevitable significant DC power consumption [42]. On the other hand, the complementary inverter (Fig. 8.1f), which comprises both p- and n-channel OTFTs, operates in a different way. For a logic high input, the n-channel OTFT drives (pulls down) the output node while the p-channel OTFT is switched off; however, for a logic low input, the p-channel OTFT drives (pulls up) the output node while the n-channel OTFT acts is switched off. A direct flow of current from the power supply to the ground occurs only when switching between the high- and low-level states, resulting in a negligible static power dissipation (<1 nW [34]). In the steady-state operation, only the gate leakage current (supposedly small current) between the gate electrode and the contacts is present. As a result, the VTC of the complementary inverter exhibits a full output swing between 0 V and V_{DD} with a sharp transition between the logic states.

The low power consumption in organic transistors is highly desirable for battery-powered or frequency-coupled applications. However, the integration of n-channel OTFTs is beneficial only if the electron mobility is comparable with the hole mobility [43]. In the dual-V_{TH} inverter presented in [43], the load OTFT is 15 times larger than the driver to achieve a trip voltage at $V_{DD}/2$. Therefore, an n-channel OTFT would be superior only if the ratio of the hole mobility to the electron mobility is less than 15. The ground-breaking work of our partners at MPI-SSR on the development of the performance and stability of complementary OTFTs led to considerable improvements in the past few years [52], which made the complementary OTFT technology a favorable choice for this work. The fabricated n-channel OTFTs have about ten-fold lower carrier mobility than p-channel OTFTs and there is still an ongoing research effort to improve these results [36].

A comprehensive study of all the different logic design styles is performed and the results are summarized in [52–54]. In view of the many advantages of the complementary logic, a further study of the impact of using n-channel OTFTs in a practical circuit on the dynamic performance is given in Sect. 8.4. A 3-bit unary current-steering DAC, which requires a binary-to-thermometer decoder, is used for this study as a circuit test vehicle. Furthermore, a 6-bit binary current-steering DAC using only p-channel OTFTs is built to demonstrate the high-performance capability of the used technology; a thousand-fold faster update rate than prior state of the art is achieved as presented in Sect. 8.3.

8.2 Functionality and Specifications

With the excessive development of functionality and performance of digital circuits, electronic systems tend to employ more digital than analog circuitry [4]. Nevertheless, the physical environment prevents such proliferation to cover the entire system because naturally occurring signals are analog. Accordingly, data converters (DACs and ADCs) are indispensable fundamental blocks which are used to interface the digital processors with the analog world. Data converters find many applications in electronic products such as displays and sensors. In this work, only DACs are considered.

8.2.1 Operation Principle

A DAC is a functional block that constructs analog signals from digital input by generating a multiple of a reference quantity (X_{ref}) for each digital input code. For an n-bit digital input sequence ($b_0 b_1 b_2 \ldots b_{n-1}$), the analog output can be described by [55]

$$X_{\text{out}} = \sum_{i=0}^{n-1} 2^i b_i X_{\text{ref}}. \tag{8.1}$$

Depending on the reference quantity X_{ref}, the analog output can be either charge, voltage or current. In fact, the reference quantity X_{ref} designates the least significant bit (LSB) value of the DAC.

In principle, a DAC is characterized by two main parameters, namely resolution and update rate. For an n-bit resolution, the digital signal can only have 2^n values. Furthermore, the update rate is the rate at which the input digital codes are converted. Therefore, for the realization of an accurate DAC, the resolution and update rate have to be maximized. According to Nyquist-Shannon theorem, the minimum update rate of the DAC has to be at least equal to two times the highest frequency of the signal; this is to avoid aliasing and to be able to fully construct the output analog signal. In the following, detailed description of the different characterization parameters of a DAC is presented.

8.2.2 Characterization Parameters

The performance of DACs is evaluated by several specification parameters. These help to evaluate and select the most suitable DAC architecture from the numerous available alternative designs for a given application. The parameters are classified into two categories, namely static and dynamic measures. The static parameters are

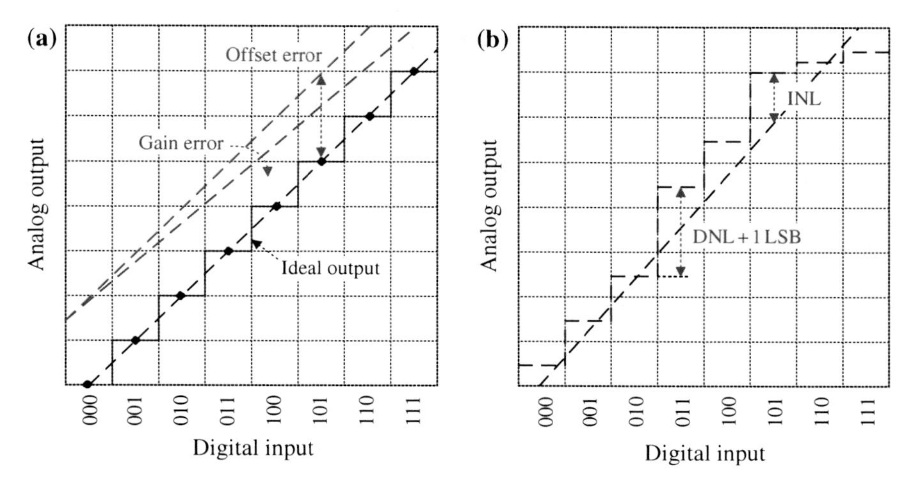

Fig. 8.2 Static specifications of a DAC. **a** Ideal transfer characteristics in addition to offset and gain errors. **b** Differential non-linearity (DNL) and integral non-linearity (INL) errors [55]. The INL shown here is extracted from the best-fit line, which is fitted to the data using the minimum mean square error (MMSE) approach. Another commonly used method is to extract the INL from the end-point line, which is a line that connects the points on the transfer function corresponding to the lowest and highest input codes. The latter approach typically yields larger INL values

offset error, gain error, differential non-linearity (DNL) and integral non-linearity (INL). Furthermore, the dynamic parameters in the time and frequency domains are maximum update rate, settling time, slew rate, glitch energy and spurious free dynamic range (SFDR). The definition of each of these nine parameters is as follows [4, 55].

8.2.2.1 Offset and Gain Errors

The offset error is the constant DC offset of the DAC transfer characteristics, while the gain error is the deviation of the characteristic slope from the ideal one (Fig. 8.2a). Note that these errors do not introduce any non-linearity and has no effect on the frequency domain.

8.2.2.2 Differential Non-linearity

The maximum DNL error is the worst case deviation of the actual to the ideal step size of 1 LSB between two adjacent input codes (Fig. 8.2b). The maximum DNL is expressed as

$$\mathrm{DNL}_{\mathrm{max}} = \max\left(\frac{V_{\mathrm{out}}(i+1) - V_{\mathrm{out}}(i) - \Delta V_{\mathrm{LSB}}}{\Delta V_{\mathrm{LSB}}}\right), \tag{8.2}$$

where $i = 0, 1, \ldots, 2^n - 2$ and ΔV_{LSB} is given by

$$\Delta V_{\mathrm{LSB}} = \frac{V_{\mathrm{out}}(2^n - 1) - V_{\mathrm{out}}(0)}{2^n - 1}. \tag{8.3}$$

In general, it is preferable to have a DAC with a DNL error that is less than ±1 LSB to guarantee proper operation.

8.2.2.3 Integral Non-linearity

The maximum INL error is the worst case deviation of the actual to the ideal linear output curve. After nullifying the gain and offset errors, the ideal output curve is either an *end-point* line connecting the minimum and full scale points or a *best-fit* line that is based on the minimum mean square error (MMSE) distance from line to samples. The latter approach—as shown in Fig. 8.2b—is used in this work.

For the end-point line approach, however, the maximum INL of a DAC is represented by the following expression:

$$\mathrm{INL}_{\mathrm{max}} = \max[V_{\mathrm{out}}(i) - V_{\mathrm{out}}(0) - \Delta V_{\mathrm{LSB}} \cdot i], \tag{8.4}$$

where $i = 0, 1, \ldots, 2^n - 1$, and ΔV_{LSB} is the slope of the line through the end points and is given by (8.3). The resulting INL using this expression is typically larger than when using the best-fit line.

By definition, a DAC is called monotonic when its output never decreases with an increasing digital input, i.e., the analog output always increases or remains constant as the digital input increases. For guaranteed monotonicity, the INL error should be less than or equals to ±0.5 LSB.

8.2.2.4 Update Rate

The update rate is the rate at which the output is sampled. According to Nyquist-Shannon theorem, the minimum update rate of the DAC has to be at least equal to two times the maximum frequency of the signal. The update rate is given in unit samples per second (S/s).

8.2.2.5 Settling Time

The settling time is defined to be the time needed by the DAC to experience a full scale transition and settle around its final value (Fig. 8.3a).

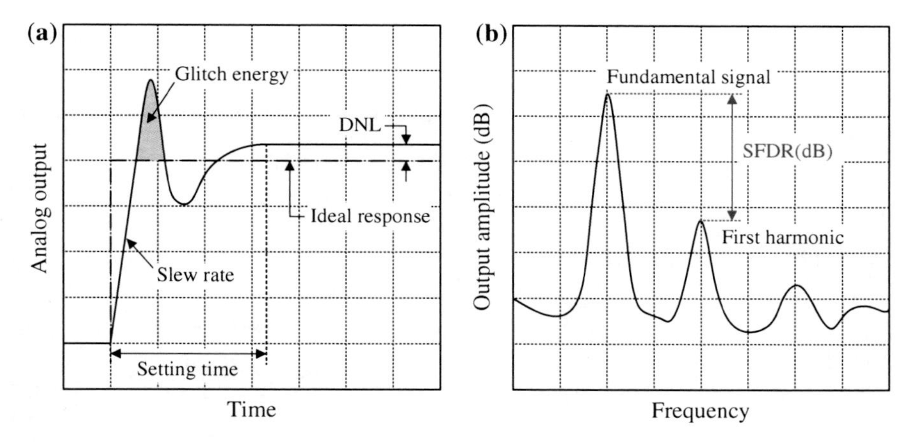

Fig. 8.3 Dynamic specifications of a DAC. **a** Analog output of a DAC as a function of time, showing the settling time, slew rate and glitch energy. **b** Frequency spectrum of a DAC output signal, illustrating the spurious free dynamic range (SFDR) [55]

8.2.2.6 Slew Rate

The slew rate is defined to be the maximal rate at which the output of the DAC can change with the varying input (Fig. 8.3a).

8.2.2.7 Glitch Energy

The glitch energy is the area under the transient response when the DAC switches between two consecutive input codes (Fig. 8.3a). An acceptable quantitative approximation of the glitch energy can be evaluated by assuming a square-shaped glitch impulse (worst case calculation). Therefore, the glitch energy is given by $E_{\mathrm{gl}} = t_{\mathrm{gl}} \times A_{\mathrm{gl}}$, where t_{gl} and A_{gl} are the timing error and the amplitude of the glitch, respectively. In practice, the glitch energy has to be adequately smaller than the LSB energy, which is given by $E_{\mathrm{LSB}} = t_{\mathrm{LSB}} \times A_{\mathrm{LSB}}$. Therefore, for a good dynamic performance, the ratio $E_{\mathrm{gl}}/E_{\mathrm{LSB}}$ has to be minimized. A simple way to minimize E_{gl} is to reduce t_{gl} by using well-placed synchronized blocks in the DAC.

8.2.2.8 Spurious Free Dynamic Range

The SFDR is defined to be the ratio between the fundamental signal power (P_{sig}) and the largest distortion component power ($P_{\text{max-dist}}$) within a specific measured frequency window (Fig. 8.3b). Accordingly, the SFDR is expressed as follows:

$$\mathrm{SFDR} = \frac{P_{\mathrm{sig}}}{P_{\mathrm{max\text{-}dist}}}. \tag{8.5}$$

Note that the SFDR depends on the operating conditions and the update rate of the DAC.

8.2.3 Design Architectures

This section gives an overview and comparison of different DAC architectures existing today [4]. Emphasis is made on the area, complexity and dynamic specifications of the DACs. For more detailed study of the operation of each architecture, the reader is referred to [56].

The DACs are mainly classified into three main categories, namely resistive, capacitive and current-steering DACs. This classification is recognized from the reference quantity (X_{ref}) used. Correspondingly, the DAC output signal can either be voltage, charge or current, respectively. Note that there are also other DAC topologies such as hybrid resistive-capacitive and pulse-width-modulation-based DACs; however, these architectures are not comprised in the focus of this work. For each of the DAC architectures, various implementations ranging from high-speed to high-accuracy designs are proposed in literature. This gives the designer the freedom to choose the most suitable implementation for a given application; nevertheless, there is always a tradeoff and a compromise has to be made.

In general, chip area and process limitations rule out the resistive-based DACs for our OTFT technology. This is because accurate and compact resistors cannot be yet realized. On the other hand, the capacitive-based DACs can be implemented using charge-divider, multiplying-capacitive or switched-capacitor designs [56]. The very large chip area needed by the charge-divider design makes it not suitable for high resolution DACs. Furthermore, the complexity of the multiplying-capacitive design makes it not a favorable choice for an OTFT technology. The switched-capacitor design, however, is suitable for OTFTs. In fact, all the capacitive-based DACs are known to have better matching properties than the resistive and current-steering DACs due to the more controllable capacitor matching.

Because of the large device-to-device variations resulting from older OTFT fabrication processes, the first approach to realize an organic-based DAC used the capacitive (switched-capacitor C-2C) architecture [29], which is less sensitive to the poor transistor matching. Despite the better capacitor matching property, the design proposed in [29] suffers from many limitations:

- A reset phase is needed every conversion clock cycle to prevent accumulating leakage errors, which are expected to be up to 0.6 LSB at 100 Hz. This degrades the speed capability of the converter.
- The leakage error of the capacitors perturb the static performance and constricts the lower speed limit of the DAC to be around 10 Hz.

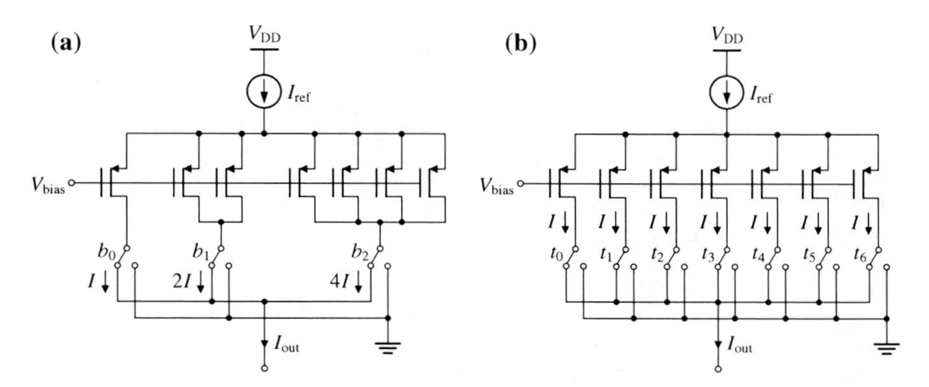

Fig. 8.4 Schematic diagram of 3-bit current-steering DACs with current-division topology. **a** Binary architecture driven by a binary-coded input. **b** Unary architecture driven by a thermometer-coded input. The reference current in both circuits is given by $I_{\mathrm{ref}} = (2^n - 1)I$, where $I = I_{\mathrm{LSB}}$

- The use of n-channel OTFTs that feature low carrier mobility of $0.005\,\mathrm{cm}^2/\mathrm{Vs}$ limits the maximum update rate to 100 Hz and results in a relatively large DAC area of $28 \times 14\,\mathrm{mm}^2$.

Therefore, for high-speed, high-accuracy and compact applications, the capacitive DACs are no longer the preferable choice for OTFT technologies.

The last category is the current-steering DACs. In this case, the current is used as the reference quantity for generating the output. The main advantage of the current-steering DAC over the resistive and capacitive DACs is the combination of high speed and compactness. This is because the transistors used to implement the current sources consume much smaller area than the resistors or capacitors. However, excellent device uniformity in this architecture is an essential necessity for the DAC linearity. Referring to Fig. 3.8, the comparison between the conventional polyimide shadow masks (used for the fabrication of the OTFT C-2C DAC) and the high-resolution silicon stencil masks (used in this work) show that the silicon stencil masks provide much improved transistor matching even for smaller device geometries, thus allowing the realization of the current-steering architecture at considerably higher sampling rate and smaller chip area.

There are basically two main design alternatives for the current-steering DAC, namely current-division and current-mirror topologies. Figure 8.4 shows a schematic diagram of a 3-bit current-steering DAC based on the current-division topology. As depicted in the figure, a reference current (I_{ref}) is divided into n-equal or binary-weighted currents depending on the architecture used. The n-equal currents are useful in case of thermometer-coded input, whereas the binary-weighted currents are useful in the case of binary-coded input. In both architectures, $2^n - 1$ transistors are required to divide the reference current. This implementation, however, has two main drawbacks: (i) the stack of the current dividing transistors on top of I_{ref} limits the output voltage swing; and (ii) the current I_{ref} must be as large as $2^n - 1$ times the

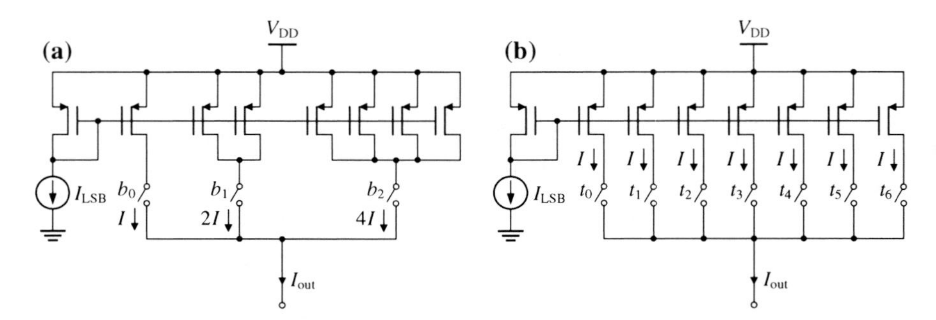

Fig. 8.5 Schematic diagram of 3-bit current-steering DACs with current-mirror topology. **a** Binary architecture driven by a binary-coded input. **b** Unary architecture driven by a thermometer-coded input. The reference current in both circuits is given by $I_{ref} = I_{LSB} = I$

unity current I_{LSB}, which results in a very large device for generating the reference current. On the other hand, the current-mirror design shown in Fig. 8.5 maximizes the output swing and requires small reference current, leading to a smaller biasing transistor. This gives rise to the current-mirror topology to be the most favorable choice for our OTFT technology.

In the binary current-steering DAC based on current mirrors (Fig. 8.5a), the current sources are organized in a binary-weighted structure and controlled by a binary-coded input. The advantages of this design includes simplicity of the architecture, small area consumption and low power dissipation. However, this design suffers from large glitches. For example, during a transition from "100" to "011", the transition may go through an intermediate value of "000" if the most significant bit (MSB) turns off before the remaining bits turn on, hence a large glitch occur. As for the unary architecture (Fig. 8.5b), $2^n - 1$ identical current sources are controlled by thermometer-coded input. Therefore, for a binary-coded input, a binary-to-thermometer decoder is used. This decoding logic adds a finite area and power to the complete design. In general, this design inherently exhibits a monotonic behavior because any increment at the digital input results in an additional current at the analog output. To combine the advantages of both unary and binary architectures and also to compensate for their limitations, segmented current-steering DACs are proposed in literature [55]; for a resolution of $n = k + m$, the n-MSBs are converted to thermometer code and driven to a unary current-steering sub-converter, while the remaining m-LSBs are driven directly to a binary current-steering sub-converter.

The use of silicon stencil masks in this work made it possible to fabricate top-contact OTFTs with high resolution and excellent uniformity. Such precise matching between the transistors is highly desirable in the current-steering DAC, where device matching is paramount. Accordingly, a 6-bit binary current-steering DAC using only p-channel OTFTs is implemented to benchmark the recently published 6-bit switched-capacitor DAC that uses complementary OTFTs [1, 29]. By comparing the results of both circuits, the presented current-steering DAC offers a dramatic performance boost (1000 times faster) and greatly reduced area consumption

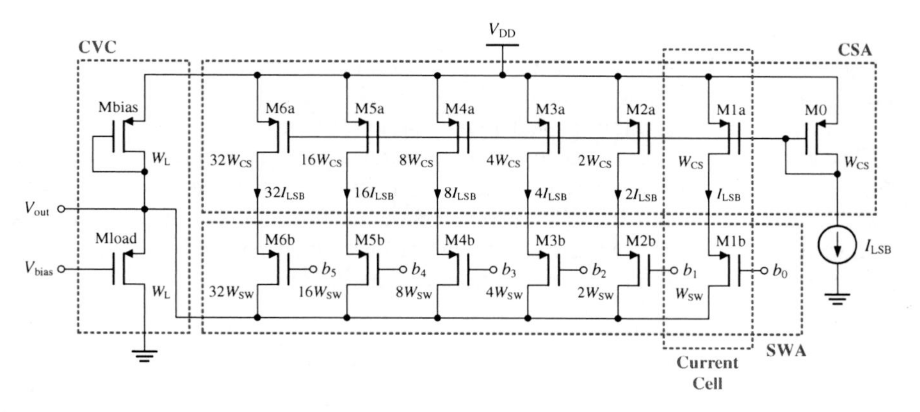

Fig. 8.6 Schematic diagram of the 6-bit binary-weighted current-steering DAC [1, 2, 4]. The circuit is composed of a current-source array (CSA) with $(W/L)_{cs} = 100/15\,\mu m$, a switch array (SWA) with $(W/L)_{sw} = 40/4\ \mu m$ and a current-to-voltage converter (CVC) with $(W/L)_L = 100/5\ \mu m$

(30 times smaller). Furthermore, a 3-bit unary current-steering DAC is designed and fabricated, allowing to demonstrate the feasibility to realize a fully custom complementary logic block, i.e., a binary-to-thermometer decoder. The functionality of this DAC is validated and a study of the impact of device and interconnect parasitics as well as the use of relatively slow n-channel OTFTs on the circuit performance is carried out. The DAC designs and measurement results are presented in the following sections. Given the successful realization of both designs, it is proposed to combine in the future both implementations in a high-resolution segmented current-steering DAC.

8.3 A High-Performance Unipolar Design

8.3.1 A 6-Bit Binary Current-Steering DAC

The schematic diagram of the designed 6-bit current-steering DAC is shown in Fig. 8.6 [1, 2, 4]. The binary-weighted current-steering architecture is employed here because of its lower complexity, smaller area consumption and lower power dissipation since no decoding logic is necessary compared with its unary current-steering counterpart. The converter consists of three main blocks: (i) a current-source array (CSA) that contains the binary-weighted current sources and an input current mirror, (ii) a switch array (SWA) that contains the binary-weighted analog switches, and (iii) a current-to-voltage converter (CVC) at the output. For the current sources M_i, 2^{i-1} identical parallel transistors are used to provide $2^{i-1}I_{LSB}$. This similarly applies for all the binary-weighted analog switches; this results in a total of 129 transistors. This type of current-steering can be designed by exclusively n- or p-channel OTFTs, hence not requiring complementary OTFT technology. Accordingly, the converter is designed with only p-channel OTFTs because of their faster operational speed and smaller area; this is mainly owed to the considerably higher carrier mobility in the p-channel ($\mu_{\mathrm{p}} \geq 10\,\mu_{\mathrm{n}}$), thus proportionally larger drive current.

The current cell comprises a current-source transistor and an analog switch. The dimensions of the current-source transistor typically depend on the technology with which the circuit is fabricated [55]. Furthermore, excellent device matching in this architecture is an essential necessity for the DAC linearity; the matching requirements are evaluated for an n-bit binary-weighted current-steering DAC by the following [29, 57]:

$$\sigma_{\mathrm{INL}} = \frac{\mathrm{INL}}{\sqrt{2}(\mathrm{erfinv}(Y))} = \frac{\sqrt{2^n}}{2} \cdot \frac{\sigma_I}{I}, \tag{8.6}$$

$$\sigma_{\mathrm{DNL}} = \frac{\mathrm{DNL}}{\sqrt{2}(\mathrm{erfinv}(Y))} = \sqrt{2^n - 1} \cdot \frac{\sigma_I}{I}, \tag{8.7}$$

where σ_I/I is the current relative standard deviation (transistor's current mismatch). For $n = 6$ bits, INL ≤ 1 LSB, DNL ≤ 1 LSB and $Y = 0.95$, which corresponds to a confidence interval of 2-σ, the required current mismatch should be $\sigma_I/I \leq 6.4\,\%$, which is the minimum value resulting from Eqs. (8.6) and (8.7).

Referring to Fig. 3.8, the 16 OTFTs fabricated by the conventional polyimide shadow masks with a channel length of 30 µm have $\sigma_I/I = 11\,\%$, which exceeds already the upper limit calculated above, whereas the 16 OTFTs fabricated by the high-resolution silicon stencil masks with a channel length of only 2 µm have $\sigma_I/I = 4\,\%$ [1, 37]. Because of the large device-to-device variations resulting from the older OTFT fabrication processes, the first approach to realize an organic-based DAC used the capacitive (switched-capacitor C-2C) architecture which is less sensitive to the poor transistor matching [29]. However, the improved transistor matching even for smaller dimensions offered by the silicon stencil masks allow the realization of the current-steering architecture at considerably higher sampling rate and smaller area consumption. Therefore, for the design of the current-source transistors in our DAC, a sufficiently large channel length of 15 µm and a channel width of 100 µm are utilized to ensure even better matching than the value displayed in Fig. 8.8b and to guarantee that $\sigma_I/I \leq 6.4\,\%$.

As long as the voltage swing at the drain nodes of the current-source transistors is reduced during switching, the considerable capacitance of the current sources would not be charged and discharged, thus the dynamic performance would not be degraded. This would have been achieved by using differential analog switches with reduced cross-point voltage at their control signals [58]. By doing so, the interconnect crossings, featuring very high parasitic capacitances ($\sim$800 nF/cm^2) and degrading the circuit performance, would have been indispensable. Therefore, single-ended analog switches are used mainly to avoid interconnect crossings and also to minimize the area consumption. Furthermore, the analog switches are designed to be operating in the saturation mode during the full output voltage swing. This approach avoids the use of cascode current sources, which require larger area and limit the output swing, while maintaining the shielding effect of the cascode structure by isolating the drain nodes of the current source transistors from the variant output node. In addition, the non-linearities of the output current-voltage characteristics of the OTFTs at low drain bias due to the non-ohmic contacts can be avoided.

These specific operating conditions put a constraint on the switch input ON voltage ($V_{sw,on}$) to keep both the current source and analog switch transistors in the saturation mode. For proper circuit operation and in close agreement with simulations, the analog switch transistors are designed with a channel length of 4 μm and a channel width of 40 μm and the switch input ON voltage is set to be 1 V.

8.3.2 Current-to-Voltage Converter

The aim of the current-to-voltage converter (CVC) is to convert linearly the output DAC current to a voltage. The linearity of the CVC is essential as it directly affects the DAC static performance. An ideal CVC would have been implemented by using a passive resistor element; however, resistors are so far difficult to be realized in the organic technology. For this reason, an OTFT-based transimpedance circuit is designed.

As shown in the schematic diagram of the DAC (Fig. 8.6), the CVC consists of only two p-channel OTFTs with $W = 100$ and $L = 5\,\mu\text{m}$; one transistor is biased in the linear regime when $V_{bias} = -2\,\text{V}$, while the other transistor is diode-connected (saturation regime) to serve as a variable current source [1, 4]. This design followed series of comparison steps which are elucidated by the simulations shown in Fig. 8.7. Two main design aspects have to be considered during the design of the CVC; the CVC has to be very simple to avoid poor yield and it must tolerate device mismatches.

The ideal case with the passive resistor element (Fig. 8.7a) experiences very low linearity error, taking into account that device mismatch is not considered in the simulations. The second case (Fig. 8.7b; when $V_{bias} = 0\,\text{V}$ and I_{bias} is constant) involves a p-channel load OTFT that is diode-connected, thus operating in the saturation mode ($V_{DG} < |V_{TH}|$). In this case, the CVC relation is controlled by the OTFT saturation equation given by the following:

$$I_{D,sat} = -\mu C_I \frac{W}{2L}(-V_{GS} + V_{TH})^2, \tag{8.8}$$

$$V_{out} = \sqrt{I_{out}\frac{2L}{\mu C_I W}} - V_{TH}. \tag{8.9}$$

This non-linear relation directly reflects the high output linearity error obtained by the simulation. In the third case (Fig. 8.7b; when $V_{bias} = -2\,\text{V}$ and I_{bias} is constant), a linearly operating load is used. Given that the DAC output voltage does not exceed 2.5 V, a bias voltage of -2 V is set in a way to ensure that the gate-source voltage does

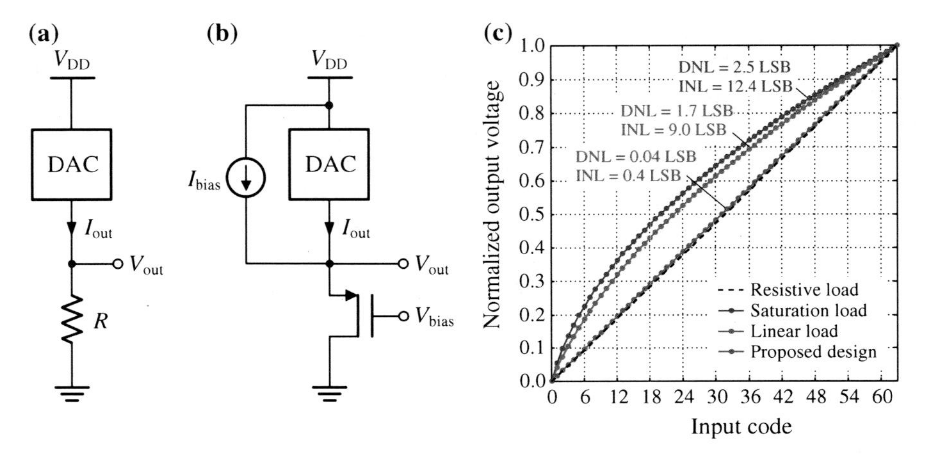

Fig. 8.7 Schematic and simulation results of different current-to-voltage converter (CVC) designs [1]. **a** Ideal passive resistive load. **b** OTFT-based transimpedance circuit. **c** Simulation results of three active designs in comparison to the ideal resistive load: (i) saturation load when $V_{bias} = 0\,V$ (diode-connected) and a constant I_{bias}, (ii) linear load when $V_{bias} = -2\,V$ and a constant I_{bias}, and (iii) a proposed design using a linearly operating load ($V_{bias} = -2\,V$) and a voltage-dependent bias current (I_{bias}) which decreases at higher DAC output voltage

not exceed the breakdown voltage ($\sim$6 V) during the complete output swing. Owing to the field-dependence in the OTFT contact resistance, which decreases as the gate-source voltage increases, the DAC linearity is perturbed. Therefore, it is proposed in the last case (Fig. 8.7b; when $V_{bias} = -2\,V$ and I_{bias} is voltage dependent) to use a variable driver current source, which decreases at higher DAC output voltage, to compensate for the non-linearity of the load. Using a diode-connected OTFT, as shown previously in the schematic (Fig. 8.6), the variable current source is realized. In this case, the gate-source voltage of the diode-connected OTFT is given by $V_{GS} = V_{out} - V_{DD}$. Therefore, the absolute value of the saturation current $I_{D,sat}$ passing through the transistor given by (8.8) decreases as the DAC output voltage (V_{out}) increases. The proposed CVC circuit proved to provide optimized DAC linearity.

8.3.3 Layout and Floorplan

Device matching and data interference have a strong impact on the DAC performance. Consequently, strict requirements apply to the design of the DAC layout and floorplan due to the relatively large number of transistors employed, especially for an organic-based analog circuit [1, 4]. Figure 8.8 shows two possible approaches for the floorplan, namely merged SWA-CSA and separated SWA-CSA [55, 59]. In the merged SWA-CSA, each unit cell comprises both the current source and the analog switch transistors. This approach has a simplified layout; however, it increases the

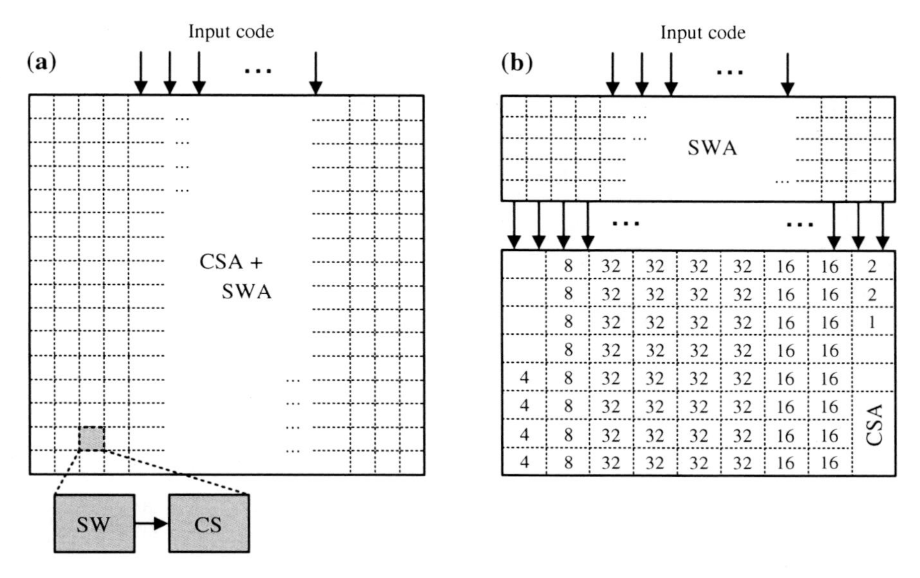

Fig. 8.8 Floorplan of current-steering DACs [1, 59]. **a** Merged SWA-CSA. **b** Separated SWA-CSA. In the separated SWA-CSA, identical transistors are placed in closer proximity, allowing better control of device matching. The numbers given in the CSA of the separated floorplan designate the arrangement of the binary-weighted OTFTs used in our 6-bit binary current-steering DAC

distances between the current source transistors and so the layout would be potentially more susceptible to the systematical gradient effects. As for the separated SWA-CSA, the identical transistors are placed in separate arrays, thus ensuring better device matching, easing power routing and minimizing signal interference due to the division of the digital and the analog parts. In spite of the increased layout complexity in the latter approach (Fig. 8.8b), it is used as it offers a substantially better conversion performance.

Furthermore, the transistors are placed in a matrix-like orientation in each array to keep the identical transistors in closer proximity. This makes it possible to exploit the excellent feature size control enabled by the stencil mask technology regardless of the non-uniformity introduced by material deposition through evaporation. Figure 8.8b shows clearly the arrangement of the transistors in the CSA by designating the binary-weights. In this way, the complexity of the routing matrix between the SWA and the CSA is minimized. With this careful layout and floorplan design, no interconnect crossings that feature high parasitic capacitances are comprised. The layout of the DAC is presented in Appendix A. Using the OTFT process technology presented in Chap. 4, the DAC is fabricated on a glass substrate. The total area of the DAC including the contact pads is only $2.6 \times 4.6\,\text{mm}^2$.

8.3.4 Calibration

After the chip is fabricated, multiple DACs are measured and a noticeable non-linearity error is observed in the transfer characteristics only during the transition of the input bit stream from "011111" to "100000" [1, 4]. Therefore, to guarantee monotonicity, the DAC is calibrated by tuning manually the switch input ON voltage ($V_{sw,on}$) for the most significant bit (MSB) to be 0 V instead of 1 V. As a result, a slightly larger current for the MSB is enabled to the output, hence the DAC linearity is optimized. The calibration method used here has a lower complexity compared to the one used in [29], where a 2-bit off-chip calibration circuit is employed.

8.3.5 Measurement Setup

By revisiting the layout of the 6-bit binary current-steering DAC, the converter has 12 input/output connections, namely the 6 digital binary-coded input bit stream (V_{sw1}, V_{sw2}, ..., V_{sw5}), the power supplies (V_{DD} and V_{GND}), the load biasing voltage (V_{bias}), the biasing LSB current (I_{LSB}), the output voltage (V_{out}), and finally an optional testing connection (V_X) that is connected to the LSB line in the routing matrix between the CSA and the SWA [1, 4]. The fabricated samples are measured in a manual probe station (SUSS PM5 analytical probe system from Karl Suss) with a careful connection between the probeheads and the 55-nm-thick gold pads. The 6-bit digital input stream is provided by a data generator (HP 8180A data generator from Hewlett Packard), while the remaining four inputs (V_{DD}, V_{GND}, V_{bias} and I_{LSB}) are supplied by a source measure unit (SMU; E5270B 8-slot precision measurement mainstream from Agilent Technologies). During DC measurements, the output node (V_{out}) is captured by the SMU. During AC measurements, however, the output node is connected to an oscilloscope (TDS3000B oscilloscope from Tektronix). An IC-CAP PEL code is written to control the equipment and automate the process by synchronizing the signals from the data generator and the SMU (see Appendix D.2). All measurements are performed in ambient air, at room temperature, at a supply voltage (V_{DD}) of 3.3 V and at relatively dimmed light environment.

8.3.6 Static and Dynamic Performances

The measured static and dynamic characteristic parameters, which fully evaluate the performance of the 6-bit binary current-steering DAC, are presented herein [1, 4]. Figure 8.9 shows the measured DC transfer function together with the corresponding DNL and INL errors. Before calibration, an undesired abrupt step during the transition of the input bit stream from "011111" to "100000" is observed, resulting in a degraded static linearity; the maximum DNL and INL are −0.97 and −1.64 LSB, respectively.

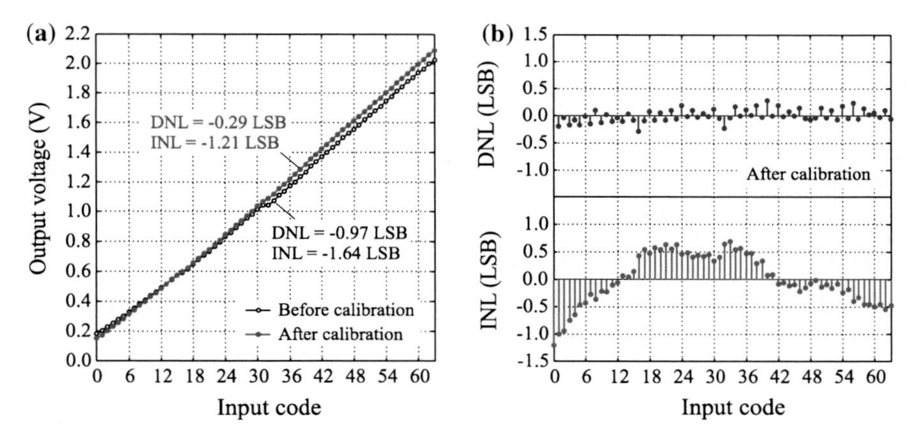

Fig. 8.9 Measured DC characteristics of the 6-bit binary current-steering DAC [1]. **a** Transfer characteristics before and after calibration. **b** Calculated DNL and INL errors after calibration

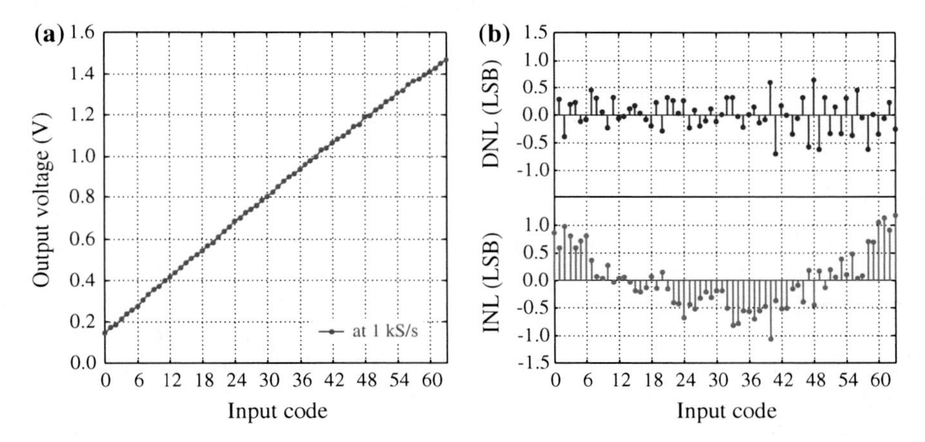

Fig. 8.10 Measured characteristics of the 6-bit binary current-steering DAC at an update rate of 1 S/s, after calibration, and at an output voltage swing of 1.3 V [1]. **a** Transfer characteristic. **b** Calculated DNL and INL errors

After calibration, however, the transfer function shows a monotonic behavior, where the maximum output voltage swing (V_{swing}) is 1.94 V with a power dissipation of 260 μW and an improved static linearity is attained. The maximum DNL and INL are −0.29 and −1.21 LSB, respectively.

Furthermore, Fig. 8.10 depicts the measured transfer function along with the calculated DNL and INL errors after calibration, at an output voltage swing of 1.3 V (at a reduced I_{LSB}), and at an update rate of 1 kS/s; the maximum DNL and INL are −0.69 and 1.16 LSB, respectively. Figure 8.11 illustrates the upper speed limit of the DAC in worst case conditions, i.e., the DAC is operated before calibration and slightly above the maximum output voltage swing ($\gtrsim$2 V). As shown in the figure,

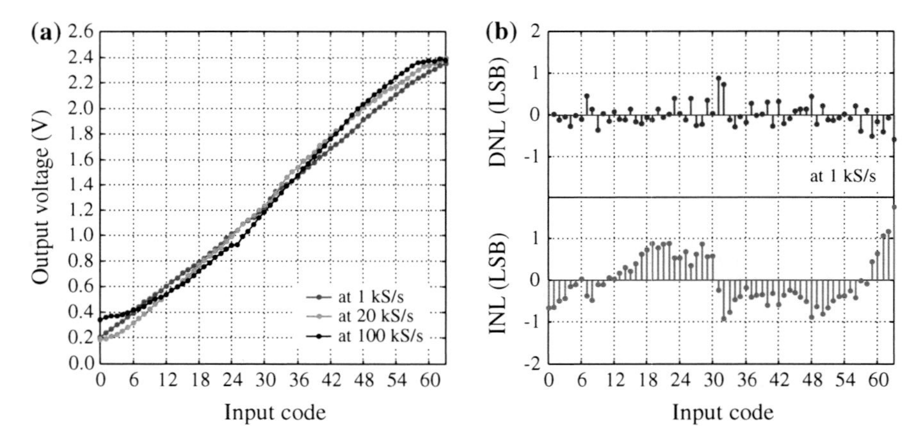

Fig. 8.11 Measured characteristics of the 6-bit binary current-steering DAC at an update rate of 1, 20 and 100 S/s [1]. The DAC is operating in worst case conditions: before calibration and slightly above the maximum output voltage swing ($\gtrsim$ 2 V). **a** Transfer characteristics. **b** Calculated DNL and INL errors at 1 S/s

as the update rate increases above ~20 kS/s, the measured transfer function exhibits higher distortion. At a swing of only 1 V and after calibration, the DAC would be able to operate at a maximum update rate as high as 100 kS/s without strong deterioration.

To measure the SFDR, 6-bit input streams with periods of 32 and 64 codes are prepared to generate sinusoidal signals by the DAC at update rates of 1, 10 and 100 kS/s. The maximum SFDR is found to be 34 dB at a signal frequency of 15.625 Hz (1 kS/s; 64 codes). Moreover, the minimum SFDR is found to be 19 dB at a signal frequency of 312.5 Hz (10 kS/s; 32 codes). Figure 8.12a shows the frequency spectrum of the measured sinusoidal signal at 3.125 kHz (100 kS/s; 32 codes); the SFDR is found to be 32 dB. Finally, Fig. 8.12b depicts the step response of the DAC at 10 S/s. From this data, the setting time, slew rate and glitch energy are found to be around 300 Hz, 0.029 V/μs and 300 μVs, respectively.

8.3.7 Benchmarking

Table 8.1 benchmarks the 6-bit binary current-steering DAC presented in this work against the 6-bit organic-based switched-capacitor (C-2C) DAC that was recently published in [29]. The comparison clearly shows that the current-steering DAC consumes 30 times smaller chip area and achieves 1000 times higher maximum update rate than the state of the art [1, 4]. These considerable improvements result from the OTFT fabrication process based on high-resolution silicon stencil masks that makes it possible to use transistors with smaller dimensions and far improved matching. In addition, the current-steering DAC achieves a maximum output voltage swing of

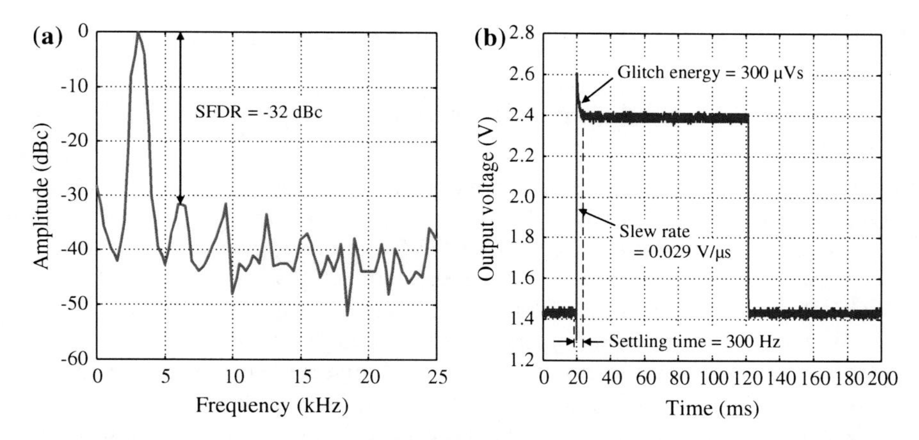

Fig. 8.12 **a** Measured output frequency spectrum of the 6-bit binary current-steering DAC at 3.125 kHz (update rate of 100 kS/s) [1]. **b** Measured step response at an update rate of 10 S/s

Table 8.1 Characteristics summary of the 6-bit binary current-steering DAC and benchmarking against the recently published organic-based 6-bit switched-capacitor (C-2C) DAC [1, 2, 4, 29]

Design parameter	[29]	This work
Architecture	Switched capacitance	Current steering
Technology	Complementary OTFTs	P-channel OTFTs
Resolution	6-bit	6-bit
Minimum feature	20 μm	4 μm
Transistor count	26	129
Capacitor count	17	0
Area consumption	$28 \times 14\,\text{mm}^2$	$2.6 \times 4.6\,\text{mm}^2$
Supply voltage	3 V	3.3 V
Maximum output swing	~1 V	~2 V
Update rate	10–100 S/s	DC–100 kS/s
Power-delay product	7 nWs (at 1 V swing)[a]	1.8 nWs (at 1 V swing)[b]
		2.6 nWs (at 2 V swing)
Maximum DNL	−0.6 LSB (at 100 S/s)	−0.69 LSB (at 1 kS/s)
Maximum INL	−0.8 LSB (at 100 S/s)	+1.16 LSB (at 1 kS/s)
SFDR	24 dB (at 10 Hz)	19 dB (at 312.5 Hz)
	29 dB (at 45 Hz)	32 dB (at 3.125 kHz)

[a]This is deduced from [29] at 100 S/s but the input bit stream is not stated
[b]This is measured at 100 kS/s and at the input bit stream "111111"

2 V, which is as high as double the swing generated by the C-2C DAC. Owing to the non-linear leakage error generated by the capacitors, the lower speed limit of the C-2C DAC is constricted to 10 S/s, while the current-steering DAC can operate at DC. Although the power dissipation of the current-steering DAC is relatively high (260 μW), the power-delay product is still smaller than that of the C-2C DAC.

8.4 Impact of Parasitics in a Complementary Design

8.4.1 A 3-Bit Unary Current-Steering DAC

The schematic diagram of the 3-bit unary current-steering DAC, including a binary-to-thermometer decoder, is depicted in Fig. 8.13 [3, 4]. The same step-by-step design flow presented in the previous section for the 6-bit binary current-steering DAC is used to calculate the dimensions of the current sources (100/15 μm), analog switches (40/4 μm) and load transistors (100/5 μm). Identical transistor dimensions as previously designed for the 6-bit binary current-steering DAC are utilized to ensure the same operating point. Accordingly, a row of drivers (level shifters) are placed between the decoder and the DAC to drive the SWA with an input ON voltage ($V_{sw,on}$) of 1 V, allowing the transistors to be operating in the desired saturation regime. However, one should note that a unary instead of a binary architecture is employed in this design, meaning that the current mirrors are identical and not binary-weighted. The decoder logic is implemented using a fully custom design as discussed in the following.

8.4.2 Binary-to-Thermometer Decoder

The DAC performance is largely determined by the design of the binary-to-thermometer decoder; therefore, the choice of the architecture is a crucial design decision [4]. There are basically three main architectures proposed in literature for the binary-to-thermometer decoders used in DACs, namely one-dimensional, two-dimensional (row/column) and multi-dimensional decoders; the selection depends on the complexity of the DAC and the number of transistors in the CSA/SWA [56]. Since our 3-bit unary current-steering DAC comprises only seven current mirrors, the simple one-dimensional decoder is the most preferable choice.

According to our review study presented in [56], there are four possible design techniques for a one-dimensional 3-bit binary-to-thermometer decoder, namely redundant-logic, ROM-based, multiplexed-based and iterative designs. By comparing the four techniques, the iterative design is superior in speed and compactness, and thus, it is selected for our DAC. The block diagram of the designed 3-bit binary-to-thermometer decoder is shown in Fig. 8.13. The logic gates are implemented in the standard complementary (dual) design [42], where $W_n = 10W_p$, W_p of an inverter is 80 μm and all channel lengths are 4 μm. This is to examine the feasibility of fabricating functional low-power complementary logic circuit using the OTFT technology and to investigate their performance.

As shown in the block diagram of the decoder (Fig. 8.13), a row of drivers (level shifters) are placed at the output. This is to drive the SWA with an input ON voltage ($V_{sw,on}$) of 1 V, allowing the transistors in the SWA to be operating in the desired saturation regime. These drivers are implemented using inverters with ground nodes that

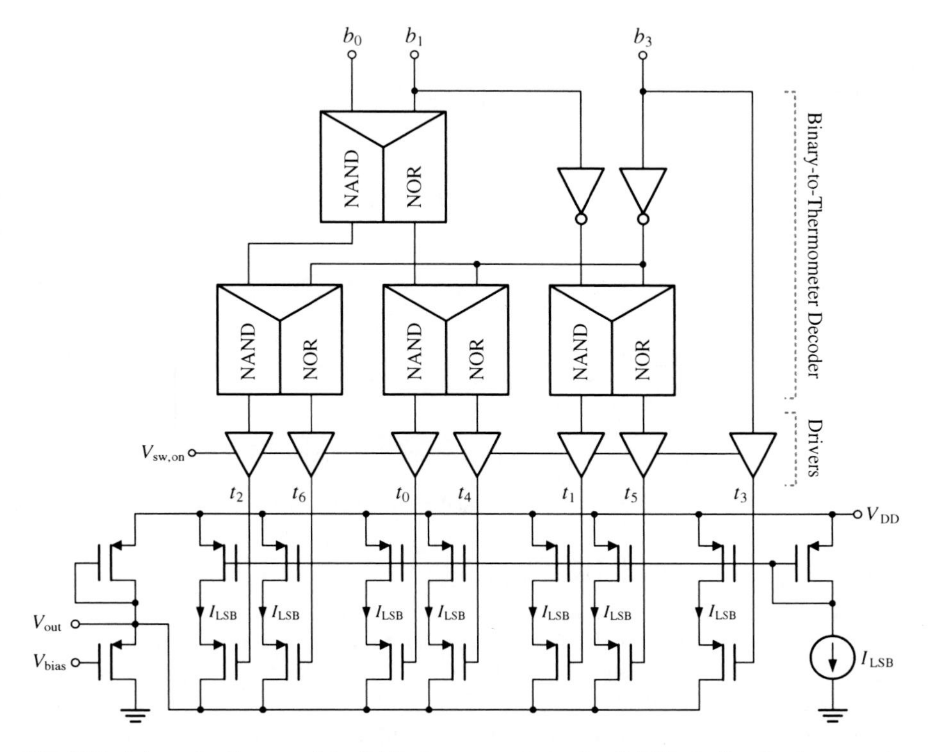

Fig. 8.13 Schematic diagram of the 3-bit unary current-steering DAC including the thermometer decoder driving complementary logic [3]

are replaced with an external supply voltage ($V_{sw,on}$). Furthermore, the output order of the decoder defines the switching scheme with which the unary current-sources are enabled to the DAC output. This switching scheme is used to compensate for errors that are introduced by systematical gradient effects (linear and/or symmetrical). A detailed analysis of the switching scheme is given in the following.

8.4.3 Switching Scheme

The switching scheme is defined to be the order of which the current sources within the CSA are switched to the DAC output [4, 55]. Conventionally, the current sources are switched in a sequential order. However, with careful layout, another switching order can be used to compensate the systematic errors that are introduced by linear and/or symmetrical gradient errors, which occur due to process, temperature or stress variations. These errors have to be considered because they contribute to the performance of the DAC. By reducing the impact of these systematic errors using a switching scheme, random errors will dominate. However, random errors are kept

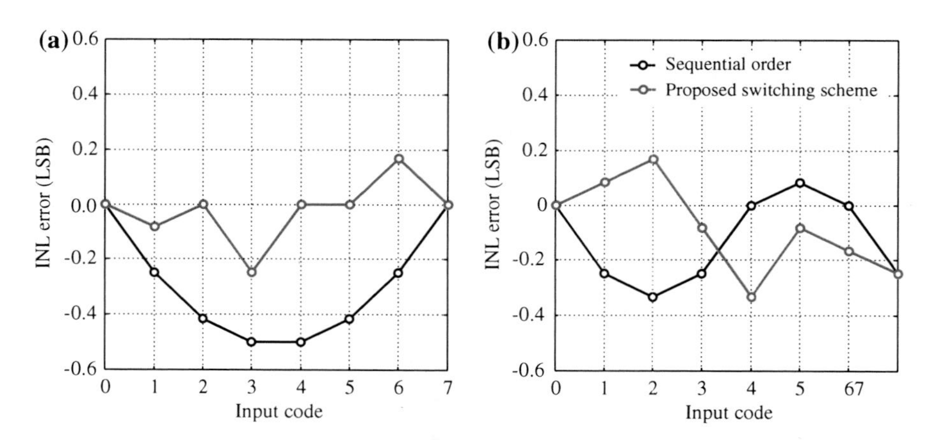

Fig. 8.14 Calculated INL error when using the sequential and the proposed switching schemes for the 3-bit unary current-steering DAC, assuming **a** linear gradient errors with 0.5 LSB peak-to-peak amplitude and **b** symmetrical gradient errors with 0.5 LSB peak-to-peak amplitude

within the desired boundary (INL and DNL < 1 LSB) by adjusting the dimensions of the active current-source transistors as discussed in the previous section.

The developed switching scheme (labelled $t_0, t_1, \ldots, t_6$) for our 3-bit unary current-steering DAC is depicted in Fig. 8.13. Given that the current-source transistors are arranged in a one-dimensional array, they are only influenced by lateral gradient errors. A comparison between the sequential order and our proposed scheme is shown in Fig. 8.14, assuming a peak-to-peak amplitude of 0.5 LSB for the linear as well as the symmetrical gradient errors. A smaller maximum INL is achieved by the proposed scheme in the case of linear gradient error, whereas the same maximum INL is provided in the case of symmetrical gradient error. The analysis presented in Sect. 4.2.4 suggests that the linear gradient error is dominant in our OTFT technology (see Figs. 4.6b and 4.10); this results from a tilt of the stencil mask during evaporation and/or the sample is outside the perpendicular projection of the evaporator source during material deposition. Nevertheless, if the linear and symmetrical gradient errors are present (by adding both components in Fig. 8.14), the proposed scheme still yields a much smaller maximum INL of -0.33 LSB compared to -0.75 LSB for the conventional sequential order.

8.4.4 Measurement and Simulation Results

The 3-bit unary current-steering DAC is fabricated on a glass substrate with a total area of $3.1 \times 6.6\,\text{mm}^2$. The area is optimized at the cost of a large number of metal crossings ($\sim$70) with about 12 pF capacitance each; this is to investigate the impact of interconnect parasitics on the circuit performance. The layout of the DAC is presented in Appendix A. Measurements are performed in ambient air and at a supply voltage

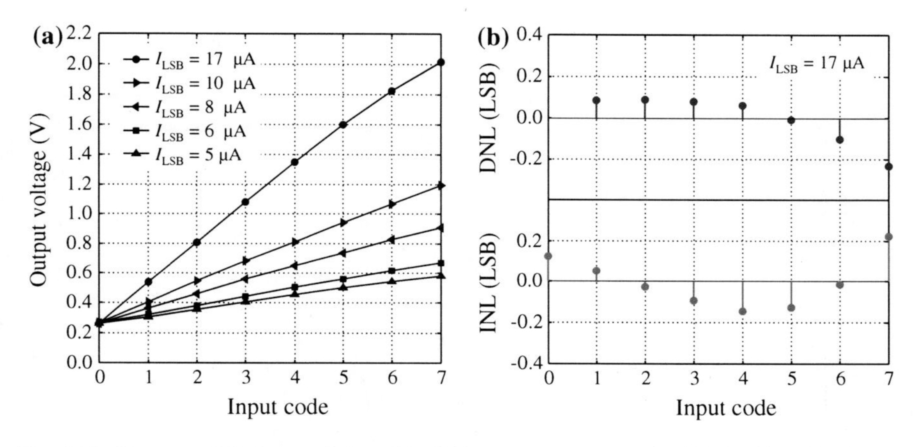

Fig. 8.15 Measured DC characteristics of the 3-bit unary current-steering DAC. **a** Transfer characteristics for different applied least-significant-bit currents (I_{LSB}). **b** Calculated DNL and INL errors when $I_{LSB} = 17\,\mu A$

of 3 V using the same test setup presented in the previous section. Figure 8.15 shows the measured DC transfer function together with the calculated DNL and INL errors. The maximum output voltage swing (V_{swing}) of the DAC is 1.75 V; in addition, the maximum DNL and INL errors are 0.23 and 0.22 LSB, respectively.

The maximum possible performance of the circuit in this case is limited by the complementary driving logic. This is attributed to the slow n-channel OTFTs and the large number of metal crossings utilized. Therefore, a case study is carried out on the binary-to-thermometer decoder to investigate the impact of the device and interconnect parasitic capacitances on the circuit performance, in particular [3].

Transient measurements and simulations of the binary-to-thermometer decoder at the output node t_4 at frequencies of 100 , 500 and 1 kHz are depicted in Figs. 8.16 and 8.17. In fact, the node t_4 is the output of the slowest logic path in the circuit (see Fig. 8.13), also known as the critical path.[1] The simulations are based on a DC/AC SPICE model for both p- and n-channel OTFTs (adopted from Chaps. 5 and 6). The frequency 1 kHz is almost the speed limit where a recognizable output can be seen. The output graphs show that the high-to-low and low-to-high signal propagation delay times are $t_{delay,HL} \cong 2$ ms and $t_{delay,LH} \cong 0.7$ ms, respectively. The output low potential is 1 V because of the offset voltage ($V_{sw,on}$) set by the drivers (see Fig. 8.13). The nearly perfect agreement of the simulated to the measured data in Fig. 8.16 shows that the DC/AC model is trustworthy. Hence, reliable prospective data for speed improvement can be deduced from the simulations.

Accordingly, Fig. 8.17 shows further transient simulation results for six different cases at 500 Hz to study the speed limiting factors of this complementary circuit. Case 1, which is similar to the simulations shown in Fig. 8.16, yielded a signal delay of 1.94 ms. In this case, the OTFTs have a gate-to-contact overlap length

[1] The critical path is the longest synthesizable path of which the largest delay of the circuit/network occurs from the input to the output.

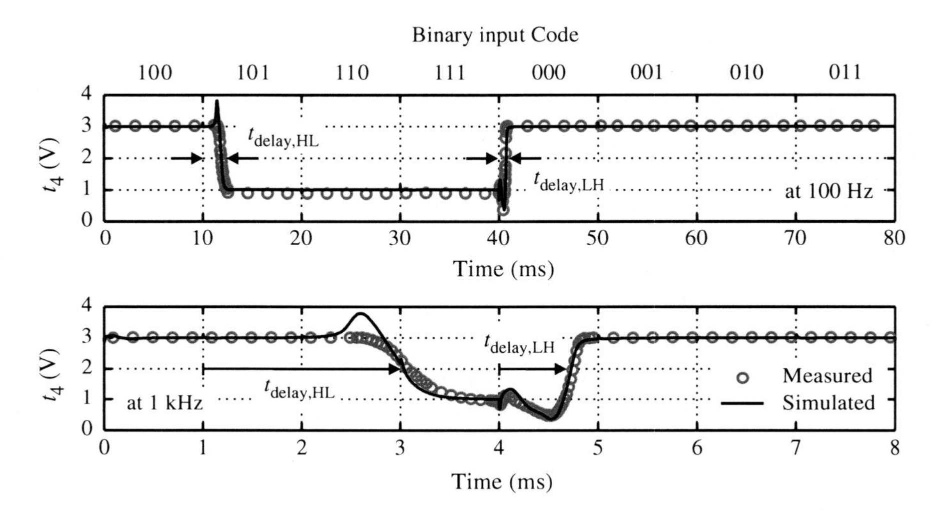

Fig. 8.16 Measured and simulated output of the binary-to-thermometer decoder circuit at node t_4 at (upper) 100 Hz and (lower) 1 kHz [3]

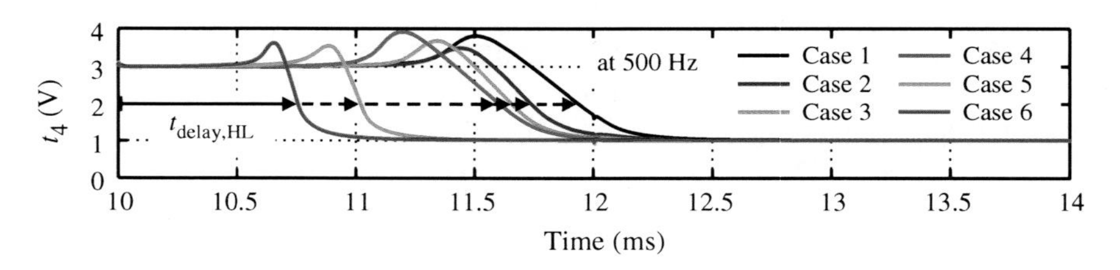

Fig. 8.17 Simulation case study to investigate the impact of device and interconnect parasitics as well as the influence of using the slow n-channel OTFTs on the performance of the binary-to-thermometer decoder circuit at 500 Hz

(L_{ov}) of 20 µm, which result in considerably large device parasitic capacitances particularly for the large n-channel OTFTs. Also limiting the circuit performance significantly is the low carrier mobility in the channel region of the n-channel OTFTs ($\mu_n = 0.02\,\text{cm}^2/\text{Vs}$). Furthermore, the parasitic capacitances of the metal crossings in this case are substantially large (about 12 pF each). By reducing only the device parasitics ($L_{ov} = 5\,\mu\text{m}$) in case 2, the signal delay is improved to 1.74 ms. In general, there is a tradeoff between the contact resistances and the overlap capacitances when reducing the gate overlap. Thus, for an optimized gate overlap ($L_{ov} = 10\,\mu\text{m}$) in case 3, the delay is further reduced to 1.65 ms. On the other hand, the reduction of only the metal parasitics by a factor of 10 in case 4 resulted in a signal delay of 1.6 ms. This indicates that the impact of metal crossings is as detrimental as the device parasitics. Furthermore, the delay could be even reduced to 1 ms, as shown in case 5, when the carrier mobility of the n-channel OTFTs is improved to $\mu_n = 1\,\text{cm}^2/\text{Vs}$ [60].

Table 8.2 Summary of the propagation delay times for the simulated case study of the binary-to-thermometer decoder [3]

Case	Description	$t_{\text{delay, HL}}$ (ms)
1	Reference case with $L_{\text{ov}} = 20\,\mu\text{m}$ and $\mu_{\text{n}} = 0.02\,\text{cm}^2/\text{Vs}$	1.94
2	Reduced device parasitics using $L_{\text{ov}} = 5\,\mu\text{m}$	1.74
3	Optimized device parasitics using $L_{\text{ov}} = 10\,\mu\text{m}$	1.65
4	Reduced parasitics of interconnect crossings by 10 times	1.60
5	Using faster n-channel OTFTs with $\mu_{\text{n}} = 1\,\text{cm}^2/\text{Vs}$	1.00
6	Combination of cases 3, 4 and 5	0.75

Finally, case 6 combines cases 3, 4 and 5 to achieve a minimum signal delay of 0.75 ms, meaning that the circuit speed would be more than doubled. The extracted propagation delay times for the six simulation cases are summarized in Table 8.2.

From this study, it can be concluded that the reduction of the parasitic capacitances of the devices and metal crossings, which is achievable by means of mask overlay and layout advancements, already leads to promising performance improvements. However, to achieve even higher speeds, the complex relationships between the channel length, gate-overlap length, contact resistance, overlap capacitance and field-effect mobility has to be carefully investigated as demonstrated in [37] and [39]. Moreover, further improvements have to come from a higher carrier mobility in the n-channel OTFTs, allowing for smaller transistors and thus even lower parasitic device capacitances.

8.5 Summary

A record in performance and compactness for organic integrated circuits is introduced in this work by using the OTFT fabrication process that is based on high-resolution silicon stencil masks. The process offers submicrometer channel lengths and superb transistor matching, thus allowing to exploit circuit concepts and topologies that would not be feasible otherwise. Accordingly, the first current-steering DACs built in an OTFT technology are demonstrated. A 6-bit binary current-steering DAC using p-channel OTFTs is realized; measurement results show that this converter is 1000 times faster, 30 times smaller and achieves 2 times higher output swing than prior state of the art. The design of the DAC includes a new highly linear current-to-voltage converter, which employs only two p-channel OTFTs and also compensates for the inherent non-linearity of the transistors. The converter has a maximum update rate of 100 kS/s, a maximum output voltage swing of 2 V and consumes an area of $2.6 \times 4.6\,\text{mm}^2$. Moreover, another 3-bit current-steering DAC, but using the unary architecture and complementary OTFTs, is fabricated on the same substrate. This converter is used as a circuit test vehicle in a simulation case study to indicate advancement opportunities for the low-voltage complementary OTFT technology. The study is based on an accurate agreement of simulated and measured data up to 1 kHz, which is the maximum operating frequency of the DAC; this is owed to the optimized DC/AC SPICE models for both p- and n-channel OTFTs. Simulation results indicate that considerable improvements in circuit performance may result from (i) optimizing drain and source overlaps with the gate electrode, (ii) avoiding metal-interconnect crossings in the circuit layout or reducing their corresponding parasitics through better dielectrics, and (iii) developing low-voltage n-channel OTFTs with a higher charge carrier mobility.

References

1. T. Zaki, F. Ante, U. Zschieschang, J. Butschke, F. Letzkus, H. Richter, H. Klauk, J.N. Burghartz, A 3.3 V 6-bit 100 kS/s current-steering digital-to-analog converter using organic p-type thin-film transistors on glass. IEEE J. Solid-State Circuits **47**(1), 292–300 (2012)
2. T. Zaki, F. Ante, U. Zschieschang, J. Butschke, F. Letzkus, H. Richter, H. Klauk, J.N. Burghartz, A 3.3V 6b 100 kS/s current-steering D/A converter using organic thin-film transistors on glass, in *IEEE International Solid-State Circuits Conference Technical Digest*, Feb 2011, pp. 324–325
3. T. Zaki, F. Ante, U. Zschieschang, J. Butschke, F. Letzkus, H. Richter, H. Klauk, J.N. Burghartz, Circuit impact of device and interconnect parasitics in a complementary low-voltage organic thin-film technology, in *Semiconductor Conference Dresden*, Sep 2011, pp. 1–4
4. T. Zaki, Design of current-steering D/A converters using organic thin-film transistors enabled by high-resolution silicon stencil masks, M.Sc. thesis, University of Stuttgart, Stuttgart, BW, Germany, Oct 2010
5. U. Zschieschang, R. Rödel, U. Kraft, K. Takimiya, T. Zaki, F. Letzkus, J. Butschke, H. Richter, J.N. Burghartz, W. Xiong, B. Murmann, H. Klauk, Low-voltage organic transistors for flexible electronics, in *Proceedings of the Conference on Design, Automation and Test in Europe*, Mar 2014, pp. 1–6
6. A. Dodabalapur, Organic and polymer transistors for electronics. Mater. Today **9**(4), 24–30 (2006)
7. G.H. Gelinck, H.E.A. Huitema, E. van Veenendaal, E. Cantatore, L. Schrijnemakers, J.B.P.H. van der Putten, T.C.T. Geuns, M. Beenhakkers, J.B. Giesbers, B.-H. Huisman, E.J. Meijer, E.M. Benito, F.J. Touwslager, A.W. Marsman, B.J.E. van Rens, D.M. de Leeuw, Flexible active-matrix displays and shift registers based on solution-processed organic transistors. Nat. Mater. **3**, 106–110 (2004)
8. I. Yagi, N. Hirari, Y. Miyamoto, M. Noda, A. Imaoka, N. Yoneya, K. Nomoto, J. Kasahara, A. Yumoto, T. Urabe, A flexible full-color AMOLED display driven by OTFTs. J. Soc. Inf. Display **16**(1), 15–20 (2008)
9. M. Noda, N. Kobayashi, M. Katsuhara, A. Yumoto, S.-I. Ushikura, R.-I. Yasuda, N. Hirai, G. Yukawa, I. Yagi, K. Nomoto, T. Urabe, A rollable AM-OLED display driven by OTFTs. Soc. Inf. Disp. Symp. Tech. Digest **41**(1), 710–713 (2010)
10. T. Sekitani, T. Yokota, U. Zschieschang, H. Klauk, S. Bauer, K. Takeuchi, M. Takamiya, T. Sakurai, T. Someya, Organic nonvolatile memory transistors for flexible sensor arrays. Science **326**(5959), 1516–1519 (2009)
11. K. Fukuda, T. Sekitani, U. Zschieschang, H. Klauk, K. Kuribara, T. Yokota, T. Sugino, K. Asaka, M. Ikeda, H. Kuwabara, T. Fukushima, T. Aida, M. Takamiya, T. Sakurai, T. Someya, A 4 V operation, flexible braille display using organic transistors, carbon nanotube actuators, and organic static random-access memory. Adv. Funct. Mater. **21**(21), 4019–4027 (2011)

12. W. Zhang, M. Ha, D. Braga, M.J. Renn, C.D. Frisbie, C.H. Kim, A 1V printed organic DRAM cell based on ion-gel gated transistors with a sub-10nW-per-cell refresh power, in *IEEE International Solid-State Circuits Conference Technical Digest*, Feb 2011, pp. 326–328
13. E. Cantatore, T.C. Geuns, A.F.A. Gruijthuijsen, G.H. Gelinck, S. Drews, D.M. de Leeuw, A 13.56MHz RFID system based on organic transponders, in *IEEE International Solid-State Circuits Conference Technical Digest*, Feb 2006, pp. 1042–1051
14. E. Cantatore, T.C.T. Geuns, G.H. Gelinck, E. van Veenendaal, A.F.A. Gruijthuijsen, L. Schrijnemakers, S. Drews, D.M. de Leeuw, A 13.56-MHz RFID system based on organic transponders. IEEE J. Solid-State Circuits **42**(1), 84–92 (2007)
15. K. Myny, S. van Winckel, S. Steudel, P. Vicca, S. de Jonge, M.J. Beenhakkers, C.W. Sele, N.A.J.M. van Aerle, G.H. Gelinck, J. Genoe, P. Heremans, An inductively-coupled 64b organic RFID tag operating at 13.56MHz with a data rate of 787b/s, in *IEEE International Solid-State Circuits Conference Technical Digest*, Feb 2008, pp. 290–614
16. K. Myny, S. Steudel, P. Vicca, M.J. Beenhakkers, N.A.J.M. van Aerle, G.H. Gelinck, J. Genoe, W. Dehaene, P. Heremans, Plastic circuits and tags for 13.56 MHz radio-frequency communication. Solid-State Electron. **53**(12), 1220–1226 (2009)
17. R. Blache, J. Krumm, W. Fix, Organic CMOS circuits for RFID applications, in *IEEE International Solid-State Circuits Conference Technical Digest*, Feb 2009, pp. 208–209
18. K. Myny, S. Steudel, S. Smout, P. Vicca, F. Furthner, B. van der Putten, A.K. Tripathi, G.H. Gelinck, J. Genoe, W. Dehaene, P. Heremans, Organic RFID transponder chip with data rate compatible with electronic product coding. Org. Electron. **11**(7), 1176–1179 (2010)
19. A.P.F. Turner, Biosensors: sense and sensibility. Chem. Soc. Rev. **42**(8), 3175–3648 (2013)
20. T. Someya, A. Dodabalapur, A. Gelperin, H.E. Katz, Z. Bao, Integration and response of organic electronics with aqueous microfluidics. Langmuir **18**(13), 5229–5302 (2002)
21. B. Crone, A. Dodabalapur, A. Gelperin, L. Torsi, H.E. Katz, A.J. Lovinger, Z. Bao, Electronic sensing of vapors with organic transistors. Appl. Phys. Lett. **78**(15), 2229–2231 (2001)
22. T. Someya, Y. Kato, T. Sekitani, S. Iba, Y. Noguchi, Y. Murase, H. Kawaguchi, T. Sakurai, Conformable, flexible, large-area networks of pressure and thermal sensors with organic transistor active matrixes. Proc. Natl. Acad. Sci. USA **102**(35), 12321–12325 (2005)
23. M. Kaltenbrunner, T. Sekitani, J. Reeder, T. Yokota, K. Kuribara, T. Tokuhara, M. Drack, R. Schwödiauer, I. Graz, S. Bauer-Gogonea, S. Bauer, T. Someya, An ultra-lightweight design for imperceptible plastic electronics. Nature **499**, 458–463 (2013)
24. H. Fuketa, K. Yoshioka, Y. Shinozuka, K. Ishida, T. Yokota, N. Matsuhisa, Y. Inoue, M. Sekino, T. Sekitani, M. Takamiya, T. Someya, T. Sakurai, 1μm-thickness 64-channel surface electromyogram measurement sheet with 2V organic transistors for prosthetic hand control, in *IEEE International Solid-State Circuits Conference Technical Digest*, Feb 2013, pp. 104–105
25. T. Someya, T. Sekitani, M. Kaltenbrunner, T. Yokota, H. Fuketa, M. Takamiya, T. Sakurai, Ultraflexible organic devices for biomedical applications, in *IEEE International Electron Devices Meeting Technical Digest*, Dec 2013, pp. 8.5.1–8.5.4
26. H. Marien, M. Steyaert, E. van Veenendaal, P. Heremans, DC-DC converter assisted two-stage amplifier in organic thin-film transistor technology on foil, in *Proceedings of the European Solid-State Circuits Conference*, Sep 2011, pp. 411–414
27. D. Raiteri, F. Torricelli, E. Cantatore, A.H.M. van Roermund, A tunable transconductor for analog amplification and filtering based on double-gate organic TFTs, in *Proceedings of the European Solid-State Circuits Conference*, Sep 2011, pp. 415–418
28. K. Fukuda, T. Sekitani, T. Yokota, K. Kuribara, T.-C. Huang, T. Sakurai, U. Zschieschang, H. Klauk, M. Ikeda, H. Kuwabara, T. Yamamoto, K. Takimiya, K.-T. Cheng, T. Someya, Organic pseudo-CMOS circuits for low-voltage large-gain high-speed operation. IEEE Electron Device Lett. **32**(10), 1448–1450 (2011)
29. W. Xiong, Y. Guo, U. Zschieschang, H. Klauk, B. Murmann, A 3-V 6-Bit C-2C digital-to-analog converter using complementary organic thin-film transistors on glass. IEEE J. Solid-State Circuits **45**(7), 1380–1388 (2010)
30. H. Marien, M.S.J. Steyaert, E. van Veenendaal, P. Heremans, A fully integrated $\Delta\Sigma$ ADC in organic thin-film transistor technology on flexible plastic foil. IEEE J. Solid-State Circuits **46**(1), 276–284 (2011)

31. W. Xiong, U. Zschieschang, H. Klauk, B. Murmann, A 3V 6b successive-approximation ADC using complementary organic thin-film transistors on glass, in IEEE *International Solid-State Circuits Conference Technical Digest*, Feb 2010, pp. 134–135
32. T.-C. Huang, K. Fukuda, C.-M. Lo, Y.-H. Yeh, T. Sekitani, T. Someya, K.-T. Cheng, Pseudo-CMOS: A design style for low-cost and robust flexible electronics. IEEE Trans. Electron Devices **58**(1), 141–150 (2011)
33. P. Heremans, W. Dehaene, M. Steyaert, K. Myny, H. Marien, J. Genoe, G. Gelinck, E. van Veenendaal, Circuit design in organic semiconductor technologies, in *Proceedings of the European Solid-State Device Research Conference*, Sep 2011, pp. 5–12
34. H. Klauk, U. Zschieschang, J. Pflaum, M. Halik, Ultralow-power organic complementary circuits. Nature **445**, 745–748 (2007)
35. U. Zschieschang, F. Ante, D. Kälblein, T. Yamamoto, K. Takimiya, H. Kuwabara, M. Ikeda, T. Sekitani, T. Someya, J. Blochwitz-Nimoth, H. Klauk, Dinaphtho[2,3-b:2',3'-f]thieno[3,2-b]thiophene (DNTT) thin-film transistors with improved performance and stability. Org. Electron. **12**(8), 1370–1375 (2011)
36. R. Rödel, F. Letzkus, T. Zaki, J.N. Burghartz, U. Kraft, U. Zschieschang, K. Kern, H. Klauk, Contact properties of high-mobility, air-stable, low-voltage organic n-channel thin-film transistors based on a naphthalene tetracarboxylic diimide. Appl. Phys. Lett. **102**(233303), 233303-1–233303-5 (2013)
37. F. Ante, D. Kälblein, T. Zaki, U. Zschieschang, K. Takimiya, M. Ikeda, T. Sekitani, T. Someya, J.N. Burghartz, K. Kern, H. Klauk, Contact resistance and megahertz operation of aggressively scaled organic transistors. Small **8**(1), 73–79 (2012)
38. F. Letzkus, T. Zaki, F. Ante, J. Butschke, H. Richter, H. Klauk, J. N. Burghartz, Si stencil masks for organic thin film transistor fabrication, in *Proceedings of the SPIE Photomask Technology*, Sep 2011, pp. 81662B-1-81662B-12
39. T. Zaki, R. Rödel, F. Letzkus, H. Richter, U. Zschieschang, H. Klauk, J.N.Burghartz, S-parameter characterization of submicrometer low-voltage organic thin-film transistors. IEEE Electron Device Lett. **34**(4), 520–522 (2013)
40. E. Cantatore, E.J. Meijer, Transistor operation and circuit performance in organic electronics, in *IEEE International Solid-State Circuits Conference Technical Digest*, Sep 2003, pp. 29–36
41. H. Klauk, M. Halik, U. Zschieschang, G. Schmid, W. Radlik, Polymer gate dielectric pentacene TFTs and circuits on flexible substrates, in *IEEE International Electron Devices Meeting Technical Digest*, Dec 2002, pp. 557–560
42. S.-M. Kang, Y. Leblebici, *CMOS Digital Integrated Circuits: Analysis and Design*, 3rd edn. (McGraw-Hill, New York, 2003)
43. I. Nausieda, K.K. Ryu, D.D. He, A.I. Akinwande, V. Bulović, Dual threshold voltage organic thin-film transistor technology. IEEE Trans. Electron Devices **57**(11), 3027–3032 (2010)
44. K.P. Pernstich, S. Haas, D. Oberhoff, C. Goldmann, D.J. Gundlach, B. Batlogg, A.N. Rashid, G. Schitter, Threshold voltage shift in organic field effect transistors by dipole monolayers on the gate insulator. J. Appl. Phys. **96**(11), 6431–6438 (2004)
45. T. Yokota, T. Nakagawa, T. Sekitani, Y. Noguchi, K. Fukuda, U. Zschieschang, H. Klauk, K. Takeuchi, M. Takamiya, T. Sakurai, T. Someya, Control of threshold voltage in low-voltage organic complementary inverter circuits with floating gate structures. Appl. Phys. Lett. **98**(19), 1933021–1933023 (2011)
46. K. Myny, M.J. Beenhakkers, N.A.J.M. van Aerle, G.H. Gelinck, J. Genoe, W. Dehaene, P. Heremans, Unipolar organic transistor circuits made robust by dual-gate technology. IEEE J. Solid-State Circuits **46**(5), 1223–1230 (2011)
47. Y. Chung, O. Johnson, M. Deal, Y. Nishi, B. Murmann, Z. Bao, Engineering the metal gate electrode for controlling the threshold voltage of organic transistors. Appl. Phys. Lett. **101**(6), 0633041–0633044 (2012)
48. K. Myny, E. van Veenendaal, G.H. Gelinck, J. Genoe, W. Dahaene, P. Heremans, An 8b organic microprocessor on plastic foil, in *IEEE International Solid-State Circuits Conference Technical Digest*, Feb 2011, pp. 322–324

49. K. Myny, E. van Veenendaal, G.H. Gelinck, J. Genoe, W. Dahaene, P. Heremans, An 8-bit, 40-instructions-per-second organic microprocessor on plastic foil. IEEE J. Solid-State Circuits **47**(1), 284–291 (2012)
50. T.-C. Huang, K. Fukuda, C.-M. Lo, Y.-H. Yeh, T. Sekitani, T. Someya, K.-T. Cheng, Pseudo-CMOS: a novel design style for flexible electronics, in *Design, Automation and Test in Europe Conference and Exhibition*, Mar 2010, pp. 154–159
51. T.-C. Huang, K.-T. Cheng, Design for low power and reliable flexible electronics: Self-tunable cell-library design. J. Display Technol. **5**(6), 206–214 (2009)
52. F. Ante, Contact effects in organic transistors, Ph.D. dissertation, Swiss Federal Institute of Technology in Lausanne, Lausanne, Switzerland, Dec 2011
53. M. Elsobky, Measurements and characterization of low-voltage organic transistors and digital logic gates, B.Sc. thesis, University of Stuttgart, Stuttgart, BW, Germany, Aug 2013
54. M. Elattar, Design of an organic-based driver circuit for smart skin applications, B.Sc. thesis, University of Stuttgart, Stuttgart, BW, Germany, Aug 2014
55. A. Van den Bosch, M. Steyaert, W. Sansen, *Static and Dynamic Performance Limitations for High Speed D/A Converters* (Kluwer Academic Publishers, Boston, 2004)
56. T. Zaki, A compact 12-bit current-steering D/A converter for HDRC camera systems, Institute for Microelectronics Stuttgart, Technical Report, 2010
57. M. Gustavsson, J.J. Wikner, N.N. Tan, *CMOS Data Converters for Communications* (Springer, New York, 2000)
58. J. Bastos, A.M. Marques, M.S.J. Steyaert, W. Sansen, A 12-bit intrinsic accuracy high-speed CMOS DAC. IEEE J. Solid-State Circuits **33**(12), 1959–1969 (1998)
59. K. O'Sullivan, C. Gorman, M. Hennessy, V. Callaghan, A 12-bit 320-MSample/s current-steering CMOS D/A converter in 0.44 mm^2. IEEE J. Solid-State Circuits **39**(7), 1064–1072 (2004)
60. R. Schmidt, J.H. Oh, Y.-S. Sun, M. Deppisch, A.-M. Krause, K. Radacki, H. Braunschweig, M. Könemann, P. Erk, Z. Bao, F. Würthner, High-performance air-stable n-channel organic thin film transistors based on halogenated perylene bisimide semiconductors. J. Am. Chem. Soc. **131**(17), 6215–6228 (2009)

Chapter 9
Conclusion

In this final chapter we draw a series of conclusions on key aspects of this thesis. We also make reference to our own publications in which parts of the results presented herein have been published previously. Organic thin-film transistors (OTFTs) are making significant inroads into flexible, large-area electronics. Together with IMS CHIPS and MPI-SSR, a new OTFT fabrication process based on high-resolution silicon stencil masks is developed, offering submicrometer channel lengths and far improved transistor matching [1–5]. This work is set out to implement a unified DC/AC compact model for the OTFTs and to demonstrate fast digital and analog organic integrated circuits. It is also sought to determine the matching capability of this new OTFT technology, to investigate the impact of parasitics on the device performance and to characterize the frequency response of individual OTFTs using a self-contained method.

Technology—The OTFTs employ air-stable, small-molecule organic semiconductors (e.g. DNTT and $F_{16}CuPc$) and exploit the inverted-staggered (bottom-gate, top-contact) configuration. The use of silicon stencil masks allows to pattern the source/drain contacts on top of the organic semiconductors without the need of solvents or elevated temperatures, which would make the small molecules prone to degradation. Furthermore, the OTFTs take full advantage of combining an oxide layer (AlO_x) with a molecular self-assembled monolayer (C14-SAM) to form a 5.3-nm-thick gate dielectric with a capacitance of the order of 1 µF/cm^2. This relatively high capacitance enables the transistors to operate at voltages as low as 3 V, at which a sufficiently high carrier concentration (10^{12}–10^{13} cm^{-2}) can be reached to create the channel. The low-voltage operation competence makes the OTFTs well-suited for battery-powered or frequency-coupled portable devices and opens the possibility to build hybrid flexible systems-in-foil (SiF), integrating large-area organic electronics with high-performance thin silicon chips.

Device Matching—The uniformity of the evaporated source/drain contacts through the stencil masks is investigated by measuring the critical dimensions (CDs) of 35 0.8-µm-wide isolated lines dispersed over the entire area of a test sample. The mean values of the CDs are found to be 0.88 ± 0.01 µm for the mask and 0.86 ± 0.19 µm for the contacts [3]. On closer inspection, a systematic variation is observed on

T. Zaki, *Short-Channel Organic Thin-Film Transistors*, Springer Theses,
DOI 10.1007/978-3-319-18896-6_9

the sample, indicating a tilt of the mask during evaporation and/or a displacement of the sample from the perpendicular projection of the evaporator source. This can be alleviated by installing an automated handling system in the future. Nevertheless, electrical characterization of 16 OTFTs fabricated by the silicon stencil masks with a channel length of 2 μm yields a standard deviation of only 4 % for the drain current, which is less than half the value measured for 16 OTFTs fabricated by conventional polyimide shadow masks but with a much larger channel length of 30 μm [6]. This eminent improvement enables to exploit circuit concepts, such as the current-steering DACs and ADCs that are known to be superior in speed and compactness, for which device matching is paramount.

Static Model—For the design of organic integrated circuits, a reliable compact model describing the steady-state characteristics of the OTFTs is highly desired. It is reported in literature that some of the OTFTs experience mobility enhancement at higher gate voltages. Conversely, the measured OTFTs display a fairly constant mobility; in fact, it reduces by just 10 % at high gate voltages due to bias stress. This makes the ensuing models based on that special mobility-enhancement behavior not adequate. Therefore, a simple unified model of the static I–V characteristics that covers all operating regimes, namely subthreshold, linear and saturation, is derived. Effects induced by the voltage-dependent parasitic resistances are carefully considered [7–9]. A precise fit of the model to the measurements of various OTFTs is achieved. Being geometry scalable, the model typically requires a single parameter set for the simulation of OTFTs with different feature sizes. There is still, however, noticeable process variations and time-dependent characteristics of OTFTs [10], which necessitate the consideration of some variability in the model parameters.

Dynamic Model—To analyse the charge storage behavior of the OTFTs, admittance (Y) measurements have been conducted at frequencies from 100 up to 1 MHz. As anticipated, the cutoff frequency (f_T) of the OTFTs tends to increase with decreasing channel length and with increasing gate bias. Because of the distributed coupling between the gate electrode and the semiconductor, a small-signal equivalent circuit based on an RC network is developed to accurately model the frequency dispersion effects; a perfect fit to the measurements over the complete frequency range is obtained [11]. Furthermore, an in-depth analysis of the intrinsic capacitances have indicated that at $V_{DS} = 0$ V, C_{gs} and C_{gd} are both equal to $1/2 \cdot C_I WL$, while at $V_{DS} \geq V_{GS} - V_{TH}$, C_{gs} and C_{gd} are equal to $2/3 \cdot C_I WL$ and zero, respectively. Thus, it is demonstrated that the channel charge dynamics can be well described by the unified Meyer's capacitance model [11]. These results in conjunction with recently published findings in [12] establish an essential modeling support for dynamic simulations and performance optimization of organic integrated circuits.

Record Performance—The ultimate frequency performance of OTFTs relies heavily on the combination of material properties, device configuration and dimensions. The propagation delay per stage of ring oscillators is commonly used as a practical benchmark for the frequency response of organic integrated circuits [5, 8, 9]. Given the variability of most OTFTs, ring oscillators provide only an average delay figure. Hence, the cutoff frequency (f_T) of stand-alone OTFTs are also evaluated in this work for the first time using S-parameter measurements.

The relation between the cutoff frequency and the channel length as well as the gate overlap is physically modeled and a good agreement to the experimental data is obtained. A record cutoff frequency as high as 3.7 MHz has been measured at $V_{GS} = -3$ V for a DNTT OTFT with $L = 0.6$ µm and $L_{ov} = 5$ µm [13]. These results justify that the OTFT technology can be well extended to the submicrometer range and that high-frequency characterization of OTFTs by means of S-parameter measurements is feasible.

Asymmetric OTFTs—Although the use of fully-patterned OTFTs minimizes the gate overlaps, the usual lack of a self-alignment process makes the overlap capacitances inevitable. It is likewise very difficult to avoid a mismatch between the source/drain contact overlaps because of the limited alignment capability of most OTFT technologies. For these reasons, the frequency response of asymmetric OTFTs, particularly those with small dimensions ($L = 1$ and $L_{ov} = 5$ µm), is examined. The intentional misalignment is facilitated by the high fidelity of the silicon stencil masks. It is thereby revealed that underestimation of the cutoff frequency is probable when assumptions, such as $C_{gd} \ll C_{gs}$ for a typical FET device, are incorporated [14].

Integrated Circuits—Owing to the remarkable uniformity and reproducibility of the fabricated OTFTs, the first organic current-steering DACs are reported in this work. A 6-bit binary DAC, comprising as many as 129 p-channel DNTT OTFTs, is implemented. The converter has a maximum update rate of 100 kS/s, a maximum output swing of 2 V, and consumes an area of $2.6 \times 4.6\,\text{mm}^2$, establishing the state of the art for organic circuits [6, 15]. Lately, an OTFT-driven flexible AM-OLED display using air-stable DNTT has been demonstrated in [16]. The adoption of the same organic semiconductor enables further integration of the DAC and the display backplane on the same foil, resulting in potential cost savings and improved reliability. The exceptional OTFT properties and DAC performance commensurate with the ever-increasing demands of having AM-OLED displays with flexible form factor, higher resolution and faster refresh rate.

While for the 6-bit DAC the use of both n-channel OTFTs and metal-interconnect crossings could be avoided, this is quite often not possible as low-power circuit design calls for a complementary technology and complex interconnect schemes. Therefore, a study of the impact of device and interconnect parasitics on the circuit performance is carried out, for which a 3-bit unary DAC as a circuit test vehicle is utilized. The converter has a maximum update rate of 1 kS/s, consumes an area of 3.1×6.6 mm^2, and incorporates 42 p-channel and 25 n-channel OTFTs [7]. The key findings of the study are as follows. First, significant reduction of the gate-overlap lengths can be counterproductive as there is a tradeoff between the overlap capacitances and the contact resistances. Second, the influence of the interconnect crossings is as detrimental as the device parasitics because of using the same gate-dielectric layer at the crossings. Finally, the performance of complementary organic circuits is strongly limited by the slow n-channel OTFTs. Thus, it is desired to introduce at least another properly isolated metal layer and improve the performance of air-stable n-channel OTFTs without sacrificing the general process compatibility.

Outlook—The OTFTs require simultaneously low-voltage operation, air stability and high bandwidth to find use in practical applications. To meet these necessary criteria, there are ongoing research efforts to develop new organic semiconductors. In view of the many advantages of complementary technologies, it is expected that this type of circuits will be further advanced. However, the performance and stability of n-channel OTFTs remain more problematic. As a result, there is an increasing interest of using oxide-based semiconductors, which mostly exhibit electron transport characteristics, in high-performance TFTs as a plausible replacement of the slow n-channel OTFTs. Although the achieved carrier mobilities are rather high, the intrinsic performance of the materials cannot be fully utilized in short-channel (O)TFTs because of the contact resistances. Accordingly, doping of the contact areas as well as using device structures with split-gate electrodes to reduce this effect are worth considering.

Besides process variations, the OTFTs are generally sensitive to atmospheric oxygen and water, bias stress, temperature and light. Effective encapsulation is thus needed to ensure sufficient lifetime and increase robustness to environmental conditions. Albeit the increased area and complexity, it is encouraging to employ circuit styles and/or special devices that allow built-in post-fabrication tunability to compensate for these effects. In addition, the OTFT models have to be modified to account for the time-dependent change of parameters, without which misleading characteristics may be predicted. Moreover, it is envisaged to exploit the stencil masks in a reel-to-reel manufacturing process to increase the throughput and fabricate commercially viable products. The first products using OTFTs are already introduced into the marketplace, yet many new innovative and promising applications will continue to appear. These are outcomes of organic chemistry and as such can be varied in a near infinite number of ways [17]:

> *If atoms were musical notes, and macromolecules music, the range of structures attainable would be beyond anything as yet heard by humankind.*
> —André Moliton and Roger C. Hirons (2012)

References

1. F. Ante, F. Letzkus, J. Butschke, U. Zschieschang, J.N. Burghartz, K. Kern, H. Klauk, Top-contact organic transistors and complementary circuits fabricated using high-resolution silicon stencil masks, in *Device Research Conference*, Jun 2010, pp. 175–176
2. F. Ante, F. Letzkus, J. Butschke, U. Zschieschang, K. Kern, J.N. Burghartz, H. Klauk, Sub-micron low-voltage organic transistors and circuits enabled by high-resolution silicon stencil masks, in *IEEE International Solid-State Circuits Conference Technical Digest*, pp. 21.6.1–21.6.4 (2010)
3. F. Letzkus, T. Zaki, F. Ante, J. Butschke, H. Richter, H. Klauk, J.N. Burghartz, Si stencil masks for organic thin film transistor fabrication, in *Proceedings of the SPIE Photomask Technology*, Sep. 2011, pp. 81662B-1–81662B-12
4. F. Letzkus, T. Zaki, F. Ante, J. Butschke, H. Richter, H. Klauk, J.N. Burghartz, Si Stencil-Masken für die Herstellung organischer Dünnschichttransistoren, in *MikroSystemTechnik Kongress*, pp. 181–184 (2011) (in German)

5. U. Zschieschang, R. Hofmockel, R. Rödel, U. Kraft, M.J. Kang, K. Takimiya, T. Zaki, F. Letzkus, J. Butschke, H. Richter, J.N. Burghartz, H. Klauk, Megahertz operation of flexible low-voltage organic thin-film transistors. Org. Electron. **14**(6), 1516–1520 (2013)
6. T. Zaki, F. Ante, U. Zschieschang, J. Butschke, F. Letzkus, H. Richter, H. Klauk, J.N. Burghartz, A 3.3 V 6-Bit 100 kS/s current-steering digital-to-analog converter using organic p-type thin-film transistors on glass. IEEE J. Solid-State Circuits **47**(1), 292–300 (2012)
7. T. Zaki, F. Ante, U. Zschieschang, J. Butschke, F. Letzkus, H. Richter, H. Klauk, J.N. Burghartz, Circuit impact of device and interconnect parasitics in a complementary low-voltage organic thin-film technology, in *Semiconductor Conference Dresden*, pp. 1–4 (2011)
8. F. Ante, D. Kälblein, T. Zaki, U. Zschieschang, K. Takimiya, M. Ikeda, T. Sekitani, T. Someya, J.N. Burghartz, K. Kern, H. Klauk, Contact resistance and megahertz operation of aggressively scaled organic transistors. Small **8**(1), 73–79 (2012)
9. R. Rödel, F. Letzkus, T. Zaki, J.N. Burghartz, U. Kraft, U. Zschieschang, K. Kern, H. Klauk, Contact properties of high-mobility, air-stable, low-voltage organic n-channel thin-film transistors based on a naphthalene tetracarboxylic diimide. Appl. Phys. Lett. **102**(233303), 233303-1–233303-5 (2013)
10. U. Kraft, U. Zschieschang, M.J. Kang, K. Takimiya, T. Zaki, F. Letzkus, J.N. Burghartz, E. Weber, H. Klauk, Evolution of the field-effect mobility and the contact resistance of low-voltage organic thin-film transistors based on air-stable, high-mobility thioacenes, in *Materials Research Society Spring Meeting and Exhibit*, poster (2013)
11. T. Zaki, S. Scheinert, I. Hörselmann, R. Rödel, F. Letzkus, H. Richter, U. Zschieschang, H. Klauk, J.N. Burghartz, Accurate capacitance modeling and characterization of organic thin-film transistors. IEEE Trans. Electron Devices **61**(1), 98–104 (2014)
12. S. Scheinert, T. Zaki, R. Rödel, I. Hörselmann, H. Klauk, J.N. Burghartz, Numerical analysis of capacitance compact models for organic thin-film transistors. Org. Electron. **15**(7), 1503–1508 (2014)
13. T. Zaki, R. Rödel, F. Letzkus, H. Richter, U. Zschieschang, H. Klauk, J.N. Burghartz, S-parameter characterization of submicrometer low-voltage organic thin-film transistors. IEEE Electron Device Lett. **34**(4), 520–522 (2013)
14. T. Zaki, R. Rödel, F. Letzkus, H. Richter, U. Zschieschang, H. Klauk, J.N. Burghartz, AC characterization of organic thin-film transistors with asymmetric gate-to-source and gate-to-drain overlaps. Org. Electron. **14**(5), 1318–1322 (2013)
15. T. Zaki, F. Ante, U. Zschieschang, J. Butschke, F. Letzkus, H. Richter, H. Klauk, J.N. Burghartz, A 3.3V 6b 100kS/s current-steering D/A converter using organic thin-film transistors on glass, in *IEEE International Solid-State Circuits Conference Technical Digest*, pp. 324–325 (2011)
16. Y. Fujisaki, Y. Nakajima, T. Takei, H. Fukagawa, T. Yamamoto, H. Fujikake, Flexible active-matrix organic light-emitting diode display using air-stable organic semiconductor of dinaphtho[2,3-b:2',3'-f]thieno[3,2-b]thiophene. IEEE Trans. Electron Devices **59**(12), 3442–3449 (2012)
17. A. Moliton, R.C. Hiorns, The origin and development of (plastic) organic electronics. Polym. Int. **61**(3), 337–341 (2012)

Appendix A
Physical Layout Design

This appendix presents the physical layout design of three stencil mask sets, which are developed for this study. Each layout has several test structures and/or circuits; however, only the components that are relevant to this study are discussed herein. The three mask sets are designated as *Die 1*, *Die 2* and *Die 3*. The designs are not arranged in a chronological order, they are rather ordered depending on the presentation flow of this work. Finally, labels and alignment structures, which are generally included in all the designs, are illustrated.

A.1 Design of Die 1

The first design, Die 1, includes test structures and OTFTs mainly for static and dynamic characterization. Figure A.1 depicts a simplified diagram showing the dimensions of the test OTFTs on Die 1. For static characterization, there are several OTFTs with channel widths of 200 μm but with various channel lengths (100–1 μm) and gate overlaps (100–2 μm). By measuring these OTFTs, one can extract the intrinsic mobility in the organic semiconductor and the contact resistance of the transistors using the transmission-line-method (TLM). As for the dynamic characterization, there are various metal-insulator-metal (MIM), metal-insulator-semiconductor (MIS) and relatively large OTFTs for admittance measurements. The layout of the MIM and MIS along with the short and open calibration structures are shown in Fig. A.2.

A.2 Design of Die 2

The second design, Die 2, includes aggressively-scaled OTFTs for S-parameter measurements. All the transistors have channel widths of 100 μm. But for the sake of the stencil masks stability, the OTFTs are dissected into four parallel transistors

T. Zaki, *Short-Channel Organic Thin-Film Transistors*, Springer Theses,
DOI 10.1007/978-3-319-18896-6

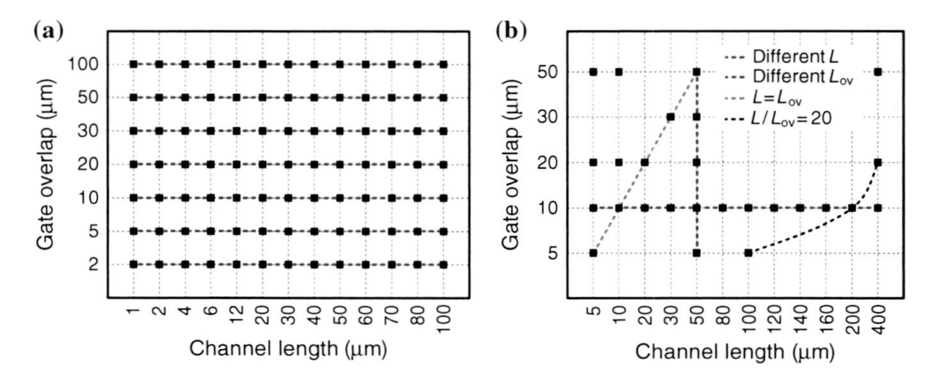

Fig. A.1 Simplified diagram (not to scale) showing the dimensions of the test OTFTs on Die 1. **a** OTFTs with channel widths of 200 μm for static characterization using the transmission line method (TLM). **b** OTFTs with channel widths of 400 and 1000 μm for dynamic characterization using admittance measurements

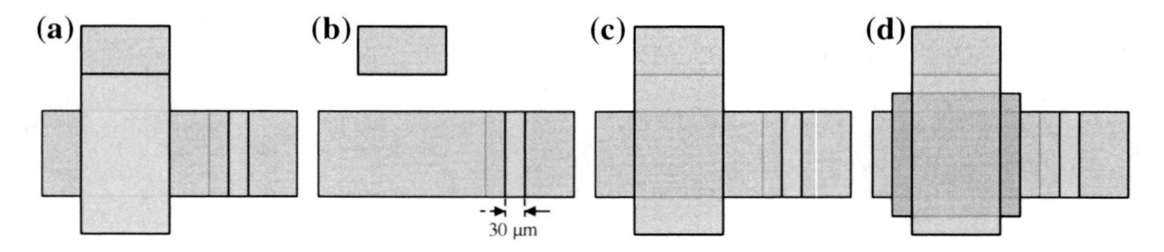

Fig. A.2 Layout of test structures on Die 1 dedicated for admittance measurements. **a** Short and **b** open calibration structures. **c** Metal-insulator-metal (MIM). **d** Metal-insulator-semiconductor (MIS). The die includes several of these structures but with different top-to-gate *overlap area*, i.e., 200×200, 100×100, 50×50, 10×10, 500×20 and $1000 \times 40\,\mu m^2$. In all of these test structures, the enclosure of the top metal inside the semiconductor as well as the non-related clearances are all set to 30 μm

($4 \times 25\,\mu m$) sharing the same semiconductor. Figure A.3 shows the layout of an OTFT and the calibration (standard short, open and through) structures. The ground-signal-ground (GSG) pads configuration is used for the special microwave wafer probes that are used during the measurements. The OTFTs on the die have channel lengths of 10, 2, 1, 0.8, 0.6 and 0.4 μm and gate overlaps of 20 and 5 μm. Moreover, the organic semiconductor layer extends beyond the periphery of the intrinsic OTFTs by 30 μm on each side (also called fringe regions). To study the impact of misalignment between the gate-source (L_{gs}) and gate-drain (L_{gd}) overlaps, where $L_{gs} + L_{gd} = 2L_{ov}$, the OTFTs with the gate-overlap lengths (L_{ov}) of 5 μm and the channel lengths (L) of 1 and 0.6 μm are duplicated and intentionally misaligned on the mask level. A total of 19 OTFTs with intentional misalignment of $\pm 1, \pm 2, \ldots, \pm 9$ are realized. This is to guarantees the fabrication of OTFTs with well-defined $L_{gs} = 1, 2, \ldots, 9\,\mu m$, while keeping $L_{gs} + L_{gd} = 10\,\mu m$. Vernier structures are employed to measure the particular degree of misalignment.

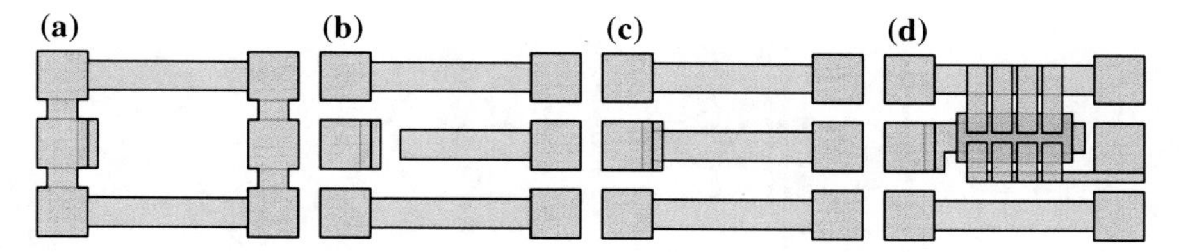

Fig. A.3 Layout of OTFT and calibration structures on Die 2 dedicated for S-parameter measurements. **a** Short, **b** open and **c** through calibration structures. **d** OTFT that is dissected into four parallel transistors sharing the same semiconductor. This is for stencil mask stability in the case of using very small channel lengths. The ground-signal-ground (GSG) pads configuration is used for the special microwave wafer probes that are used during the measurements

Table A.1 Geometries of the OTFTs used by the unipolar and the complementary (bipolar) ring oscillators on Die 3

Type	L (μm)	$W_{s,p}$ (μm)	$W_{l,n}$ (μm)	L_{ov} (μm)	Remark
Unipolar	4	110	24	20	Includes one crossing
	4	110	24	20	No interconnect crossings
	2	110	24	10	No interconnect crossings
	1	110	24	5	No interconnect crossings
	1	72	24	5	No interconnect crossings
Bipolar	4	40	80	20	No interconnect crossings
	1	40	80	5	No interconnect crossings

A.3 Design of Die 3

The third design, Die 3, includes fast and compact organic ICs, namely ring oscillators and current-steering digital-to-analog converters (DACs). The electrical results for both organic ICs achieved a world record in performance with respect to low-voltage OTFTs. The layouts of both are explained herein.

The design includes five unipolar (only p-channel OTFTs) and two complementary (both p- and n-channel OTFTs) 11-stage ring oscillators. The ring oscillators employ OTFTs with different channel lengths (L), gate overlaps (L_{ov}), and channel widths for the source (W_s) and the load (W_l) transistors. In the case of the complementary design, the width of the n-channel OTFTs (W_n) is twice as big as the width of the p-channel OTFTs (W_p). All the dimensions are summarized in Table A.1. Figure A.4 shows the layout of the ring oscillators. A folded configuration is employed for performance optimization. The designs exploit a relatively large output inverter to drive the load, i.e., the oscilloscope, and provide a large output swing.

Furthermore, the design also includes two different DACs, namely a 6-bit binary current-steering DAC and a 3-bit unary current-steering DAC. Strict requirements apply here to the design of the layout and floorplan due to the relatively large number of transistors employed, particularly for an organic-based analog circuit [1]. Device matching and data interference have a strong impact on the DAC performance.

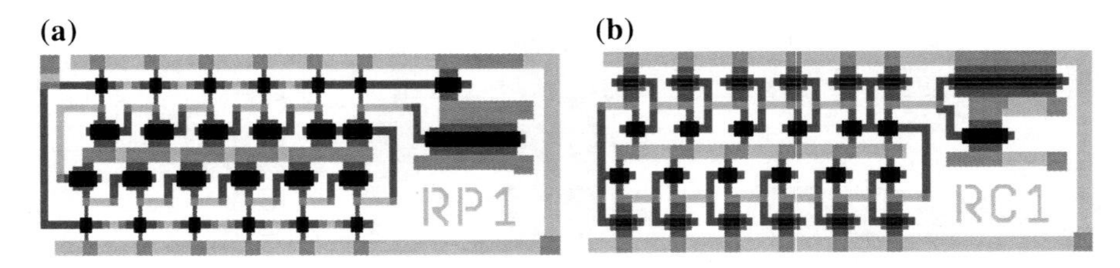

Fig. A.4 Layout of **a** unipolar (only p-channel OTFTs) and **b** complementary (both p- and n-channel OTFTs) 11-stage ring oscillators [1]

In principle, there are two methods for arranging the transistors in a current-steering DAC, i.e., either by merging the switches and the current-source transistors in one array or by splitting them into two separate arrays, namely switch array (SWA) and current-source array (CSA). In the merged approach, each unit cell comprises both the current-source and the switch transistors. This method has a simplified layout; however, it increases the distance between the current-source transistors and so the layout will be potentially more susceptible to systematical gradient effects. As for the separated SWA-CSA, the identical transistors are placed in separate arrays, thus ensuring better device matching, easing power routing, and minimizing signal interference as a result of keeping the digital and the analog blocks apart. In spite of the increased layout complexity in the latter approach, it is selected here as it offers a substantially better conversion performance.

For simplicity, the layout design of the DACs is described here by making use of the so called stick diagrams. These are just simple line-based illustrations that allow layout engineering to handle the routing without the consideration of spacings and geometries.

With respect to the 6-bit binary current-steering DAC, 63 current-source and 63 switch transistors are needed [2]. Therefore, a matrix-like orientation of the OTFTs in the SWA and the CSA is preferable to bring the transistors in close proximity. This makes it possible to exploit the excellent feature size control enabled by the stencil mask technology regardless of the non-uniformity introduced by material deposition through evaporation. With careful layout and floorplan design, no interconnect crossings that feature high parasitic capacitances are comprised. A simplified version of the layout of the 6-bit binary DAC is shown in Fig. A.5a, noting that the binary input stream is directly driven to the switches. A minimum channel length of 4 μm and a minimum gate overlap of 20 μm are used by the OTFTs. The total area of the 6-bit binary DAC including the contact pads is $2.6 \times 4.6\,\text{mm}^2$.

On the other hand, the 3-bit unary current-steering DAC comprises 7 current-source and 7 switch transistors [3]. As only few number of transistors are used for the conversion, higher complexity in the layout would counteract the desired

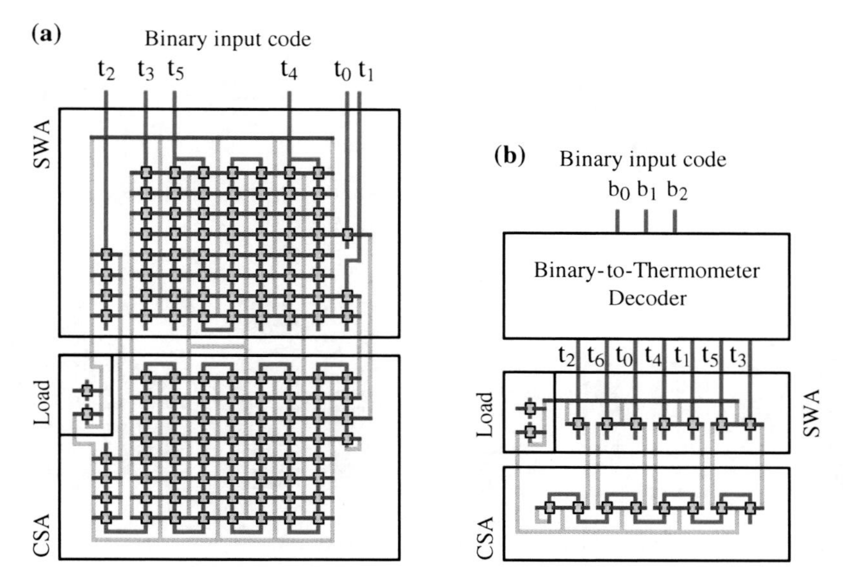

Fig. A.5 Simplified layout (*stick diagram*) of **a** the 6-bit binary and **b** the 3-bit unary current-steering DACs. The load in both circuits is a current-to-voltage converter (CVC). In the 6-bit binary DAC, the binary input code is arranged in a way to minimize the complexity of the routing matrix between the SWA and the CSA. Note that the crossings of the input lines (that use the aluminum-gate layer) with the *horizontal top*-metal line (that is supposed to be for the maximum supply voltage) are avoided in the final layout. In the 3-bit unary DAC, the output order of the thermometer decoder defines the switching scheme, through which the unary current-sources are enabled to the output. This switching scheme is used instead of the conventional sequential order to compensate for errors that are introduced by systematical gradient effects (linear or symmetrical) [1–3]

performance. For this reason, the simple one-dimensional array orientation for the CSA and the SWA is preferable. However, a clear separation between the analog and the digital parts should still be maintained. A simplified version of the layout of the 3-bit unary DAC is shown in Fig. A.5b. In this case, a 3-bit binary-to-thermometer decoder is needed. A minimum channel length of 4 μm and a minimum gate overlap of 20 μm are used by the OTFTs. The total area of the 3-bit unary DAC including the contact pads and the decoder circuit is $3.1 \times 6.6\,\text{mm}^2$.

A.4 Labels and Alignment Structures

For each die design, a total of five (in some cases four) masks are required. As a result, all the designs include labels for the five physical layers, which are entitled bottom-metal (B), gate (G), organic semiconductors for n-channel OTFTs (N) and p-channel OTFTs (P), and top-metal (T). Figure A.6a depicts the layout of the labels, which are typically placed at the top of each die design. The purpose of these labels is

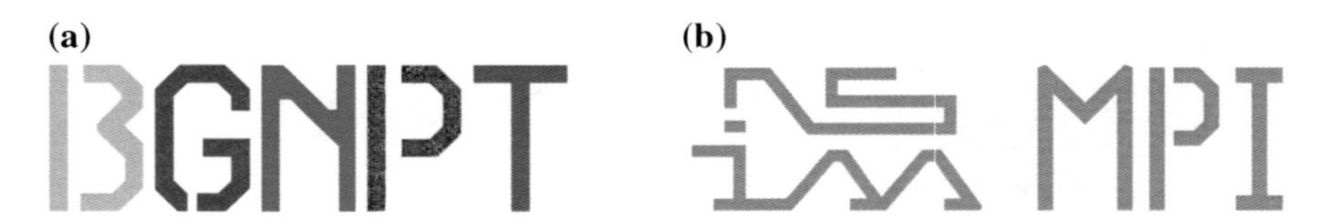

Fig. A.6 **a** Layout of the physical layer labels. **b** Layout of IMS CHIPS and MPI-SSR logos/initials [1]

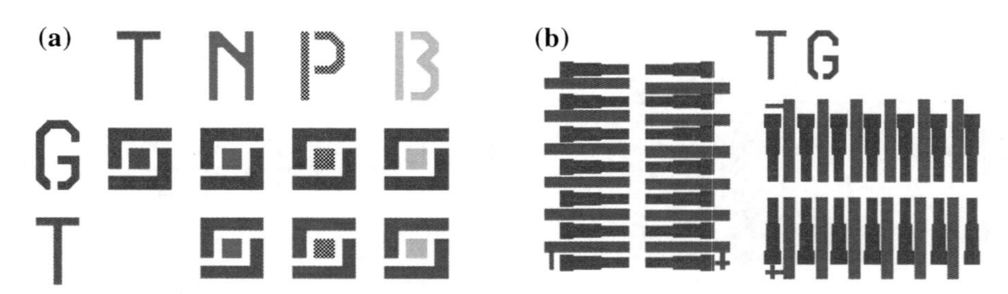

Fig. A.7 Layout of **a** box-in-box and **b** *vertical/horizontal* vernier-scale alignment structures [1]

to designate clearly the five different layers in the layout and the stencil masks after fabrication. For stencil mask stability, donut-shaped isolated features that could fall out during stencil etching are completely avoided. As shown for labels *G* and *P*, the features are mechanically and firmly attached by the use of more than one anchor. Furthermore, the layout of IMS CHIPS and MPI-SSR logos/initials are implemented and placed at the bottom of each die design (Fig. A.6b).

Moreover, several alignment structures are placed at the four corners of each die design to measure and evaluate visually the degree of misalignment between the layout layers after the fabrication. Box-in-box structures are used to measure the degree of misalignment between almost all the layers with an accuracy of about 1 μm (Fig. A.7a). Since the alignment between the top metal and the gate is the most critical as it defines the dimensions of the gate-source and gate-drain overlaps, vernier scales that offer higher precision (about 0.1 μm) are used for those two layers (Fig. A.7b).

Appendix B
De-embedding Procedure

Full two-port S-parameters of OTFTs are measured in this work using an HP 3577A vector network analyzer (VNA). Systematical errors, which are mostly introduced by the test setup, are observed during the measurements [4]. Accordingly, a de-embedding process is employed to remove the undesired parasitic effects of the physical network placed between the VNA and the device under test (DUT), i.e., an OTFT. In principle, the de-embedding process uses a model of the test fixture and mathematically removes the parasitics from the overall measurements. There are many different approaches that have been developed for modeling and removing these errors such as 3-term, 8-term, 12-term and 16-term error correction models [4–6]. In this work, the 12-term error correction model, which requires standard short, open, through and matched terminations for a two-port network calibration, is utilized; the method produced quite promising results. This appendix describes the error model and the calibration method used.

Figure B.1 shows a generic block diagram of a two-port network with an error adapter, which separates the DUT (an OTFT in our case) from an ideal measurement system [7]. Furthermore, Fig. B.2 depicts the twelve system errors involved when measuring a two-port device/network in the forward and reverse directions [4]. The error model and the S-parameters of the DUT are represented with a signal flow graph, which is a simple way to represent and helps to analyze the transmitted and reflected signals from a device/network [4]. Moreover, it is useful for cascaded network topologies, since flow graphs of adjacent networks can be connected in series. Directed lines in the flow graph represent the dependency of the signal flow through the network; for example, the signal flow from node a_1 to b_1 is defined as the input reflection coefficient (S_{11i}). At this point, there has to be a clear distinction between the intrinsic (actual de-embedded) and the extrinsic (measured) S-parameters, which are given by

T. Zaki, *Short-Channel Organic Thin-Film Transistors*, Springer Theses,
DOI 10.1007/978-3-319-18896-6

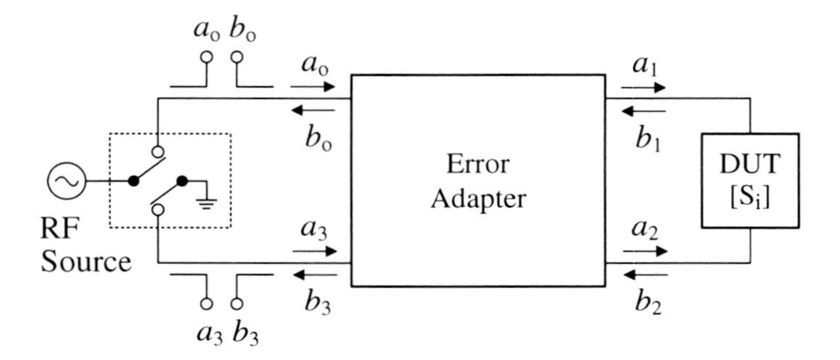

Fig. B.1 Generic block diagram of a two-port network with an error adapter, which separates the DUT from an ideal measurement system [7]

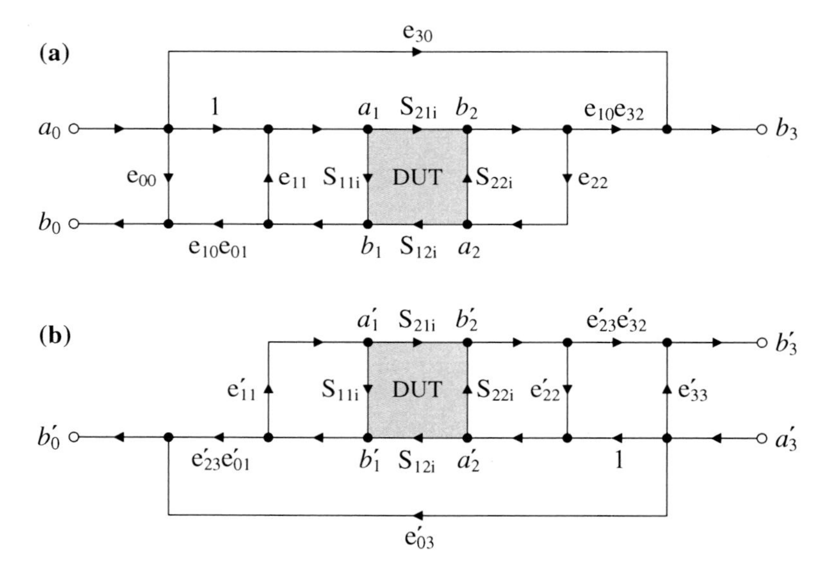

Fig. B.2 Signal flow diagram of the twelve-term error model with the S-parameters of a two-port DUT in the **a** forward and **b** reverse directions [6]

$$\begin{aligned} b_1 &= S_{11i} \cdot a_1 + S_{12i} \cdot a_2 \\ b_2 &= S_{21i} \cdot a_1 + S_{22i} \cdot a_2 \end{aligned} \qquad \begin{bmatrix} b_1 \\ b_2 \end{bmatrix} = \begin{bmatrix} S_{11i} & S_{12i} \\ S_{21i} & S_{22i} \end{bmatrix} \begin{bmatrix} a_1 \\ a_2 \end{bmatrix}, \tag{B.1}$$

$$\begin{aligned} b_0 &= S_{11e} \cdot a_0 + S_{12e} \cdot a_3 \\ b_3 &= S_{21e} \cdot a_0 + S_{22e} \cdot a_3 \end{aligned} \qquad \begin{bmatrix} b_0 \\ b_3 \end{bmatrix} = \begin{bmatrix} S_{11e} & S_{12e} \\ S_{21e} & S_{22e} \end{bmatrix} \begin{bmatrix} a_0 \\ a_3 \end{bmatrix}. \tag{B.2}$$

The subscripts i and e denote the intrinsic (of the DUT) and extrinsic (including parasitics) S-parameters, respectively.

Referring to Fig. B.2, e_{00} and e'_{33} are the directivity error terms resulting form signal leakage through the directional couplers on the corresponding ports, $e_{10}e_{01}$ and $e'_{23}e'_{32}$ are the reflection tracking terms resulting from the path differences between the test and reference paths, e_{11} and e'_{22} are the source match terms resulting from the VNA's test port impedance not being perfectly matched to the source impedance, $e_{10}e_{32}$ and $e'_{23}e'_{01}$ are the transmission tracking terms resulting from the path differences between the load and reference paths, e_{22} and e'_{11} are the load match terms resulting from the VNA's test port impedance not being perfectly matched to the load impedance, and finally e_{30} and e'_{03} are the leakage error terms resulting from the crosstalk between the corresponding ports. By applying Mason's gain formula to the forward and reverse flow graphs (Fig. B.2) and using the definitions of S_i and S_e given by (B.2) and (B.1), respectively, a relationship between S_i, S_e and the twelve error adapter coefficients can be derived [6]. Then by solving for S_i using linear algebra operations, we obtain [4–6]

$$S_{11i} = \frac{\left(\frac{S_{11e}-e_{00}}{e_{10}e_{01}}\right)\left[1+\left(\frac{S_{22e}-e'_{33}}{e'_{23}e'_{32}}\right)e'_{22}\right] - e_{22}\left(\frac{S_{21e}-e_{30}}{e_{10}e_{32}}\right)\left(\frac{S_{12e}-e'_{03}}{e'_{23}e'_{01}}\right)}{\left[1+\left(\frac{S_{11e}-e_{00}}{e_{10}e_{01}}\right)e_{11}\right]\left[1+\left(\frac{S_{22e}-e'_{33}}{e'_{23}e'_{32}}\right)e'_{22}\right] - \left(\frac{S_{21e}-e_{30}}{e_{10}e_{32}}\right)\left(\frac{S_{12e}-e'_{03}}{e'_{23}e'_{01}}\right)e_{22}e'_{11}}, \tag{B.3}$$

$$S_{21i} = \frac{\left(\frac{S_{21e}-e_{30}}{e_{10}e_{32}}\right)\left[1+\left(\frac{S_{22e}-e'_{33}}{e'_{23}e'_{32}}\right)(e'_{22}-e_{22})\right]}{\left[1+\left(\frac{S_{11e}-e_{00}}{e_{10}e_{01}}\right)e_{11}\right]\left[1+\left(\frac{S_{22e}-e'_{33}}{e'_{23}e'_{32}}\right)e'_{22}\right] - \left(\frac{S_{21e}-e_{30}}{e_{10}e_{32}}\right)\left(\frac{S_{12e}-e'_{03}}{e'_{23}e'_{01}}\right)e_{22}e'_{11}}, \tag{B.4}$$

$$S_{22i} = \frac{\left(\frac{S_{22e}-e'_{33}}{e'_{23}e'_{32}}\right)\left[1+\left(\frac{S_{11e}-e_{00}}{e_{10}e_{01}}\right)e_{11}\right] - e'_{11}\left(\frac{S_{21e}-e_{30}}{e_{10}e_{32}}\right)\left(\frac{S_{12e}-e'_{03}}{e'_{23}e'_{01}}\right)}{\left[1+\left(\frac{S_{11e}-e_{00}}{e_{10}e_{01}}\right)e_{11}\right]\left[1+\left(\frac{S_{22e}-e'_{33}}{e'_{23}e'_{32}}\right)e'_{22}\right] - \left(\frac{S_{21e}-e_{30}}{e_{10}e_{32}}\right)\left(\frac{S_{12e}-e'_{03}}{e'_{23}e'_{01}}\right)e_{22}e'_{11}}, \tag{B.5}$$

$$S_{12i} = \frac{\left(\frac{S_{12e}-e'_{03}}{e'_{23}e'_{01}}\right)\left[1+\left(\frac{S_{11e}-e_{00}}{e_{10}e_{01}}\right)(e_{11}-e'_{11})\right]}{\left[1+\left(\frac{S_{11e}-e_{00}}{e_{10}e_{01}}\right)e_{11}\right]\left[1+\left(\frac{S_{22e}-e'_{33}}{e'_{23}e'_{32}}\right)e'_{22}\right] - \left(\frac{S_{21e}-e_{30}}{e_{10}e_{32}}\right)\left(\frac{S_{12e}-e'_{03}}{e'_{23}e'_{01}}\right)e_{22}e'_{11}}. \tag{B.6}$$

Note that each of the intrinsic S-parameters is a function of all the four extrinsic S-parameters and the twelve error adapter terms. Therefore, the VNA must perform the complete forward as well as the reverse sweeps in order to update any of the intrinsic S-parameters.

When a one-port device is being measured, the 12-term model presented in Fig. B.2 simplifies to just the terms describing the directivity, port match and tracking errors

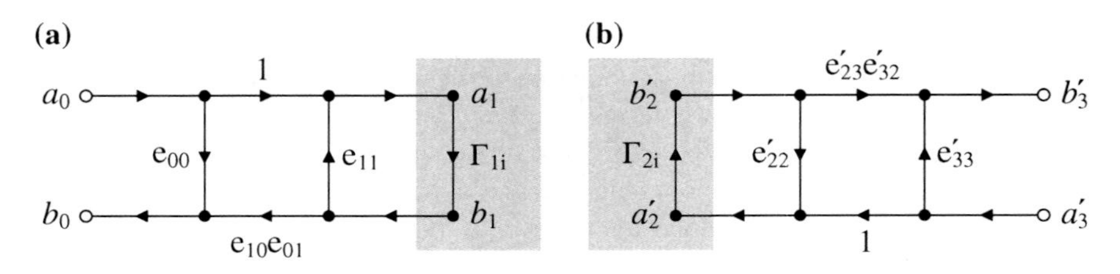

Fig. B.3 Signal flow diagram of the three-term error model with the reflection coefficient of a one-port DUT in the **a** forward and **b** reverse directions [6]

at each port, as shown in Fig. B.3. Solving for the forward flow graph using Mason's gain formula yields the following expressions [6]:

$$\Gamma_{1e} = \frac{b_0}{a_0} = \frac{e_{00} - (e_{00}e_{11} - e_{10}e_{01})\Gamma_{1i}}{1 - e_{11}\Gamma_{1i}}, \tag{B.7}$$

$$\Gamma_{1i} = \frac{b_1}{a_1} = \frac{\Gamma_{1e} - e_{00}}{e_{11}\Gamma_{1e} - (e_{00}e_{11} - e_{10}e_{01})}, \tag{B.8}$$

where Γ_{1e} and Γ_{1i} are the extrinsic (measured) and intrinsic (actual) reflection coefficients at port 1, respectively. In principle, the reflection coefficient Γ_1 is equivalent to S_{11} but for a one-port device. By letting $\Delta e = e_{00}e_{11} - e_{10}e_{01}$, equations (B.7) and (B.8) can be rewritten as

$$e_{00} + \Gamma_{1i}\Gamma_{1e}e_{11} - \Gamma_{1i}\Delta e = \Gamma_{1e}. \tag{B.9}$$

Thus, the three error terms can be determined by measuring three known standards such as open ($\Gamma_{1i} = 1$), short ($\Gamma_{1i} = -1$) and matched ($\Gamma_{1i} = 0$) loads, that yield a system of three linear equations. These three equations can then be solved simultaneously to calculate e_{00}, e_{11} and $e_{10}e_{01}$. The exact same approach applies for the reverse flow graph to extract e'_{33}, e'_{22} and $e'_{23}e'_{32}$.

Once the twelve error terms in Fig. B.2 are known, the effects of the test fixture can be de-embedded from the measurements using (B.3)–(B.6). The two-port calibration steps to extract these twelve error terms are as follows. Considering initially the forward flow graph, the first step is to calibrate port 1 using the procedure described above for the simplified one-port model. This is performed using the short, open and matched loads to determine e_{00}, e_{11} and $e_{10}e_{01}$. In the second step, the leakage error term e_{30} from port 1 to port 2 is directly obtained by measuring S_{21e} while connecting matched loads to both ports. The third step is to extract e_{22} and $e_{10}e_{32}$ while connecting port 1 and port 2 together (through load) using the following expressions [6]:

$$e_{22} = \frac{S_{11e} - e_{00}}{S_{11e}e_{11} - (e_{00}e_{11} - e_{10}e_{01})}, \tag{B.10}$$

$$e_{10}e_{32} = (S_{21e} - e_{30})(1 - e_{11}e_{22}). \tag{B.11}$$

This concludes the extraction of the six error terms in the forward direction. Then the same procedure can be used to solve for the remaining six error terms in the reverse direction.

To achieve a high fidelity in the de-embedding process, the steps above are performed on a standard calibration substrate with well-characterized terminations having identical pad configuration and dimensions with that of the OTFT. This guarantees that any undesired effects introduced by the cables, probe holders and on-wafer pads are successfully removed.

Appendix C
Analytical Derivations

Evaluation of the static as well as the dynamic performance of the organic thin-film transistors (OTFTs) requires an understanding of the device physics to be able to accurately model their behavior. A small-signal model, consisting of a distributed resistance-capacitance equivalent circuit, is used to characterize both the resistive and the reactive parts of the intrinsic OTFT. Furthermore, a distributed resistance network is used to analyze the current injection mechanism through the contact-semiconductor interfaces. Accordingly, the expressions for the channel capacitance and the contact resistance of the OTFTs are derived in this appendix.

C.1 Derivation of the Channel Capacitance

The frequency response of OTFTs by means of admittance measurements is investigated in this work. As shown in Fig. C.1, evaluation of the intrinsic part of an OTFT requires the use of a distributed resistance-capacitance (RC) equivalent circuit. For this setup, the source and drain contacts are electrically shorted and connected to the low terminal (virtual ground) of an LCR meter, while the gate electrode is connected to the high terminal. At voltages below the threshold voltage, no channel exists and the capacitance measured is just the sum of the relatively small (assuming $L \gg L_{\text{ov}}$) gate-source and gate-drain overlap capacitances. However, as the gate bias exceeds the threshold voltage, an accumulation channel is created; consequently, the channel resistance reduces and an additional geometrical gate-to-channel capacitance is measured. Since there is no potential difference between the drain and source electrodes, the channel resistance is assumed to be uniform along the whole channel. The expression of the gate-to-channel capacitance is derived herein (adopted from [8]).

Referring to Fig. C.1, a particular section of the distributed RC circuit of length Δx has the following resistance and capacitance:

T. Zaki, *Short-Channel Organic Thin-Film Transistors*, Springer Theses,
DOI 10.1007/978-3-319-18896-6

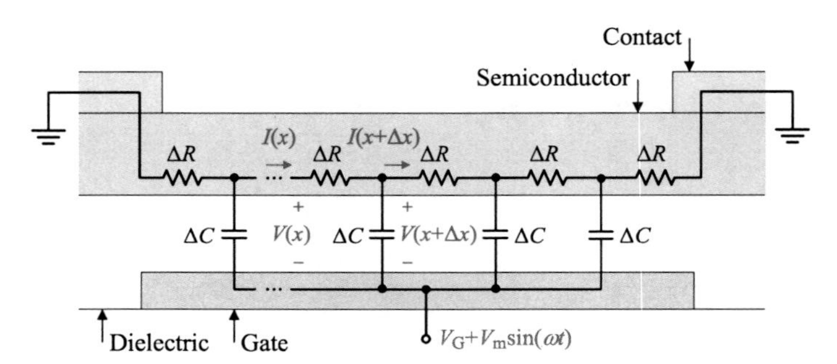

Fig. C.1 Schematic cross section of an inverted-staggered (*bottom*-gate, *top*-contact) OTFT (not to scale) with a channel length that is much larger than the gate-overlap length ($L \gg L_{ov}$). The small-signal model of the intrinsic part of the OTFT using a distributed RC equivalent circuit, assuming the transistor is operated in the accumulation regime, is depicted *inside* the schematic. In the shown setup, the source and drain contacts are electrically shorted and connected to the low terminal (*virtual ground*) of an LCR meter, while the gate electrode is connected to the high terminal to bias the transistor with a DC potential (V_G) along with a superimposed AC potential ($V_m \sin \omega t$)

$$\Delta R = \frac{\Delta x R_{\text{sheet}}}{W}, \tag{C.1}$$

$$\Delta C = C_I W \Delta x, \tag{C.2}$$

where W (m) is the width of the transistor, R_{sheet} ($\Omega/\square$) is the voltage dependent sheet resistance within the channel and C_I (F/cm^2) is the gate dielectric capacitance per unit area. By applying the Kirchhoff's current law and then substituting for ΔC using (C.2), we get

$$\begin{aligned}
I(x + \Delta x) - I(x) &= -j\omega \Delta C\, V(x + \Delta x) \\
I(x + \Delta x) - I(x) &= -j\omega C_I W \Delta x\, V(x + \Delta x) \\
\frac{I(x + \Delta x) - I(x)}{\Delta x} &= -j\omega C_I W [V(x) - \Delta R\, I(x)],
\end{aligned}$$

where ω (rad/s) is the angular frequency at which the capacitance is measured. By substituting for ΔR using (C.1) and then taking the limit $\Delta x \to 0$ at both sides, we obtain

$$\lim_{\Delta x \to 0} \left(\frac{I(x + \Delta x) - I(x)}{\Delta x} \right) = \lim_{\Delta x \to 0} \left(-j\omega C_I W \left[V(x) - \frac{\Delta x R_{\text{sheet}} I(x)}{W} \right] \right)$$

$$\frac{\partial I}{\partial x} = -j\omega C_I W V(x). \tag{C.3}$$

On the other hand, applying Kirchhoff's voltage law yields the following:

$$V(x + \Delta x) - V(x) = -\Delta R\, I(x).$$

Again by substituting for ΔR using (C.1) and then taking the limit $\Delta x \to 0$ at both sides, we get

$$\lim_{\Delta x \to 0} \left(\frac{V(x + \Delta x) - V(x)}{\Delta x} \right) = \lim_{\Delta x \to 0} \left(- \frac{R_{\text{sheet}}}{W} I(x) \right)$$

$$\frac{\partial V}{\partial x} = -\frac{R_{\text{sheet}}}{W} I(x). \tag{C.4}$$

Solving (C.3) and (C.4) together gives the following second-order linear differential equation:

$$\frac{\partial^2 V}{\partial x^2} = j\omega R_{\text{sheet}} C_{\text{I}} V(x). \tag{C.5}$$

Let $1/\lambda^2 = j\omega R_{\text{sheet}} C_{\text{I}}$, equation (C.5) has the general solution $V(x) = Ae^{x/\lambda} + Be^{-x/\lambda}$ in which λ is a complex decay length. To evaluate the terminal capacitance, we impose the boundary conditions $V(x = 0) = V(x = L) = V_{\text{m}}$ to obtain A and B:

$$A = V_{\text{m}} \left(\frac{1 - e^{-L/\lambda}}{e^{L/\lambda} - e^{-L/\lambda}} \right),$$

$$B = V_{\text{m}} \left(\frac{e^{L/\lambda} - 1}{e^{L/\lambda} - e^{-L/\lambda}} \right).$$

Thus, the function $V(x)$ can now be expressed as

$$V(x) = \frac{V_{\text{m}}}{e^{L/\lambda} - e^{-L/\lambda}} \left[(1 - e^{-L/\lambda}) e^{x/\lambda} - (1 - e^{L/\lambda}) e^{-x/\lambda} \right]. \tag{C.6}$$

The current $I(x)$ can then be evaluated by substituting $V(x)$ in (C.4):

$$I(x) = -\frac{W}{\lambda R_{\text{sheet}}} \frac{V_{\text{m}}}{e^{L/\lambda} - e^{-L/\lambda}} \left[(1 - e^{-L/\lambda}) e^{x/\lambda} + (1 - e^{L/\lambda}) e^{-x/\lambda} \right]. \tag{C.7}$$

We can deduce from this expression that $I(x = 0) = -I(x = L)$ as expected from the setup shown in Fig. C.1.

The measured admittance by the LCR meter is given by $Y_{\text{ch}} = G_{\text{ch}} + j\omega C_{\text{ch}}$, where G_{ch} and C_{ch} are the measured channel conductance and capacitance. From the setup shown in Fig. C.1, the measured capacitance C_{ch} is expressed as follows:

$$C_{\text{ch}} = \frac{1}{\omega} \text{Im} \left[\frac{2I(x = 0)}{V(x = 0)} \right] = -\frac{1}{\omega} \text{Im} \left[\frac{2I(x = L)}{V(x = L)} \right]. \tag{C.8}$$

Substituting (C.7) in (C.8) results in

$$C_{\text{ch}} = \text{Im}\left[\underbrace{\frac{2W}{\omega\lambda R_{\text{sheet}}}}_{f_1} \overbrace{\frac{1}{e^{L/\lambda} - e^{-L/\lambda}} (\underbrace{e^{L/\lambda}}_{f_2} + e^{-L/\lambda} - 2)}^{f_3} \right]. \tag{C.9}$$

Given that $1/\lambda^2 = j\omega R_{\text{sheet}} C_{\text{I}}$, the terms given by f_1 and f_2 in (C.9) can be written as

$$f_1 = \sqrt{\frac{2C_{\text{I}}W^2}{\omega R_{\text{sheet}}}}(1 + j) = \frac{C_{\text{I}}WL}{\alpha}(1 + j), \tag{C.10}$$

$$f_2 = e^{\sqrt{\frac{\omega R_{\text{sheet}} C_{\text{I}} L^2}{2}}(1+j)} = e^{\alpha(1+j)}, \tag{C.11}$$

where $\alpha = \sqrt{\omega R_{\text{sheet}} C_{\text{I}} L^2/2}$. Substituting (C.11) into the term given by f_3 in (C.9) yields

$$\begin{aligned} f_3 &= \frac{e^{\alpha(1+j)} + e^{-\alpha(1+j)} - 2}{e^{\alpha(1+j)} - e^{-\alpha(1+j)}} \\ &= \frac{(e^{\alpha} + e^{-\alpha})\cos\alpha + j(e^{\alpha} - e^{-\alpha})\sin\alpha - 2}{(e^{\alpha} - e^{-\alpha})\cos\alpha + j(e^{\alpha} + e^{-\alpha})\sin\alpha}. \end{aligned} \tag{C.12}$$

The measured capacitance C_{ch} can therefore be finally obtained by substituting (C.10) and (C.12) in (C.9) and applying some simplification to the trigonometric expression:

$$C_{\text{ch}} = \frac{C_{\text{I}}WL}{\alpha}\left(\frac{\sinh\alpha + \sin\alpha}{\cosh\alpha + \cos\alpha}\right). \tag{C.13}$$

Following a similar procedure to that used to derive the capacitance, the measured loss (G_{ch}/ω) can be expressed as

$$\frac{G_{\text{ch}}}{\omega} = \frac{C_{\text{I}}WL}{\alpha}\left(\frac{\sinh\alpha - \sin\alpha}{\cosh\alpha + \cos\alpha}\right). \tag{C.14}$$

C.2 Derivation of the Contact Resistance

The contact resistance has a considerable impact on the performance of the OTFTs. When the channel length is reduced, the relative contribution of the contact resistance to the total device resistance increases, and thus, the effectivemobility decreases [9].

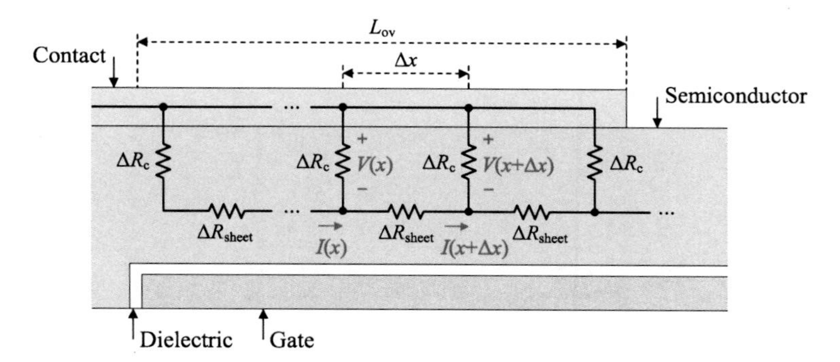

Fig. C.2 Schematic cross section of the contact part of an inverted-staggered (*bottom*-gate, *top*-contact) OTFT (not to scale). A distributed resistance network, which models the current injection mechanism, is depicted *inside* the schematic. For simplification, the contact resistance is considered ohmic, even though this assumption is not reflecting the true nature of the contact/semiconductor interface

In a general approach, the transmission line method (TLM) is used to extract the contact resistance of the OTFTs [10]. Evaluation of the contact resistance of our inverted-staggered (bottom-gate, top-contact) OTFTs, in particular, requires the use of a distributed resistance network as shown in Fig. C.2. The distributed resistance network is a simplified equivalent circuit that is used to electrically describe and model the current injection mechanism through the contact interfaces between the metallic source/drain contacts and the organic semiconductor. In this section, the equivalent circuit is analyzed and accordingly the expression of the equivalent contact resistance R_C is derived (adopted from [10]).

By considering a section of the distributed resistance network shown in Fig. C.2 of length Δx, the vertical ΔR_C (Ω) and horizontal ΔR_{sheet} (Ω) components can be expressed as

$$\Delta R_C = \frac{\rho_c}{W \Delta x}, \tag{C.15}$$

$$\Delta R_{sheet} = \frac{R_{sheet} \Delta x}{W}, \tag{C.16}$$

where W (m) is the width of the transistor, ρ_c (Ωm^2) is the contact resistivity multiplied by the equivalent distance between the source/drain contacts and the channel, and R_{sheet} ($\Omega/\Box$) is the voltage dependent sheet resistance within the channel. By applying the Kirchhoff's current law, we obtain

$$I(x + \Delta x) - I(x) = \frac{V(x)}{\Delta R_C}.$$

Then by substituting for ΔR_{C} using (C.15) and then taking the limit $\Delta x \to 0$ at both sides, we find

$$\lim_{\Delta x \to 0}\left(\frac{I(x+\Delta x)-I(x)}{\Delta x}\right) = \lim_{\Delta x \to 0}\left(\frac{W}{\rho_{\mathrm{c}}}V(x)\right)$$

$$\frac{\partial I}{\partial x} = \frac{W}{\rho_{\mathrm{c}}}V(x). \tag{C.17}$$

Applying now the Kirchhoff's voltage law and then substituting for $\Delta R_{\mathrm{sheet}}$ using (C.16) result in the following:

$$V(x+\Delta x)-V(x) = \Delta R_{\mathrm{sheet}} I(x+\Delta x)$$

$$V(x+\Delta x)-V(x) = \frac{R_{\mathrm{sheet}}\Delta x}{W} I(x+\Delta x)$$

$$\frac{V(x+\Delta x)-V(x)}{\Delta x} = \frac{R_{\mathrm{sheet}}}{W}\left[I(x)+\frac{V(x)}{\Delta R_{\mathrm{C}}}\right].$$

By substituting for ΔR_{C} using (C.15) and then taking the limit $\Delta x \to 0$ at both sides, we obtain

$$\lim_{\Delta x \to 0}\left(\frac{V(x+\Delta x)-V(x)}{\Delta x}\right) = \lim_{\Delta x \to 0}\left(\frac{R_{\mathrm{sheet}}}{W}\left[I(x)+\frac{W\Delta x}{\rho_{\mathrm{c}}}V(x)\right]\right)$$

$$\frac{\partial V}{\partial x} = \frac{R_{\mathrm{sheet}}}{W} I(x). \tag{C.18}$$

Solving (C.17) and (C.18) together gives the following second-order linear differential equation:

$$\frac{\partial^2 I}{\partial x^2} = \frac{R_{\mathrm{sheet}}}{\rho_{\mathrm{c}}} I(x). \tag{C.19}$$

Let $1/L_{\mathrm{T}}^2 = R_{\mathrm{sheet}}/\rho_{\mathrm{c}}$, equation (C.19) has the general solution $I(x) = Ae^{x/L_{\mathrm{T}}} + Be^{-x/L_{\mathrm{T}}}$, where L_{T} is a characteristic transfer length over which 63 % of the charge flow between the source/drain contacts and the semiconductor transfers [11]. To evaluate the equivalent contact resistance, we impose the boundary conditions $I(x = 0) = 0$ and $I(x = L_{\mathrm{ov}}) = I_{\mathrm{m}}$, where L_{ov} (m) is the gate-overlap length, to obtain A and B:

$$A = -B = \frac{I_{\mathrm{m}}}{e^{L_{\mathrm{ov}}/L_{\mathrm{T}}} - e^{-L_{\mathrm{ov}}/L_{\mathrm{T}}}}.$$

Thus, the function $I(x)$ can now be expressed as

$$I(x) = \frac{I_{\mathrm{m}}}{e^{L_{\mathrm{ov}}/L_{\mathrm{T}}} - e^{-L_{\mathrm{ov}}/L_{\mathrm{T}}}} \left(e^{x/L_{\mathrm{T}}} - e^{-x/L_{\mathrm{T}}}\right). \tag{C.20}$$

Moreover, the function $V(x)$ can then be evaluated by substituting $I(x)$ in (C.17):

$$V(x) = \frac{\rho_{\mathrm{c}}}{W L_{\mathrm{T}}} \frac{I_{\mathrm{m}}}{e^{L_{\mathrm{ov}}/L_{\mathrm{T}}} - e^{-L_{\mathrm{ov}}/L_{\mathrm{T}}}} \left(e^{x/L_{\mathrm{T}}} + e^{-x/L_{\mathrm{T}}}\right). \tag{C.21}$$

Accordingly, the equivalent contact resistance R_{C} can be calculated as follows:

$$R_{\mathrm{C}} = \frac{V(x = L_{\mathrm{ov}})}{I(x = L_{\mathrm{ov}})} = \frac{\rho_{\mathrm{c}}}{W L_{\mathrm{T}}} \frac{\left(e^{L_{\mathrm{ov}}/L_{\mathrm{T}}} + e^{-L_{\mathrm{ov}}/L_{\mathrm{T}}}\right)}{\left(e^{L_{\mathrm{ov}}/L_{\mathrm{T}}} - e^{-L_{\mathrm{ov}}/L_{\mathrm{T}}}\right)}. \tag{C.22}$$

Given that $1/L_{\mathrm{T}}^2 = R_{\mathrm{sheet}}/\rho_{\mathrm{c}}$, the contact resistance (only at one side of the OTFT, i.e., either the source or the drain side) can be finally reduced to the following expression:

$$R_{\mathrm{C}} = \frac{R_{\mathrm{sheet}} L_{\mathrm{T}}}{W} \coth\left(\frac{L_{\mathrm{ov}}}{L_{\mathrm{T}}}\right). \tag{C.23}$$

This expression describes the relationship between the transfer length L_{T} and the gate-overlap length L_{ov} [11]. If the overlap length is much larger than the transfer length ($L_{\mathrm{ov}} \gg L_{\mathrm{T}}$), eqs. (C.22) and (C.23) simplify to $R_{\mathrm{C}} \cong \rho_{\mathrm{c}}/(W L_{\mathrm{T}})$, which indicates that the contact resistance is independent on the overlap length. This implies that, in this case, the contact region is partially active with an equivalent overlap area of $W \cdot L_{\mathrm{T}}$. On the other hand, if the contact length is much smaller than the transfer length ($L_{\mathrm{ov}} \ll L_{\mathrm{T}}$), eqs. (C.22) and (C.23) simplify to $R_{\mathrm{C}} \cong \rho_{\mathrm{c}}/(W L_{\mathrm{ov}})$. This means that, in this case, the contact resistance is inversely proportional to the contact length and the entire contact region with an area of $W \cdot L_{\mathrm{ov}}$ is active.

Appendix D
Code Listings

In this appendix, the Cadence SKILL codes for layout design rule checking and for creating an OTFT parameterized cell are listed. Furthermore, the IC-CAP PEL codes for static and dynamic measurement routines are presented. Finally, the MATLAB code for S-parameter characterization is given herein.

D.1 Cadence SKILL Codes for Layout Design

The software suites from Cadence Design Systems, namely Cadence Virtuoso and Cadence Assura Physical Verification, are used during the physical layout design of the organic integrated circuits. Accordingly, the scripting language SKILL is used to define the design rules that are necessary for the physical verification and also to describe parameterized cells (PCells) for automated layout design. In fact, SKILL is not an acronym; however, it is the name of the language.

D.1.1 Design Rule Check

The SKILL code written for the design rule check (DRC) is presented in Listing D.1, noting that some syntax specific for Cadence Assura is also used within the code, for example, for layer processing (e.g. geomAnd, geomOr, et cetera). After the initialization of layers and the creation of new procedures, the code is divided into two parts to define the design rules of both the organic integrated circuits and the silicon stencil masks. As given in Tables 4.6 and 4.7 and Fig. 4.13, the output error/warning messages of the code are set to refer to the same designated design rule numbers. Some messages, however, do not have numbers, which indicate that they are general rules (e.g. the error message which specifies that a certain layout layer is off-grid).

T. Zaki, *Short-Channel Organic Thin-Film Transistors*, Springer Theses,
DOI 10.1007/978-3-319-18896-6

In the beginning of the code, the five layers of the organic thin-film transistors (OTFTs) are generated and they are assigned to their corresponding GDSII numbers (Table 4.5). Within the design framework, the OTFT layers are assigned to five different drawing layers, namely *NDIFF*, *PDIFF*, *POLY*, *METAL1* and *METAL2*; these are just preset drawing layers in the technology file of IMS CHIPS, which is used in our layout designs. Furthermore, the PCells are initially ignored in the code to minimize the DRC runtime. This is because it is ensured already during the description of the PCells that they satisfy the recommended design rules.

The listed code for the DRC is restricted to the geometric and connectivity configuration shown in Fig. 4.13. However, a more sophisticated version of the code (not listed herein) is also implemented to take into consideration other issues, for example, other transistor configurations (e.g. fingers and multipliers of OTFTs). The simplified version is given herein just to demonstrate the concept. Moreover, design rules 10 and 11 (Table 4.7), which set maximum aspect ratios for the mask bridges attached at both ends and at one end, respectively, are checked in the given version of the code only for the top-metal layer. This is because the top-metal layer includes the most critical and finest structures.

Listing D.1 SKILL code for layout design rule checking (DRC)

```
1   ;--- Start of design rule checking code
2   drcExtractRules(
3
4   ;--- Ignoring PCells from the DRC
5   avParameters(
6      ?ignoreCell(cell("PCELLS"))
7   )
8
9   ;--- Generation of commonly used derived layers
10  layerDefs("df2"
11     nsemi      = layer("NDIFF"       type("drawing"))
12     psemi      = layer("PDIFF"       type("drawing"))
13     gate       = layer("POLY"        type("drawing"))
14     top        = layer("METAL1"      type("drawing"))
15     bottom     = layer("METAL2"      type("drawing"))
16     chipsize   = layer("prBoundary" type("drawing"))
17     dataextent = cellBoundary(root)
18  )
19
20  layerDefs("gds2"
21     nsemi  = layer(40)
22     psemi  = layer(39)
23     gate   = layer(46)
24     top    = layer(49)
25     bottom = layer(51)
26  )
27
28  ;--- Layer derivation rules
29  ngate   = geomAnd(nsemi gate)
30  pgate   = geomAnd(psemi gate)
31  npgate  = geomCat(ngate pgate)
32  npsemi  = geomAnd(nsemi psemi)
```

```
33  semi     = geomCat(nsemi psemi)
34  gatef    = geomAndNot(gate semi)
35  all_lay  = geomOr(bottom gate nsemi psemi top)
36
37  ;--- Contacts
38  gatecont  = geomAnd(gate bottom)
39  metalcont = geomAnd(top bottom)
40
41  ;--- Define new procedures
42  procedure(
43     procWidthCheck(layer1 value comment)
44     drc(layer1 width < value comment withCornerTouch)
45  )
46
47  procedure(
48     procSpCkSameLay(layer1 value comment)
49     drc(layer1 sepNotch < value comment)
50  )
51
52  procedure(
53     procSpCkDifLay(layer1 layer2 value comment)
54     drc(layer1 layer2 sep < value comment)
55  )
56
57  procedure(
58     procEnclCheck(layer1 layer2 value comment)
59     drc(layer1 layer2 enc < value comment shielded(layer2) opposite)
60  )
61
62  procedure(
63     procOverlapCheck(layer1 layer2 value comment)
64     drc(layer1 layer2 ovlp < value comment)
65  )
66
67  procedure(
68     procMinSizeCk(layer1 value comment)
69     checkerr1 = drc(layer1 width < value)
70     checkerr2 = drc(layer1 area < value*value)
71     checkerr  = geomOr(checkerr1 checkerr2)
72     errorLayer(checkerr comment)
73  )
74
75  procedure(
76     procAreaCheck(layer1 value comment)
77     drc(layer1 area > value comment)
78  )
79
80  ;-----------------------------
81  ;--- CHECK ORGANIC IC RULES ---
82  ;-----------------------------
83
84  ;--- Check off-grid errors
85  offGrid(nsemi  0.1 "original NSemi grid should be 0.1")
86  offGrid(psemi  0.1 "original PSemi grid should be 0.1")
87  offGrid(gate   0.1 "original Gate grid should be 0.1")
88  offGrid(top    0.1 "original Top grid should be 0.1")
89  offGrid(bottom 0.1 "original Bottom grid should be 0.1")
90
91  ;--- Check non-90-degree errors
92  non_90 = geomOr(
93     geomGetNon90(nsemi)
94     geomGetNon90(psemi)
95     geomGetNon90(gate)
96  )
97  errorLayer(non_90 "only horizontal or vertical edges are allowed")
98
```

```
99  ;--- Check non-45-degree errors
100 non_45 = geomOr(
101     geomGetNon45(top)
102     geomGetNon45(bottom)
103 )
104 errorLayer(non_45 "only edges with a multiple of 45 degrees are allowed")
105
106 ;--- Check channel length
107 gate_and_semi   = geomAnd(gate semi)
108 trans_active    = geomOverlap(gate_and_semi top keep >= 2)
109 metal_for_trans = geomOverlap(top trans_active)
110 grown_semi      = geomGrow(semi 20 20 20 20)
111 metal_for_trans = geomAnd(metal_for_trans grown_semi)
112 procSpCkSameLay(metal_for_trans 2.0 "1.1, min. channel length is 2")
113
114 ;--- Check gate-overlap
115 procOverlapCheck(top trans_active 20.0 "2.1, min. gate-overlap is 20")
116
117 ;--- Check enclosures
118 procEnclCheck(top  semi 20.0 "3.1, min. Top-enclose-Semi is 20")
119 procEnclCheck(semi top  20.0 "3.2, min. Semi-enclose-Top is 20")
120 procEnclCheck(semi gate 20.0 "3.3, min. Semi-enclose-Gate is 20")
121 procEnclCheck(gate semi 20.0 "3.4, min. Gate-enclose-Semi is 20")
122
123 ;--- Check top-metal spacing outside transistor active regions
124 top_wo_trans = geomAndNot(top metal_for_trans)
125 procSpCkSameLay(top_wo_trans 20.0 "4.1, min. Top-to-Top space is 20")
126
127 ;--- Check related clearances
128 procSpCkSameLay(gate   20.0 "4.2, min. Gate-to-Gate space is 20")
129 procSpCkSameLay(psemi  20.0 "4.3, min. PSemi-to-PSemi space is 20")
130 procSpCkSameLay(nsemi  20.0 "4.4, min. NSemi-to-NSemi space is 20")
131 procSpCkSameLay(bottom 20.0 "4.5, min. Bottom-to-Bottom space is 20")
132
133 ;--- Check non-related layer spacings
134 procSpCkDifLay(top   gate   20.0 "5.1, min. Top-to-Gate space is 20")
135 procSpCkDifLay(top   semi   20.0 "5.2 & 5.3, min. Top-to-Semi space is 20")
136 procSpCkDifLay(top   bottom 20.0 "5.4, min. Top-to-Bottom space is 20")
137 procSpCkDifLay(gate  semi   20.0 "5.5 & 5.6, min. Gate-to-Semi space is 20")
138 procSpCkDifLay(gate  bottom 20.0 "5.7, min. Gate-to-Bottom space is 20")
139 procSpCkDifLay(psemi nsemi  20.0 "5.8, min. PSemi-to-NSemi space is 20")
140 procSpCkDifLay(semi  bottom 20.0 "5.9 & 5.10, min. Semi-to-Bot space is 20")
141
142 ;--- Check layer widths
143 procWidthCheck(top    20.0 "6.1, min. Top width is 20.0")
144 procWidthCheck(gate   20.0 "6.2, min. Gate width is 20.0")
145 procWidthCheck(bottom 20.0 "6.3, min. Bottom width is 20.0")
146
147 ;--- Check direct contact areas
148 procMinSizeCk(metalcont 20.0 "7.1, min. Top-to-Bottom cont. area is 20x20")
149 procMinSizeCk(gatecont  20.0 "7.2, min. Gate-to-Bottom cont. area is 20x20")
150
151 ;--- Check interconnect crossings
152 top_and_gate    = geomAnd(top gate)
153 intcon_crossing = geomAndNot(top_and_gate trans_active)
154 procMinSizeCk(intcon_crossing 20.0 "8.1a, min. crossing area is 20x20")
155 errorLayer(intcon_crossing "8.1b, warning: crossings should be avoided")
156
157 ;--------------------------------
158 ;--- CHECK STENCIL MASK RULES ---
159 ;--------------------------------
160
```

```
161 ;--- Check maximum active chip area
162 size_x_1 = geomShift(dataextent 0 1.0)
163 size_x_2 = geomAndNot(size_x_1 dataextent)
164 size_x_3 = drc(size_x_2 area > 1.0*16000.0)
165 errorLayer(size_x_3 "max. recom. horiz. active chip area is 16 mm")
166
167 size_y_1 = geomShift(dataextent 1.0 0)
168 size_y_2 = geomAndNot( size_y_1 dataextent)
169 size_y_3 = drc(size_y_2 area > 1.0*16000.0)
170 errorLayer(size_y_3 "max. recom. vert. active chip area is 16 mm")
171
172 ;--- Check donut-shaped hole structures
173 hol_bottom = geomGetHoled(bottom)
174 hol_gate   = geomGetHoled(gate)
175 hol_nsemi  = geomGetHoled(nsemi)
176 hol_psemi  = geomGetHoled(psemi)
177 hol_top    = geomGetHoled(top)
178 hol_struc  = geomCat(hol_bottom hol_gate hol_nsemi hol_psemi hol_top)
179 errorLayer(hol_structures "donut-shaped holes are not acceptable")
180
181 ;--- Check remaining mask silicon area
182 drc(all_lay coverage >= 0.5 "9, remain. mask silicon area should be >=50%")
183
184 ;--- Check aspect ratio of beams attached at both ends
185 ;--- Checking only for top-metal layer with separation <= 20
186 ;--- Checking only at the transistor active regions
187 brdg = drc(metal_for_trans sep <= 20 app > 0.0 opposite parallel shielded)
188 bad_bridges  = geomGetRectangle(brdg aspectRatio > 50)
189 errorLayer(bad_bridges "10, aspect ratio of bridges should be <= 50")
190
191 ;--- Check aspect ratio of beams attached at one end
192 ;--- Checking only for top-metal layer with separation <= 20
193 notch_shielded = drc(top notch <= 20 app > 0.0 opposite parallel shielded)
194 bad_notches    = geomGetRectangle(notch_shielded aspectRatio > 15)
195 errorLayer(bad_notches "11, aspect ratio of notches should be <= 15")
196
197 ;--- End of design rule checking code
198 )
```

D.1.2 Parameterized Cells

The SKILL code written to create an OTFT PCell is presented in Listing D.2. In principle, a PCell represents a circuit component that depends on its governing parameters and is automatically generated by the code. Other PCells are also implemented to create more components, such as labels, testing modules, pads and alignment structures, all with the desired drawing layers. However, only the OTFT PCell is given herein for reference. The same drawing layers mentioned above are used again in this code.

The OTFT PCell has five input parameters, namely channel type (*Type*), top-metal width (*Top_width*), channel length (*Ch_length*), gate overlap (*Gate_ovlp*), enclosure of the top metal inside the semiconductor (*Semi_enc_Top*) and finally the amount of intentional vertical misalignment of the top metal with the bottom gate (*Vernier*).

The latter parameter, which quantifies the amount of intentional misalignment, is called *Vernier* because it is very similar to the vernier scale that is typically placed on chip to resolve the alignment errors. This parameter is needed, especially in this study, to characterize the impact of unintentional process misalignment on the device performance.

It is important to note that the channel width of the OTFT depends on the value of the parameter *Semi_enc_Top*. If *Semi_enc_Top* is greater than zero, then the channel width is defined by the top-metal width; but if *Semi_enc_Top* is smaller than zero, then the channel width is defined by the semiconductor width.

Moreover, a set of local variables, namely enclosure of the semiconductor inside the top-metal (*Top_enc_Sem*), enclosure of the gate inside the semiconductor (*Sem_enc_Gate*) and enclosure of the semiconductor inside the gate (*Gate_enc_Sem*), are defined within the code. These parameters are set in this way to guarantee that the generated OTFT complies with the layout design rules (Table 4.6).

Listing D.2 SKILL code for OTFT layout parameterized cell (PCell)

```
1  ;--- Define parameterized cell for the OTFT
2  pcDefinePCell(
3
4  ;--- Initialize PCell name and parameters
5  list(ddGetObj("PCELLS") "OTFT" "layout")
6     ((Type string "p")
7      (Top_width float 100.0)
8      (Ch_length float 10.0)
9      (Gate_ovlp float 20.0)
10     (Sem_enc_Top float 20.0)
11     (Vernier float 0.0)
12    ) ; end of pcell parameters
13
14 ;--- Body of code
15 let(
16    ;--- List of local variables extracted from DRC
17    ((Top_enc_Sem 20.0)
18     (Sem_enc_Gate 20.0)
19     (Gate_enc_Sem 20.0)
20    ) ; end of local variables
21
22    ;--- Draw first top-metal electrode
23    rodCreateRect(
24       ?width  Top_width
25       ?length Top_enc_Sem+Sem_enc_Gate+Gate_ovlp+Vernier
26       ?layer  list("METAL1","drawing")
27       ?origin list(0 0)
28    ) ; end of rodCreateRect
29
30    ;--- Draw second top-metal electrode
31    rodCreateRect(
32       ?width  Top_width
33       ?length Top_enc_Sem+Sem_enc_Gate+Gate_ovlp-Vernier
34       ?layer  list("METAL1","drawing")
35       ?origin list(0 Top_enc_Sem+Sem_enc_Gate+Gate_ovlp+Vernier+Ch_length)
36    ) ; end of rodCreateRect
37
```

```
38     ;--- Draw bottom-gate electrode
39     if( Sem_enc_Top>0
40     then
41        rodCreateRect(
42           ?width  2*Gate_enc_Sem+2*Sem_enc_Top+Top_width
43           ?length 2*Gate_ovlp+Ch_length
44           ?layer  list("POLY","drawing")
45           ?origin list(-1*Sem_enc_Top-Gat_enc_Sem Top_enc_Sem+Sem_enc_Gate)
46        ) ; end of rodCreateRect
47     else
48        rodCreateRect(
49           ?width  2*Gate_enc_Sem+Top_width
50           ?length 2*Gate_ovlp+Ch_length
51           ?layer  list("POLY","drawing")
52           ?origin list(-1*Gate_enc_Sem Top_enc_Sem+Sem_enc_Gate)
53        ) ; end of rodCreateRect
54     ) ; end of if
55
56     ;--- Draw organic Semconductor
57     case( Type
58        ( "p"
59           rodCreateRect(
60              ?width  2*Sem_enc_Top+Top_width
61              ?length 2*Sem_enc_Gate+2*Gate_ovlp+Ch_length
62              ?layer  list("PDIFF","drawing")
63              ?origin list(-1*Sem_enc_Top Top_enc_Sem)
64           ) ; end of rodCreateRect
65        )
66        ( "n"
67           rodCreateRect(
68              ?width  2*Sem_enc_Top+Top_width
69              ?length 2*Sem_enc_Gate+2*Gate_ovlp+Ch_length
70              ?layer  list("NDIFF","drawing")
71              ?origin list(-1*Sem_enc_Top Top_enc_Sem)
72           ) ; end of rodCreateRect
73        )
74        ( t
75           rodCreateRect(
76              ?width  2*Sem_enc_Top+Top_width
77              ?length 2*Sem_enc_Gate+2*Gate_ovlp+Ch_length
78              ?layer  list("PDIFF","drawing")
79              ?origin list(-1*Sem_enc_Top Top_enc_Sem)
80           ) ; end of rodCreateRect
81        )
82     ) ; end of case
83  ) ; end of let
84  ) ; end of pcDefinePCell
```

D.2 IC-CAP PEL Codes for Measurement Routines

The Integrated Circuits Characterization and Analysis Program (IC-CAP) is used to control our measurement equipment and to automate data collection. Using IC-CAP functions and the parameter extraction language (PEL), several measurement routines are defined to characterize the organic transistors and circuits. This section presents some samples of the IC-CAP PEL codes written for the static or dynamic characterization of the organic thin-film transistors (OTFTs) and the organic digital-to-analog converters (DACs).

D.2.1 Static Characterization of the Organic Transistors

Depending on the availability of the measurement equipment, the static current–voltage (I–V) characterization of the organic thin-film transistors (OTFTs) are carried out using an Agilent E5270B 8-slot precision measurement mainframe, an HP 4145A semiconductor parameter analyzer, or an HP 4141B DC source/monitor unit. In this measurement routine, the linear and saturation input characteristics as well as the output characteristics of the OTFTs are measured consecutively. The IC-CAP PEL code written to automate this process is given in Listing D.3, noting that all the input and output parameters are adjusted in advance using the graphical user interface of IC-CAP. The written code runs the different measurement setups and saves the results to a certain directory. It is found that the OTFTs are photosensitive; therefore, before taking any real measurements, the transistors are typically measured several times in advance using less number of input samples to remove any hysteresis that could be present as a result of using the light source of the microscope. To perform special measurements, such as up/down staircase sweep, other codes are written thereto, but only the simplest version of the codes is listed herein for reference.

Listing D.3 IC-CAP PEL code for static characterization of OTFTs

```
1  !--- Start of DC measurements
2  update_explicit
3
4  !--- Global variables declaration
5  global_var file_name
6
7  !--- Setting names and labels
8  Sample     = "IMS2"
9  Trans_type = "P"
10 Width      = "100"  !in micrometers
11 Length     = "4"    !in micrometers
12 GOV        = "20"   !in micrometers
13
14 !--- Getting the date from the system
15 hour       = system$("date +%H")
16 minute     = system$("date +%M")
17 second     = system$("date +%S")
18 time       = val$(hour)&val$(minute)&val$(second)
19 file_name  = val$(Sample)&"_"&val$(Trans_type)&"_W"& ...
20              val$(Width)&"L"&val$(Length)&"G"& ...
21              val$(GOV) !&@"_"@&val$(time)
22
23 !--- Set output path
24 output_path= "C:\Agilent\user\Zaki\Test_Transistors\"&val$(file_name)
25
```

```
26  !--- Remove hysteresis
27  iccap_func(".","Measure")
28  iccap_func(".","Measure")
29  iccap_func(".","Measure")
30  iccap_func(".","Measure")
31
32  !--- Static characterisation of the OTFT
33  iccap_func("../Test_Transfer_Linear","Measure")
34  iccap_func("../Test_Transfer_Linear","Export Data Measured", ...
35             output_path&"_tLin")
36
37  iccap_func("../Test_Transfer_Sat","Measure")
38  iccap_func("../Test_Transfer_Sat","Export Data Measured", ...
39             output_path&"_tSat")
40
41  iccap_func("../Test_Out","Measure")
42  iccap_func("../Test_Out","Export Data Measured", ...
43             output_path&"_out")
44
45  !--- End of DC measurements
46  return_value 0
```

D.2.2 Dynamic Characterization of the Organic Transistors

The dynamic characterization of the OTFTs by means of admittance measurements are carried out using an HP 4284A LCR meter. The capacitance (C) and the conductance (G) of OTFTs, metal-insulator-metal (MIM) and/or metal-insulator-semiconductor (MIS) devices are measured as a function of voltage and frequency. Listing D.4 presents the IC-CAP PEL code written to perform the measurements as a function of the voltage, noting that the process to conduct the measurements as a function of the frequency is quite similar. Before running this code, it is important to calibrate first the LCR meter using standard short and open structures to remove any parasitics and to obtain the desired characteristics at the terminals of the device under test (DUT). As mentioned before in the static characterization, it is also necessary to measure the DUT several times in the beginning to remove any hysteresis resulted from the photosensitivity of the organic semiconductors.

If an OTFT is to be measured in this setup, the source and drain contacts are electrically shorted and connected to the low terminal (virtual ground) of the LCR meter, while the gate electrode is connected to the high terminal. However, the application of a drain-source bias to the OTFT results in a nonuniform distribution of charges along the channel. Therefore, a different setup is also built to investigate the variations in the stored charges of the OTFT at different drain-source and gate-source voltages. Accordingly, the gate electrode is connected to the low terminal of the LCR meter, while the source and drain contacts are connected alternatively to the high terminal of the LCR meter and a DC voltage source (Keithley 230 programmable DC power supply). In this case, four measurements are carried out, i.e., gate-source and gate-drain capacitances as a function of drain-source and gate-source voltages. 894/ing D.5 presents the IC-CAP PEL code written to perform the gate-drain capacitance measurements as a function of the drain-source bias, noting that the other processes are in fact not much different from this one.

Listing D.4 IC-CAP PEL code for admittance-voltage measurements

```
1  !--- Start of CV measurements
2  update_explicit
3
4  !--- Global variables declaration
5  global_var gpib_name
6  global_var ins_address
7  global_var fd_date
8
9  !--- Initializing the address of HP LCR-4284A
10 gpib_name = "gpib0"
11 ins_address = "14"
12
13 !--- Getting the date from the system
14 year    = system$("date +%Y")
15 month   = system$("date +%m")
16 day     = system$("date +%d")
17 hour    = system$("date +%H")
18 minute  = system$("date +%M")
19 second  = system$("date +%S")
20 fn_date = val$(day)&val$(month)&val$(year)
21 fn_time = val$(hour)&val$(minute)&val$(second)
22
23 !--- Setting names and labels
24 sample = "IMS16_W200L200_MIM" ! Sample name
25 Prim_name = "CP" ! Parallel capacitance
26 Sec_name  = "G" ! Conductance
27 printer is CRT
28
29 !--- Initializing the communication to the HP LCR-4284A
30 hpib_num = HPIB_open(val$(gpib_name))
31 status   = HPIB_clear(hpib_num, ins_address)
32
33 !--- Setting the frequency samples
34 freq_samples = 9
35 Freq = "iccap_array["&val$(freq_samples)&"]"
36 Freq[0] = "500Hz" ! duplicate to remove hyst
37 Freq[1] = "500Hz"
38 Freq[2] = "5kHz"
39 Freq[3] = "10kHz"
40 Freq[4] = "40kHz"
41 Freq[5] = "100kHz"
42 Freq[6] = "300kHz"
43 Freq[7] = "600kHz"
44 Freq[8] = "1MHz"
45
46 !--- Setting the DC input voltages
47 Vbias_first = 0
48 Vbias_last = -3
49 Vbias_step = -0.2
50 volt_samples = ((Vbias_last-Vbias_first)//Vbias_step)+1
51 Vbias = "iccap_array["&val$(volt_samples)&"]"
52 Vbias[0] = 0
53 for i = 0 to volt_samples-1
54    Vbias[i] = Vbias_first+(i*Vbias_step)
55 next i
56 Vbias_rate = 6 ! DC bias rate in seconds
57
```

```
58  !--- Sending commands to the LCR meter
59  status = hpib_write(hpib_num,ins_address,"*Rst")
60  status = hpib_write(hpib_num,ins_address,"FUNC:IMP "& ...
61                     val$(Prim_name)&val$(Sec_name))
62  status = hpib_write(hpib_num,ins_address,"VOLT:LEV 100mV")
63  status = hpib_write(hpib_num,ins_address,"OUTP:HPOW ON")
64  status = hpib_write(hpib_num,ins_address,"TRIG:SOUR BUS")
65  status = hpib_write(hpib_num,ins_address,"forM:DATA ASCII")
66  status = hpib_write(hpib_num,ins_address,"APER LONG,4")
67  status = hpib_write(hpib_num,ins_address,"CORR:LENG 2M; METH SING")
68  status = hpib_write(hpib_num,ins_address,"TRIG:DEL 100ms")
69  status = hpib_write(hpib_num,ins_address,"AMPL:ALC ON")
70  status = hpib_write(hpib_num,ins_address,"OUTP:DC:ISOL ON")
71  status = hpib_write(hpib_num,ins_address,"DISP:PAGE MEAS")
72
73  !--- Initialization of the output arrays
74  prim_meas = "iccap_array["&val$(volt_samples*freq_samples)&"]"
75  sec_meas = "iccap_array["&val$(volt_samples*freq_samples)&"]"
76  prim_meas[0] = ""
77  sec_meas[0] = ""
78  dummy=0
79
80  !--- Loop to measure at each frequency sample
81  for freq_idx = 0 to freq_samples-1
82     ! Print header
83     print ""
84     print "------------------"
85     print "Freq["&val$(freq_idx)&"] = "&val$(Freq[freq_idx])
86     print "------------------"
87     print ""
88
89     ! Define buffer
90     status = hpib_write(hpib_num,ins_address,"MEM:DIM DBUF,"& ...
91                        val$(freq_samples*volt_samples))
92     status = hpib_write(hpib_num,ins_address,"MEM:FILL DBUF")
93     status = hpib_write(hpib_num,ins_address,"FREQ "&val$(Freq[freq_idx]))
94
95     ! Loop to measure at each DC voltage
96     for volt_idx = 0 to volt_samples-1
97        print "Measuring at Vbias["&val$(volt_idx)&"] = "& ...
98              val$(Vbias[volt_idx])&"V"
99        status = hpib_write(hpib_num,ins_address,"BIAS:STATE OFF")
100       status = hpib_write(hpib_num,ins_address,"BIAS:VOLT "& ...
101                          val$(Vbias[volt_idx]))
102       status = hpib_write(hpib_num,ins_address,"BIAS:STATE ON")
103       status = hpib_write(hpib_num,ins_address,"trig")
104       dummy  = wait(Vbias_rate)
105    next volt_idx
106
107    ! Ask LCR meter for measured data
108    status = hpib_write(hpib_num,ins_address,"MEM:READ? DBUF")
109
```

```
110       ! Loop to read the measurements at each DC voltage
111       for volt_idx = 0 to (volt_samples -1)
112          status = hpib_read(hpib_num, ins_address ,12 ,"A", ...
113                         prim_meas[freq_idx*volt_samples+volt_idx])
114          status = hpib_read(hpib_num, ins_address ,1 ,"A",dummy)
115          status = hpib_read(hpib_num, ins_address ,12 ,"A", ...
116                         sec_meas[freq_idx*volt_samples+volt_idx])
117          status = hpib_read(hpib_num, ins_address ,7 ,"A",dummy)
118
119          if prim_meas[freq_idx*volt_samples+volt_idx] == +9.90000E+37 then
120             prim_meas[freq_idx*volt_samples+volt_idx] = 0
121          end if
122          if sec_meas[freq_idx*volt_samples+volt_idx] == +9.90000E+37 then
123             sec_meas[freq_idx*volt_samples+volt_idx] = 0
124          end if
125
126          print ""
127          print "Vbias["&val$(volt_idx)&"] = "&val$(Vbias[volt_idx])
128          print "","PRIM("&val$(freq_idx*volt_samples+volt_idx)&") = "& ...
129                val$(prim_meas[freq_idx*volt_samples+volt_idx])
130          print "","SEC("&val$(freq_idx*volt_samples+volt_idx)&") = "& ...
131                val$(sec_meas[freq_idx*volt_samples+volt_idx])
132       next volt_idx
133
134       ! Clear data buffer
135       status = hpib_write(hpib_num, ins_address ,"MEM:CLE DBUF")
136    next freq_idx
137
138    !--- Set DC bias state to off
139    status = hpib_write(hpib_num, ins_address ,"BIAS:STATE OFF")
140
141    !--- Print measured data to a file
142    path = "C:/Agilent/user/Zaki/AC_Measurements/"& ...
143           val$(sample)&"_CV_Meas_"&val$(fn_date)
144    op_dir = sys_path$(val$(path))
145    dummy = system("mkdir -p "&val$(path))
146
147    !--- Create output file directory for the primary parameter
148    prim_op_dir = path&"/"&val$(sample)&"_CV_"&val$(Prim_name)& ...
149                  "_V"&val$(volt_samples)&"F"&val$(freq_samples)& ...
150                  "_"&val$(fn_time)&".txt"
151    prim_op_dir = sys_path$(val$(prim_op_dir))
152    dummy = system("rm -f "&val$(prim_op_dir)&" 2> /dev/null")
153    dummy = system("touch "&val$(prim_op_dir))
154
155    !--- Create output file directory for the secondary parameter
156    sec_op_dir = path&"/"&val$(sample)&"_CV_"&val$(Sec_name)& ...
157                 "_V"&val$(volt_samples)&"F"&val$(freq_samples)& ...
158                 "_"&val$(fn_time)&".txt"
159    sec_op_dir = sys_path$(val$(sec_op_dir))
160    dummy = system("rm -f "&val$(sec_op_dir)&" 2> /dev/null")
161    dummy = system("touch "&val$(sec_op_dir))
162
```

```
163 !--- Writing headers to the output files
164 Header1 = "Vbias [V]"
165 Header2 = "Vbias [V]"
166 Subhead = ""
167 for freq_idx = 0 to freq_samples-1
168    Header1 = Header1&"  "&val$(Prim_name)
169    Header2 = Header2&"  "&val$(Sec_name)
170    Subhead = Subhead&"  freq="&val$(Freq[freq_idx])
171 next freq_idx
172 printer is val$(prim_op_dir)
173 print Header1
174 print Subhead
175 printer is val$(sec_op_dir)
176 print Header2
177 print Subhead
178 printer is CRT
179
180 !--- Write measured data to the output files
181 for i = 0 to volt_samples-1
182    prim_textbody = val$(Vbias[i])
183    sec_textbody = val$(Vbias[i])
184    for k = 0 to freq_samples-1
185       prim_textbody = prim_textbody&"  "&val$(prim_meas[i+k*volt_samples])
186       sec_textbody = sec_textbody&" "&val$(sec_meas[i+k*volt_samples])
187    next k
188    printer is val$(prim_op_dir)
189    print prim_textbody
190    printer is val$(sec_op_dir)
191    print sec_textbody
192 next i
193 printer is CRT
194
195
196 !--- Plot the data in IC-CAP
197 for i = 0 to freq_samples-1
198    ! Trim the unit from the value
199    Freq[i] = trim$(val$(Freq[i]),"Hz")
200    leng = strlen(val$(Freq[i]))
201    ! In IC-CAP 1e6 is represented by MEG and not M
202    if substr$(val$(Freq[i]),leng) == "M" then
203       Freq[i] = substr$(val$(Freq[i]),1,leng-1)&"MEG"
204    end if
205 next i
206
207 !--- Copy input DC voltages to IC-CAP input
208 ip_name = "Vdc"
209 iccap_func(".","New Input",ip_name)
210 iccap_func(ip_name,"SetTableFieldValue","Mode","V")
211 iccap_func(ip_name,"SetTableFieldValue","+ Node","A")
212 iccap_func(ip_name,"SetTableFieldValue","- Node","GROUND")
213 iccap_func(ip_name,"SetTableFieldValue","Unit","SMU1")
214 iccap_func(ip_name,"SetTableFieldValue","Compliance","0")
215 iccap_func(ip_name,"SetTableFieldValue","Sweep Type","LIN")
216 iccap_func(ip_name,"SetTableFieldValue","Sweep Order","1")
217 iccap_func(ip_name,"SetTableFieldValue","Start",val$(Vbias_first))
218 iccap_func(ip_name,"SetTableFieldValue","Stop",val$(Vbias_last))
219 iccap_func(ip_name,"SetTableFieldValue","# of Points",val$(volt_samples))
220 iccap_func(ip_name,"SetTableFieldValue","Step Size",val$(Vbias_step))
221
222 !--- Set subplot for the primary parameter
223 prim_plot_name = "Primary_plot"
224 iccap_func(".","New Plot",prim_plot_name)
225 iccap_func(prim_plot_name,"SetTableFieldValue","Report Type","XY GRAPH")
226 iccap_func(prim_plot_name,"SetTableFieldValue","Header","Capacitance (C)")
227 iccap_func(prim_plot_name,"SetTableFieldValue","X Data",ip_name)
228 iccap_func(prim_plot_name,"SetTableFieldValue","X Axis Type","LINEAR")
```

```
229
230 !--- Set subplot for the secondary parameter
231 sec_plot_name = "Secondary_Plot"
232 iccap_func(".","New Plot",sec_plot_name)
233 iccap_func(sec_plot_name,"SetTableFieldValue","Report Type","XY GRAPH")
234 iccap_func(sec_plot_name,"SetTableFieldValue","Header","Conductance (G)")
235 iccap_func(sec_plot_name,"SetTableFieldValue","X Data",ip_name)
236 iccap_func(sec_plot_name,"SetTableFieldValue","X Axis Type","LINEAR")
237
238 !--- Set multi-plot for both parameters
239 main_plot_name = "Plot_CV_Measurement"
240 iccap_func(".","New Plot",main_plot_name)
241 iccap_func(main_plot_name,"SetTableFieldValue","Report Type","MULTI PLOT")
242 iccap_func(main_plot_name,"SetTableFieldValue","# of Plots","2")
243 iccap_func(main_plot_name,"SetTableFieldValue","Plot 0",prim_plot_name)
244 iccap_func(main_plot_name,"SetTableFieldValue","Plot 1",sec_plot_name)
245 iccap_func(main_plot_name,"SetTableFieldValue","# of GUIs","0")
246 iccap_func(main_plot_name,"SetTableFieldValue","Orientation","V")
247 iccap_func(main_plot_name,"SetTableFieldValue","Plots Per Row","1")
248 iccap_func(main_plot_name,"SetTableFieldValue","Header", ...
249            "CV Measurement Curves")
250
251 !--- Copy measurement results to IC-CAP output
252 for i = 0 to freq_samples-1
253    ! Output of the primary parameter
254    prim_op_name = "PRIM"&val$(i)
255    iccap_func(".","New Output",prim_op_name)
256    iccap_func(prim_op_name,"SetTableFieldValue","Mode","C")
257    iccap_func(prim_op_name,"SetTableFieldValue","High Node","A")
258    iccap_func(prim_op_name,"SetTableFieldValue","Low Node","GROUND")
259    iccap_func(prim_op_name,"SetTableFieldValue","Unit","SMU1")
260    iccap_func(prim_op_name,"SetTableFieldValue","Type","M")
261
262    ! Output of the secondary parameter
263    sec_op_name = "SEC"&val$(i)
264    iccap_func(".","New Output",sec_op_name)
265    iccap_func(sec_op_name,"SetTableFieldValue","Mode","G")
266    iccap_func(sec_op_name,"SetTableFieldValue","High Node","A")
267    iccap_func(sec_op_name,"SetTableFieldValue","Low Node","GROUND")
268    iccap_func(sec_op_name,"SetTableFieldValue","Unit","SMU1")
269    iccap_func(sec_op_name,"SetTableFieldValue","Type","M")
270
271    ! Set plot table field values
272    iccap_func(prim_plot_name,"SetTableFieldValue", ...
273               "Y Data "&val$(i),"PRIM"&val$(i))
274    iccap_func(sec_plot_name,"SetTableFieldValue", ...
275               "Y Data "&val$(i),"SEC"&val$(i))
276
277    ! Copy IC-CAP global array to local array for copy2output command
278    m = 0
279    complex Primcopy[volt_samples]
280    complex Seccopy[volt_samples]
281    for k = i*volt_samples to ((i+1)*volt_samples)-1
282       Primcopy[m]=VAL(prim_meas[k])
283       Seccopy[m]=VAL(sec_meas[k])
284       m = m+1
285    next k
286    x = copy2output(Primcopy,prim_op_name,"M")
287    x = copy2output(Seccopy,sec_op_name,"M")
288 next i
289
290 !--- Setup plot options
291 iccap_func("Plotoptions","SetTableFieldValue","White Background","Yes")
292 iccap_func("Plotoptions","SetTableFieldValue","Font Type","Calibri")
293 iccap_func("Plotoptions","SetTableFieldValue","Font Size","18")
```

```
294 iccap_func("Plotoptions","SetTableFieldValue","Symbol Size","6")
295 iccap_func("Plotoptions","SetTableFieldValue","Measured Trace", ...
296            "Solid Line with Symbols")
297 iccap_func("Plotoptions","SetTableFieldValue","Multicolor", ...
298            "Display all curves with the configured trace color")
299 iccap_func(main_plot_name,"Display Plot")
300 iccap_func(main_plot_name,"TitleOff")
301 iccap_func(main_plot_name,"AreatoolsOff")
302
303 !--- Save plot image
304 op_dir = path&"/"&val$(sample)&"_CV_Meas_"&val$(fn_time)&".png"
305 iccap_func(main_plot_name,"Save Image",op_dir)
306
307 !--- End of CV measurements
308 return_value 0
```

Listing D.5 IC-CAP PEL code for intrinsic capacitance measurements

```
1  !--- Start of intrinsic capacitance CGD vs. VDS measurements
2  update_explicit
3
4  !--- Global variables declaration
5  global_var gpib_name
6  global_var lcr_address
7  global_var dc_address
8  global_var dc_port_num
9
10 !--- Initializing the addresses of the measurement equipment
11 gpib_name = "gpib0"
12 lcr_address = "14" ! HP LCR-4284A
13 dc_address = "13" ! KEITHLEY 230 DC Power
14 dc_port_num = "4" ! Output number of the DC power supply
15
16 !--- Getting the date from the system
17 year    = system$("date +%Y")
18 month   = system$("date +%m")
19 day     = system$("date +%d")
20 hour    = system$("date +%H")
21 minute  = system$("date +%M")
22 second  = system$("date +%S")
23 fn_date = val$(day)&val$(month)&val$(year)
24 fn_time = val$(hour)&val$(minute)&val$(second)
25
26 !--- Setting names and labels
27 sample = "IMS16"
28 module = "OTFT"
29 transistor = "W400L200G10"
30 Prim_name = "CP" ! Parallel capacitance
31 Sec_name  = "G" ! Conductance
32 printer is CRT
33
34 !--- Initializing the communication to the measurement equipment
35 hpib_num = hpib_open(val$(gpib_name))
36 status   = hpib_clear(hpib_num,lcr_address)
37 status   = hpib_clear(hpib_num,dc_address)
38
39 !--- Set the sampling frequency
40 Freq = "500Hz"
41
42 !--- Set the input DC VGS voltage
43 VGS_first = -2
44 VGS_last = -3
45 VGS_step = -0.5
```

```
46  VGS_samples = ((VGS_last-VGS_first)//VGS_step)+1
47  VGS = "iccap_array["&val$(VGS_samples)&"]"
48  VGS[0] = 0
49  for i = 0 to VGS_samples-1
50     VGS[i] = VGS_first+(i*VGS_step)
51  next i
52
53  !--- Set the input DC VDS voltage
54  VDS_first = 0
55  VDS_last = -3
56  VDS_step = -0.2
57  VDS_samples = ((VDS_last-VDS_first)//VDS_step)+1
58  VDS = "iccap_array["&val$(VDS_samples)&"]"
59  VDS[0] = 0
60  for i = 0 to VDS_samples-1
61     VDS[i] = VDS_first+(i*VDS_step)
62  next i
63
64  !--- Set the DC bias rate in seconds
65  Vbias_rate = 8
66
67  !--- Sending commands to the LCR meter
68  status = hpib_write(hpib_num,lcr_address,"*Rst")
69  status = hpib_write(hpib_num,lcr_address,"FUNC:IMP "& ...
70                        val$(Prim_name)&val$(Sec_name))
71  status = hpib_write(hpib_num,lcr_address,"VOLT:LEV 100mV")
72  status = hpib_write(hpib_num,lcr_address,"OUTP:HPOW ON")
73  status = hpib_write(hpib_num,lcr_address,"TRIG:SOUR BUS")
74  status = hpib_write(hpib_num,lcr_address,"FORM:DATA ASCII")
75  status = hpib_write(hpib_num,lcr_address,"APER MED,4")
76  status = hpib_write(hpib_num,lcr_address,"CORR:LENG 2M; METH SING")
77  status = hpib_write(hpib_num,lcr_address,"TRIG:DEL 100ms")
78  status = hpib_write(hpib_num,lcr_address,"AMPL:ALC ON")
79  status = hpib_write(hpib_num,lcr_address,"OUTP:DC:ISOL ON")
80  status = hpib_write(hpib_num,lcr_address,"DISP:PAGE MEAS")
81  status = hpib_write(hpib_num,lcr_address,"FREQ "&val$(Freq))
82
83  !--- Sending commands to the DC power supply
84  status=hpib_write(hpib_num,dc_address,"DOX")
85  status=hpib_write(hpib_num,dc_address,"POX")
86  status=hpib_write(hpib_num,dc_address,"ROX")
87  status=hpib_write(hpib_num,dc_address,"M1X")
88
89  !--- Initialization of the output arrays
90  prim_meas = "iccap_array["&val$(VDS_samples*VGS_samples)&"]"
91  sec_meas  = "iccap_array["&val$(VDS_samples*VGS_samples)&"]"
92  prim_meas[0] = ""
93  sec_meas[0]  = ""
94  dummy_read   = 0
95
96  !--- Loop to measure at each VGS
97  for VGS_idx = 0 to VGS_samples-1
98     ! Print header
99     print ""
100    print "------------------"
101    print "VGS["&val$(VGS_idx)&"] = "&val$(VGS[VGS_idx])
102    print "------------------"
103    print ""
104
105    ! Set the DC voltages
106    VG = 0
107    VS = VG - VGS[VGS_idx]
108
109    ! Define buffer
110    status = hpib_write(hpib_num,lcr_address,"MEM:DIM DBUF,"& ...
111                        val$(VGS_samples*VDS_samples))
```

```
112       status = hpib_write(hpib_num, lcr_address, "MEM:FILL DBUF")
113
114       ! Loop to measure at each VDS
115       for VDS_idx = 0 to VDS_samples-1
116          ! Print header
117          print "Measuring at VDS["&val$(VDS_idx)&"] = "&val$(VDS[VDS_idx])&"V"
118
119          ! Set the DC voltages
120          VD = VDS[VDS_idx] + VS
121
122          ! Set DC power supply value
123          status = hpib_write(hpib_num, dc_address, "F0X")
124          status = hpib_write(hpib_num, dc_address, "V"&val$(VS)&"X")
125          status = wait(0.1)
126          status = hpib_write(hpib_num, dc_address, "F1X")
127          status = wait(1)
128
129          ! Set DC power supply value
130          status = hpib_write(hpib_num, lcr_address, "BIAS:STATE OFF")
131          status = hpib_write(hpib_num, lcr_address, "BIAS:VOLT "&val$(VD))
132          status = hpib_write(hpib_num, lcr_address, "BIAS:STATE ON")
133          status = hpib_write(hpib_num, lcr_address, "trig")
134          status = wait(Vbias_rate)
135       next VDS_idx
136
137       ! Set the DC power supply to stand by and DC bias of LCR meter to off
138       status = hpib_write(hpib_num, dc_address, "F0X")
139       status = hpib_write(hpib_num, lcr_address, "BIAS:STATE OFF")
140
141       ! Ask LCR meter for measured data
142       status = hpib_write(hpib_num, lcr_address, "MEM:READ? DBUF")
143
144       ! Loop to read the measurements at each DC voltage
145       for VDS_idx = 0 to VDS_samples-1
146          status = hpib_read(hpib_num, lcr_address, 12, "A", ...
147                             prim_meas[VGS_idx*VDS_samples+VDS_idx])
148          status = hpib_read(hpib_num, lcr_address, 1, "A", dummy_read)
149          status = hpib_read(hpib_num, lcr_address, 12, "A", ...
150                             sec_meas[VGS_idx*VDS_samples+VDS_idx])
151          status = hpib_read(hpib_num, lcr_address, 7, "A", dummy_read)
152
153          if prim_meas[VGS_idx*VDS_samples+VDS_idx] == +9.90000E+37 then
154             prim_meas[VGS_idx*VDS_samples+VDS_idx] = 0
155          end if
156          if sec_meas[VGS_idx*VDS_samples+VDS_idx] == +9.90000E+37 then
157             sec_meas[VGS_idx*VDS_samples+VDS_idx] = 0
158          end if
159
160          print ""
161          print "VDS["&val$(VDS_idx)&"] = "&val$(VDS[VDS_idx])
162          print "", "PRIM("&val$(VGS_idx*VDS_samples+VDS_idx)&") = "& ...
163                val$(prim_meas[VGS_idx*VDS_samples+VDS_idx])
164          print "", "SEC("&val$(VGS_idx*VDS_samples+VDS_idx)&") = "& ...
165                val$(sec_meas[VGS_idx*VDS_samples+VDS_idx])
166       next VDS_idx
167
168       ! Clear data buffer
169       status = hpib_write(hpib_num, lcr_address, "MEM:CLE DBUF")
170    next VGS_idx
171
172    ! Set to local mode and terminate HPIB communication
173    status = hpib_write(hpib_num, dc_address, "X")
174    status = hpib_command(hpib_num, 1, dc_address)
175    status = hpib_close(hpib_num, dc_address)
176    status = hpib_close(hpib_num, lcr_address)
177
```

```
178  !--- Print measured data to a file
179  path = "C:/Agilent/user/Zaki/AC_Measurements/"& ...
180          val$(sample)&"_CGD_vs_VDS_"&val$(fn_date)
181  op_dir = sys_path$(val$(path))
182  dummy = system("mkdir -p "&val$(path))
183
184  !--- Create output file directory for the primary parameter
185  prim_op_dir = path&"/"&val$(sample)&"_"&val$(module)&"_"& ...
186                val$(transistor)&"_CGD_vs_VDS_"&"VDS"& ...
187                val$(VDS_samples)&"VGS"&val$(VGS_samples)&"_"& ...
188                val$(fn_time)&".txt"
189  prim_op_dir = sys_path$(val$(prim_op_dir))
190  dummy = system("rm -f "&val$(prim_op_dir)&" 2> /dev/null")
191  dummy = system("touch "&val$(prim_op_dir))
192
193  !--- Create output file directory for the secondary parameter
194  sec_op_dir = path&"/"&val$(sample)&"_"&val$(module)&"_"& ...
195               val$(transistor)&"_GGD_vs_VDS_"&"VDS"& ...
196               val$(VDS_samples)&"VGS"&val$(VGS_samples)&"_"& ...
197               val$(fn_time)&".txt"
198  sec_op_dir = sys_path$(val$(sec_op_dir))
199  dummy = system("rm -f "&val$(sec_op_dir)&" 2> /dev/null")
200  dummy = system("touch "&val$(sec_op_dir))
201
202  !--- Writing headers to the output files
203  Header1 = "VDS [V]"
204  Header2 = "VDS [V]"
205  Subhead = ""
206  for VGS_idx = 0 TO VGS_samples-1
207     Header1 = Header1&"  "&val$(Prim_name)
208     Header2 = Header2&"  "&val$(Sec_name)
209     Subhead = Subhead&"  VGS="&val$(VGS[VGS_idx])
210  next VGS_idx
211  printer is val$(prim_op_dir)
212  print Header1
213  print Subhead
214  printer is val$(sec_op_dir)
215  print Header2
216  print Subhead
217  printer is CRT
218
219  !--- Write measured data to the output files
220  for i = 0 to VDS_samples-1
221     prim_textbody = val$(VDS[i])
222     sec_textbody = val$(VDS[i])
223     for k = 0 to VGS_samples-1
224        prim_textbody = prim_textbody&"  "&val$(prim_meas[i+k*VDS_samples])
225        sec_textbody = sec_textbody&" "&val$(sec_meas[i+k*VDS_samples])
226     next k
227     printer is val$(prim_op_dir)
228     print prim_textbody
229     printer is val$(sec_op_dir)
230     print sec_textbody
231  next i
232  printer is CRT
233
234
235  !--- Plot the data in IC-CAP
236  !--- Copy input DC voltages to IC-CAP input
237  ip_name = "VDS"
238  iccap_func(".","New Input",ip_name)
239  iccap_func(ip_name,"SetTableFieldValue","Mode","V")
240  iccap_func(ip_name,"SetTableFieldValue","+ Node","A")
241  iccap_func(ip_name,"SetTableFieldValue","- Node","GROUND")
242  iccap_func(ip_name,"SetTableFieldValue","Unit","SMU1")
243  iccap_func(ip_name,"SetTableFieldValue","Compliance","0")
244  iccap_func(ip_name,"SetTableFieldValue","Sweep Type","LIN")
245  iccap_func(ip_name,"SetTableFieldValue","Sweep Order","1")
```

```
246 iccap_func(ip_name,"SetTableFieldValue","Start",val$(VDS_first))
247 iccap_func(ip_name,"SetTableFieldValue","Stop",val$(VDS_last))
248 iccap_func(ip_name,"SetTableFieldValue","# of Points",val$(VDS_samples))
249 iccap_func(ip_name,"SetTableFieldValue","Step Size",val$(VDS_step))
250
251 !--- Set subplot for the primary parameter
252 prim_plot_name = "Primary_plot"
253 iccap_func(".","New Plot",prim_plot_name)
254 iccap_func(prim_plot_name,"SetTableFieldValue","Report Type","XY GRAPH")
255 iccap_func(prim_plot_name,"SetTableFieldValue","Header","Capacitance (C)")
256 iccap_func(prim_plot_name,"SetTableFieldValue","X Data",ip_name)
257 iccap_func(prim_plot_name,"SetTableFieldValue","X Axis Type","LINEAR")
258
259 !--- Set subplot for the secondary parameter
260 sec_plot_name = "Secondary_Plot"
261 iccap_func(".","New Plot",sec_plot_name)
262 iccap_func(sec_plot_name,"SetTableFieldValue","Report Type","XY GRAPH")
263 iccap_func(sec_plot_name,"SetTableFieldValue","Header","Conductance (G)")
264 iccap_func(sec_plot_name,"SetTableFieldValue","X Data",ip_name)
265 iccap_func(sec_plot_name,"SetTableFieldValue","X Axis Type","LINEAR")
266
267 !--- Set multi-plot for both parameters
268 main_plot_name = "Plot_CV_Measurement"
269 iccap_func(".","New Plot",main_plot_name)
270 iccap_func(main_plot_name,"SetTableFieldValue","Report Type","MULTI PLOT")
271 iccap_func(main_plot_name,"SetTableFieldValue","# of Plots","2")
272 iccap_func(main_plot_name,"SetTableFieldValue","Plot 0",prim_plot_name)
273 iccap_func(main_plot_name,"SetTableFieldValue","Plot 1",sec_plot_name)
274 iccap_func(main_plot_name,"SetTableFieldValue","# of GUIs","0")
275 iccap_func(main_plot_name,"SetTableFieldValue","Orientation","V")
276 iccap_func(main_plot_name,"SetTableFieldValue","Plots Per Row","1")
277 iccap_func(main_plot_name,"SetTableFieldValue","Header", ...
278            "Measured CGD vs. VDS")
279
280 !--- Copy measurement results to IC-CAP output
281 for i = 0 to VGS_samples-1
282    ! Output of the primary parameter
283    prim_op_name = "C_VGS"&val$(i)
284    iccap_func(".","New Output",prim_op_name)
285    iccap_func(prim_op_name,"SetTableFieldValue","Mode","C")
286    iccap_func(prim_op_name,"SetTableFieldValue","High Node","A")
287    iccap_func(prim_op_name,"SetTableFieldValue","Low Node","GROUND")
288    iccap_func(prim_op_name,"SetTableFieldValue","Unit","SMU1")
289    iccap_func(prim_op_name,"SetTableFieldValue","Type","M")
290
291    ! Output of the secondary parameter
292    sec_op_name = "G_VGS"&val$(i)
293    iccap_func(".","New Output",sec_op_name)
294    iccap_func(sec_op_name,"SetTableFieldValue","Mode","G")
295    iccap_func(sec_op_name,"SetTableFieldValue","High Node","A")
296    iccap_func(sec_op_name,"SetTableFieldValue","Low Node","GROUND")
297    iccap_func(sec_op_name,"SetTableFieldValue","Unit","SMU1")
298    iccap_func(sec_op_name,"SetTableFieldValue","Type","M")
299
300    ! Set plot table field values
301    iccap_func(prim_plot_name,"SetTableFieldValue", ...
302               "Y Data "&val$(i),"C_VGS"&val$(i))
303    iccap_func(sec_plot_name,"SetTableFieldValue", ...
304               "Y Data "&val$(i),"G_VGS"&val$(i))
305
306    ! Copy IC-CAP global array to local array for copy2output command
307    m = 0
308    complex Primcopy[VDS_samples]
309    complex Seccopy[VDS_samples]
310    for k = i*VDS_samples to ((i+1)*VDS_samples)-1
```

```
311         Primcopy[m]=VAL(prim_meas[k])
312         Seccopy[m]=VAL(sec_meas[k])
313         m = m+1
314      next k
315      x = copy2output(Primcopy,prim_op_name,"M")
316      x = copy2output(Seccopy,sec_op_name,"M")
317   next i
318
319   !--- Setup plot options
320   iccap_func("PlotOptions","SetTableFieldValue","White Background","Yes")
321   iccap_func("PlotOptions","SetTableFieldValue","Font Type","Calibri")
322   iccap_func("PlotOptions","SetTableFieldValue","Font Size","18")
323   iccap_func("PlotOptions","SetTableFieldValue","Symbol Size","6")
324   iccap_func("PlotOptions","SetTableFieldValue","Measured Trace", ...
325              "Solid Line with Symbols")
326   iccap_func("PlotOptions","SetTableFieldValue","Multicolor", ...
327              "Display all curves with the configured trace color")
328   iccap_func(".","Close All")
329   iccap_func(main_plot_name,"Display Plot")
330   iccap_func(main_plot_name,"TitleOff")
331   iccap_func(main_plot_name,"AreaToolsOff")
332
333   !--- Save plot image
334   op_dir = path&"/"&val$(sample)&"_"&val$(module)&"_"& ...
335            val$(transistor)&"_CGD_vs_VDS_"&"VDS"& ...
336            val$(VDS_samples)&"VGS"&val$(VGS_samples)&"_"& ...
337            val$(fn_time)&".png"
338   iccap_func(main_plot_name,"Save Image",op_dir)
339
340   !--- End of intrinsic capacitance CGD vs. VDS measurements
341   return_value 0
```

D.2.3 Static Characterization of the Digital-to-Analog Converters

The static characterization of the organic digital-to-analog converters (DACs) are carried out using an HP 8180A data generator and an Agilent E5270B 8-slot precision measurement mainframe. The IC-CAP PEL code written to control the measurement setup and collect the output of the 6-bit binary current-steering DAC is presented in Listing D.6. Other similar codes (not listed herein) are written during the course of this work to analyse the dynamic performance of the DAC and also to characterize the other DAC topology, i.e., the 3-bit unary current-steering DAC.

Listing D.6 IC-CAP PEL code for measuring the 6-bit binary DAC

```
1   !--- Start of 6-bit DAC measurements
2   update_explicit
3
4   !--- Global variables declaration
5   global_var gpib_name ! General purpose interface bus name
6   global_var ins_address ! Instrument address
7   global_var fd_date ! Date
8
```

```
9  !--- Initializing the address of Agilent E5270B SMU
10 gpib_name = "gpib10"
11 ins_address = "7"
12
13 !--- Getting the date from the system
14 year    = system$("date + %Y")
15 month   = system$("date + %m")
16 day     = system$("date + %d")
17 hour    = system$("date + %H")
18 minute  = system$("date + %M")
19 second  = system$("date + %S")
20 fn_date = val$(year)&val$(month)&val$(day)
21 fd_date = val$(fn_date)&val$(hour)&val$(minute)&val$(second)
22
23 !--- Initializing the file for saving measurement results
24 op_dac = "C:\Agilent\users\Zaki\Results"&"\DAC_OUT_"&val$(fd_date)&".txt"
25 op_dac = sys_path$(val$(op_dac))
26 dummy = system("rm -f "&val$(op_dac)&" 2> /dev/null")
27 dummy = system("touch "&val$(op_dac))
28 printer is val$(op_dac)
29 !        Line number
30 !        |          Binary coded input
31 !        |          |           Output voltage
32 !        |          |           |     Gate voltage
33 !        |          |           |     |    Supply current
34 !        |          |           |     |    |
35 print "Sequence Input_Code  Vout  Vg   Idd"
36 printer is CRT
37
38
39 !--- Initializing the communication to the HP 8180A data generator
40 hpib_num = hpib_open(val$(gpib_name))
41 status = hpib_clear(hpib_num,ins_address)
42
43 !--- Sending commands to the data generator instrument
44 status = hpib_write(hpib_num,ins_address,"OUT2") ! Open output ports
45 status = hpib_write(hpib_num,ins_address,"RUN") ! Run the instrument
46 dummy = wait(2) ! wait 2 seconds
47
48 !--- Measurement of the ramp signal for the 6-bit DAC
49 for i = 1 to 64   ! Outer loop: 64 input codes
50   status = hpib_write(hpib_num,ins_address,"FWD")
51   for j = 1 to 5  ! Inner loop: Five time measurement
52     iccap_func(".","Measure")
53     printer is val$(op_dac)
54     print val$(i),val$(i-1),val$(Vout[0]),val$(VGcs[0]),val$(Idd[0])
55     printer is CRT
56   next j
57 next i
58
59 !--- Sending commands to the data generator instrument
60 status = hpib_write(hpib_num,ins_address,"STP") ! Stop the instrument
61 status = hpib_write(hpib_num,ins_address,"OUT1") ! Close output ports
62 status = hpib_write(hpib_num,ins_address,"GTL") ! Go to local
63
64 !--- End of 6-bit DAC measurements
65 return_value 0
```

D.3 MATLAB Codes for S-Parameter Characterization

Besides IC-CAP, MATLAB (Matrix Laboratory) has been used to control measurement equipment and to automate data collection, particularly for S-parameter characterization. Full two-port S-parameter measurements that relate the AC currents and voltages between the drain and the gate electrodes of our OTFTs are carried out using an HP 3577A vector network analyzer (VNA) at frequencies up to 5 MHz. The MATLAB code written to automate this process is given in Listing D.7.

Furthermore, MATLAB has been exploited in this work to analyze all the measurements (e.g. to remove the impact of parasitics from the measured S-parameters using the de-embedding procedure presented in Appendix B) and also to implement compact models, which are not only related to the S-parameter characterization. These codes, however, are not not listed herein.

Listing D.7 MATLAB code for S-parameter characterization of OTFTs

```
1  %--- Start of S-parameter measurements
2  clc;
3  clear all;
4
5  %--- Disconnect and delete all instrument objects
6  Instrreset;
7
8  %--- Initialize the vector network analyzer (VNA)
9  VNA_name = 'H3577A';
10 VNA_hpib_address = 8;
11 disp(['Inititalize VNA ', VNA_name, ...
12       ' Instrument (HP-IB Address: ', num2str(VNA_hpib_address),')'])
13 VNA = visa('agilent', ['GPIB0::', num2str(VNA_hpib_address), '::0::INSTR']);
14 set(VNA, 'InputBufferSize', 50000, 'Timeout', 60);
15 fopen(VNA);
16
17 %--- Specify the input frequencies
18 start_freq = 100e3;
19 end_freq = 5e6;
20 freq_resolution = 401;
21 step_freq = (end_freq-start_freq)./(freq_resolution-1);
22 freq_values = [start_freq:step_freq:end_freq]';
23
24 %--- Configurate the VNA
25 disp(['Initialize VNA control settings'])
26 fprintf(VNA, ['IPR', ';']);
27 fprintf(VNA, ['SAM ', num2str(100), ' MV;']);
28 fprintf(VNA, ['FRA ', num2str(100), ' KHZ;']);
29 fprintf(VNA, ['FRB ', num2str(5), ' MHZ;']);
30 fprintf(VNA, ['SR4', ';']);
31 fprintf(VNA, ['I11', ';']);
32 fprintf(VNA, ['SWT ', num2str(5), ' SEC;']);
33 fprintf(VNA, ['ST1', ';']);
34 fprintf(VNA, ['SM2', ';']);
35 fprintf(VNA, ['AV7', ';']);
36 fprintf(VNA, ['LNR ', num2str(0), ' MET;']);
37 fprintf(VNA, ['LNA ', num2str(0), ' MET;']);
38 fprintf(VNA, ['LNB ', num2str(0), ' MET;']);
39 fprintf(VNA, ['REF', num2str(0), ' DBR;']);
40 fprintf(VNA, ['DIV', num2str(10), ' DBR;']);
```

```
41
42  %--- Define the measurement routine and the labels
43  Sn = {'11','21','12','22'};
44  Load = 'OTFT_W100L06G5_VG3VD2';
45  Date = '08082012';
46
47  %--- Store the measured S-parameters in registers
48  disp(['Store S-parameters in registers'])
49  fprintf(VNA,['I', char(Sn(i)), ';']);
50  pause on; pause (15); pause off;
51  fprintf(VNA, ['SD', num2str(i), ';']);
52  pause on; pause (5); pause off;
53
54  %--- Read data from registers
55  disp(['Read data from registers'])
56  temp = str2num(query(VNA,['DD', num2str(i), ';']));
57  pause on; pause (5); pause off;
58  temp = temp';
59  eval(['S',char(Sn(i)) , ...
60       ' = temp(1:2:length(temp), 1) + j*temp(2:2:length(temp), 1)']);
61
62  %--- Close and delete the VNA
63  fclose(VNA); delete (VNA);
64
65  %--- Generate output file
66  file_name_and_path = ['Measurement_Results/', Date, ...
67                        '_S',char(Sn(i)),'_',Load,'.xls'];
68  fid = fopen(file_name_and_path, 'wt');
69  fprintf(fid, ['Freq_Hz\t', 'real\t', 'imag', '\r\n']);
70  j = 1;
71  for i = 1:freq_resolution
72      fprintf(fid, [num2str(freq_values(i)), '\t', ...
73                    num2str(temp(j)), '\t', ...
74                    num2str(temp(j+1)), '\t', '\r\n']);
75      j = j+2;
76  end
77  fclose('all'); % End of S-parameter measurements
```

References

1. T. Zaki, Design of current-steering D/A converters using organic thin-film transistors enabled by high-resolution silicon stencil masks. M.Sc. thesis, University of Stuttgart, Stuttgart, BW, Germany, 2010
2. T. Zaki, F. Ante, U. Zschieschang, J. Butschke, F. Letzkus, H. Richter, H. Klauk, J.N. Burghartz, A 3.3 V 6-Bit 100 kS/s current-steering digital-to-analog converter using organic p-type thin-film transistors on glass. IEEE J. Solid-State Circuits **47**(1), 292–300 (2012)
3. T. Zaki, F. Ante, U. Zschieschang, J. Butschke, F. Letzkus, H. Richter, H. Klauk, J.N. Burghartz, Circuit impact of device and interconnect parasitics in a complementary low-voltage organic thin-film technology, in *Semiconductor Conference Dresden*, Sept 2011, pp. 1–4
4. De-embedding and embedding S-parameter networks using a vector network analyzer. Application Note 1364–1, Agilent Technologies (2001)
5. Applying error correction to network analyzer measuerements. Application Note 1287–3, Agilent Technologies (1999)
6. D.K. Rytting, Network analyzer error models and calibration methods. Presentation, Agilent Technologies (1998)
7. J.V. Butler, D.K. Rytting, M.F. Iskander, R.D. Pollard, M. Vanden,Bossche, 16-term error model and calibration procedure for on-wafer network analysis measurements. IEEE Trans. Microw. Theory Tech. **39**(12), 2211–2217 (1991)
8. D.W. Greve, V.R. Hay, Interpretation of capacitance voltage characteristics of polycrystalline silicon thin film transistors. J. Appl. Phys. **61**(3), 1176–1180 (1987)

9. F. Ante, D. Kälblein, T. Zaki, U. Zschieschang, K. Takimiya, M. Ikeda, T. Sekitani, T. Someya, J.N. Burghartz, K. Kern, H. Klauk, Contact resistance and megahertz operation of aggressively scaled organic transistors. Small **8**(1), 73–79 (2012)
10. D.K. Schroder, *Semiconductor Material and Device Characterization*, 3rd edn. (Wiley-IEEE Press, Hoboken, 2006)
11. F. Ante, Contact effects in organic transistors. Ph.D. Dissertation, Swiss Federal Institute of Technology in Lausanne, Lausanne, Switzerland, 2011